I0057477

Handbook of Microwave Engineering

Handbook of Microwave Engineering

Edited by **Alessandro Torello**

New York

Published by NY Research Press,
23 West, 55th Street, Suite 816,
New York, NY 10019, USA
www.nyresearchpress.com

Handbook of Microwave Engineering
Edited by Alessandro Torello

International Standard Book Number: 978-1-63238-519-2 (Hardback)

Printed in the United States of America.

Contents

Preface

This book discusses the fundamentals as well as modern approaches of microwave engineering. It outlines the processes and applications of this field in detail. Microwave engineering refers to that area of science and technology which studies the design and applications of microwave circuits and systems. This field of study is used in making satellites, radars, wireless radios, etc. This book presents researches and studies performed by experts across the globe. It attempts to understand the multiple branches that fall under this discipline and comprehend how such concepts have practical applications. This book studies, analyses and upholds the pillars of microwave engineering and its utmost significance in modern times. This text is appropriate for students seeking detailed information in this area as well as for experts.

The information shared in this book is based on empirical researches made by veterans in this field of study. The elaborative information provided in this book will help the readers further their scope of knowledge leading to advancements in this field.

Finally, I would like to thank my fellow researchers who gave constructive feedback and my family members who supported me at every step of my research.

Editor

1

Synthesis, Structural and Photophysical Properties of $Gd_2O_3:Eu^{3+}$ Nanostructures Prepared by a Microwave Sintering Process

Ana P. de Moura[1], Larissa H. Oliveira[1], Içamira C. Nogueira[2], Paula F. S. Pereira[1], Máximo S. Li[3], Elson Longo[1], José A. Varela[1], Ieda L. V. Rosa[4*]

[1]Chemistry Institute, State University of Sao Paulo-UNESP, Araraquara, Brazil
[2]Department of Engineering Materials, Federal University of Sao Carlos, São Carlos, Brazil
[3]Institute of Physics of São Carlos, USP, São Carlos, Brazil
[4]Department of Chemistry, Federal University of Sao Carlos, São Carlos, Brazil
Email: *ilvrosa@ufscar.br

Abstract

In this paper, we report the obtention of gadolinium oxide doped with europium ($Gd_2O_3:Eu^{+3}$) by thermal decomposition of the $Gd(OH)_3:Eu^{3+}$ precursor prepared by the microwave assisted hydrothermal method. These systems were analyzed by thermalgravimetric analyses (TGA/DTA), X-ray diffraction (XRD), structural Rietveld refinement method, fourrier transmission infrared absorbance spectroscopy (FT-IR), field emission scanning electron microscopy (FE-SEM) and photoluminescence (PL) measurement. XRD patterns, Rietveld refinement analysis and FT-IR confirmed that the $Gd(OH)_3:Eu^{3+}$ precursor crystallize in a hexagonal structure and space group *P6/m*, while the $Gd_2O_3:Eu^{3+}$ powders annealed in range of 500°C and 700°C crystallized in a cubic structure with space group Ia-3. FE-SEM images showed that $Gd(OH)_3:Eu^{3+}$ precursor and $Gd_2O_3:Eu^{3+}$ are composed by aggregated and polydispersed particles structured as nanorods-like morphology. The excitation spectra consisted of an intense broad band with a maximum at 263 nm and the Eu^{3+} ions can be excitated via matrix. The emission spectra presented the characteristics $^5D_0 \rightarrow {}^7F_{0,1,2,3 \text{ and } 4}$ transitions of the Eu^{3+} ion, whose main emission, $^5D_0 \rightarrow {}^7F_2$, is observed at 612 nm. The photophysical properties indicated that the microwave sintering treatment favored the Eu^{3+} ions connected to the O-Gd linkages in the Gd_2O_3 matrix. Also, the emission in the $Gd_2O_3:Eu^{3+}$ comes from the energy transfered from the Gd-O linkages to the $[EuO_8]^{\bullet}$ clusters in the crystalline structure.

*Corresponding author.

Keywords

Gadolinuim Oxide, Europium Luminescence, Nanorods

1. Introduction

One-dimensional nanomaterials, such as nanotubes, nanowires, nanobelts or nanoribbons have attracted much interest in the past decade due to their physical properties and potential applications in nanotechnology fields [1]-[8]. Moreover, these materials can be applied as displays, catalysts, biological sensing, and other optoelectronic devices [9]-[11].

The demand for efficiency and high resolution waveguides, lamps and other optical devices has also stimulated the discovery of new luminescent materials with superior properties. Thus, there has been a tremendous interest in the subject of materials science for the development of new luminescent materials. The improved performance of display requires high-quality phosphors for sufficient brightness and long-term stability. To enhance the luminescent characteristics of phosphors, extensive research has been carried out on rare-earth activated oxide phosphors due to their superiority in color purity, chemical and thermal stabilities [12]-[14]. In this context, lanthanide hydroxides and oxides have actively been investigated for its application in multilayered capacitors, luminescent lamps and displays, solid-laser devices, optoelectronic data storages, waveguides, and heterogeneous catalysts. Their composition, structure and particle size depend on the synthesis method. Moreover, the chemical homogeneity and morphology of the synthesized products determine the effectiveness of their properties [11] [15] [16]. When they are applied for a fluorescent labeling, for instance, there are several advantages such as sharp emission spectra, long lifetimes, and high resistance against photobleaching in comparison with conventional organic fluorophores and quantum dots [17]-[19].

In particular, the gadolinium oxide doped with Eu^{3+} (Gd_2O_3:Eu^{3+}) exhibits a strong paramagnetic behavior (S 1/4 72) as well as strong UV and cathode-rays have also been observed in the lanthanide (Sm^{3+}, Er^{3+}) doped Gd_2O_3 excited luminescence, which are useful in biological fluorescent label, contrast agent, and display applications [20]-[22]. In addition, Gd_2O_3:Eu^{3+} is a very efficient X-ray and thermo-luminescent phosphor [23].

Europium ion in a trivalent state is one of the most studied rare earth element because of the simplicity of its emission spectra and due to the wide application as red phosphor in color TV screens. Eu^{3+} *f-f* transitions are sensitive to its local environment. The monitoring of different concentrations of the Eu^{3+} content into a ceramic material is very interesting in understanding the nature of the lattice modifiers as well as the degree of order-disorder into its crystalline structure. The most intense *f-f* transition is the $^5D_0 \rightarrow {}^7F_2$ transition at 616 nm. When this ion is presented in a non-centrosymmetric site, it can be used as an activator ion with red emission which has been used in the most commercial red phosphor. Moreover, the intensity of Eu^{3+} excitations at around 394 and 465 nm is improved in these materials as compared with most other Eu^{3+} doped phosphors [24] [25]. Because of it, this ion is able to be applied as biological sensors, phosphors, electroluminescent devices, optical amplifiers or lasers when it is used as a dopant in a variety of ceramic materials [26]-[28].

A variety of preparation methods have been developed to reduce the reaction temperature and achieve a small particle size of high quality Gd_2O_3:Eu^{3+} phosphors [11] [29]-[32].

Microwave heat processing has been successfully applied for the preparation of micro or nanosized inorganic materials [33]-[38]. The microwave-assisted heating is a greener approach to synthesize materials in a shorter time (from several minutes to a few hours) and with lower power consumption (hundreds of Watts) compared to the conventional heating at the same temperatures [39]-[43]. This is a consequence of directly and uniformly heating of the components, and exchange in the reaction selectivity, which can increase the reactional rates (microwave catalysis). Consequently, microwave synthesis is becoming quite common in several material sciences areas, nanotechnology, inorganic, organic, biochemical, or pharmaceutical laboratories [44]-[50].

In the present work, we investigated the photo-physical properties of Gd_2O_3: Eu^{3+} phosphors obtained by the thermal decomposition in range of 500˚C and 700˚C of the $Gd(OH)_3$:Eu^{3+} precursor prepared by the microwave assisted hydrothermal method. These materials were structured and microstructurally analyzed by means of X-ray diffraction (XRD), Rietveld refinement method, fourrier transmission infrared absorbance spectroscopy (FT-IR), field emission scanning electron microscopy (FE-SEM). The photo-physical properties were investigated through the excitation and emission spectra of the Eu^{3+} ion as well as lifetime measurements.

2. Experimental Procedure

2.1. Synthesis of the Precursors

The synthesis of the precursors was performed using the following procedure: In a typical synthesis, 1.8 g of Gd_2O_3 and 0.018 g of Eu_2O_3 were dissolved in 3.0 mL of the HNO_3 solution. After the formation of a clear solution, this solution was kept under constant heating until complete evaporation of the acid. Then 80 mL of distilled water were added to the solution and stirred for 30 min at room temperature. After that, an aqueous KOH (2.0 M) solution was added until the pH of solution was adjusted to be in the range of 12 giving rise to a colloidal precipitates. After stirring for about 30 min, the resultant solution was transferred to a Teflon lined stainless autoclave. This autoclave was then sealed and placed into a microwave system (MH) using 2.45 GHz microwave radiation with maximum power of 800 W. The MH conditions were kept at 140°C for 1 minute. The white powders obtained ($Gd(OH)_3$:Eu^{3+}) were collected, washed with water and ethanol, and then dried at 60°C for 8 h under atmospheric air in a conventional furnace.

2.2. Synthesis of Gd_2O_3:Eu^{3+} Powders

The Gd_2O_3:Eu^{3+} powders were obtained from thermal decomposition of the $Gd(OH)_3$:Eu^{3+} precursors. These precursor powders were placed in ceramic crucibles and heated in a microwave sintering furnace at 500°C, 550°C, 600°C, 650°C and 700°C for 5 min under an ambient atmosphere using a heating rate of 5°C/min producing white powders denoted as Gd_2O_3:Eu^{3+}.

2.3. Characterization

The $Gd(OH)_3$:Eu^{3+} and Gd_2O_3:Eu^{3+} powders were structurally characterized by X-ray diffraction (XRD) in normal routine and Rietveld routine using a Rigaku-DMax/2500PC (Japan) with Cu-Kα radiation (λ = 1.5406 Å) and in the 2θ range from 10° to 130° with a scanning rate of 0.02°/min. Fourier Transmission Infrared absorbance spectroscopy (FT-IR) analysis were taken in a FT-IR Bruker model EQUINOX spectrophotometer in range of 500 and 4000 cm^{-1}. Crystals morphologies were verified using a Scanning Electron Microscope (Jeol JSM-6460LV microscope). Photoluminescence (PL) was measured with a Thermal Jarrel-Ash Monospec 27 monochromator and a Hamamatsu R446 photomultiplier. The 350.7 nm exciting wavelength of a krypton ion laser (Coherent Innova) was used, with the nominal output power of the laser power kept at 200 mW. All the measurements were taken at room temperature. The excitation and emission spectra of the Gd_2O_3:Eu^{3+} powders were measured in a Jobin Yvon-Fluorolog 3 spectrofluorometer at room temperature using a 450 W xenon lamp as excitation energy source. Lifetime data of the Eu^{3+} $^5D_0 \rightarrow {}^7F_2$ (λ_{exc} = 394 nm, λ_{em} = 612 nm) transition in the Gd_2O_3:Eu^{3+} samples were evaluated from the decay curves using the emission wavelength set at 612 nm and excitation wavelength set at 393 nm.

3. Results and Discussion

3.1. Thermogravimetric Analyses (TGA/DTA)

Figure 1 presents the TGA curve of the as-prepared $Gd(OH)_3$:Eu^{3+} powder. It can be seen in this figure, during the MH process, the thermal degradation of the $Gd(OH)_3$:Eu^{3+} powders occurs in a two-step process in accordance to observed by Chang *et al.* [51]. The first-step occurs at 436°C, where a weight loss of 8.3% is reported, while the second-step, occurs from 436°C to 700°C, a weight loss of 3.3% is revealed. From these results, it was confirmed that the MH treatments of the hydroxide precursor give rise to a stable Gd_2O_3:Eu^{3+} matrix at temperatures up to 400°C. Moreover, the two-step dehydration process indicates the presence of an intermediate phase in addition to the starting hexagonal $Gd(OH)_3$:Eu^{3+} and the final cubic Gd_2O_3:Eu^{3+}. The theoretical weight loss of each process in this work is in a good agreement with shown by the literature.

3.2. X-Ray Diffraction (XRD) and Rietveld Refinement Analyses

XRD patterns of the as-prepared $Gd(OH)_3$:Eu^{3+} are presented in **Figure 2(A)**, where the reflectance peaks of a

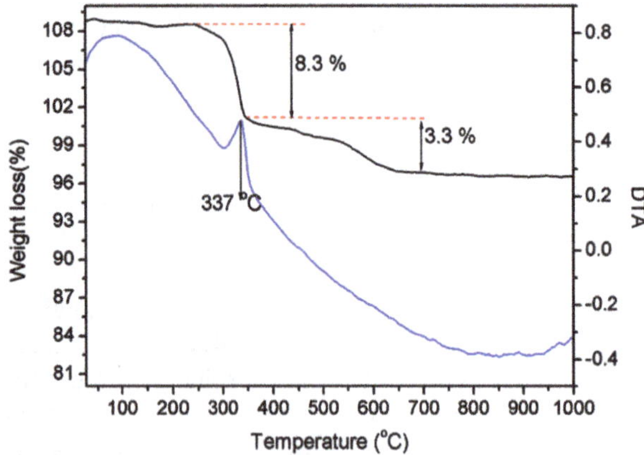

Figure 1. TGA/DTA curves of the as prepared $Gd(OH)_3:Eu^{3+}$.

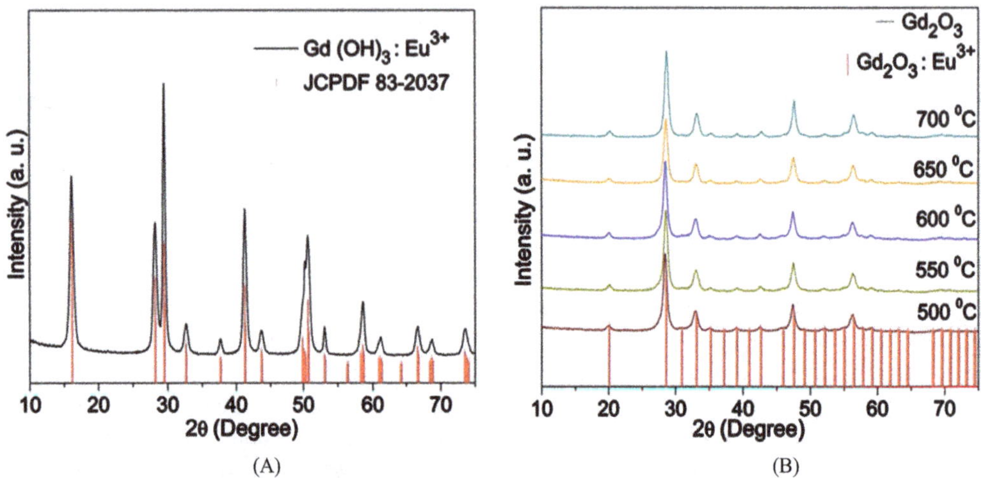

Figure 2. (A) XRD patterns of $Gd(OH)_3:Eu^{3+}$ precursor obtained by the MH conditions at 140°C for 1 min. (B) XRD patterns of $Gd_2O_3:Eu^{3+}$ powders prepared by thermal decomposition of the precursor $Gd(OH)_3:Eu^{3+}$.

pure hexagonal $Gd(OH)_3:Eu^{3+}$ phase with space group *P63/m* can be perfectly indexed in agreement with the respective Inorganic Crystal Structure Database (ICSD) card number 200,093. After the thermal decomposition of the $Gd(OH)_3:Eu^{3+}$ in range of 500°C to 700°C for 5 min (**Figure 2(B)**), it was observed that all the samples can be perfectly indexed to the cubic structure of crystalline Gd_2O_3 and space group Ia-3 (ICSD # 94892). None secondary phases were detected in these samples, indicating that all the $Gd(OH)_3:Eu^{3+}$ precursor heated from 500°C to 700°C giving rise to $Gd_2O_3:Eu^{3+}$ powders. In addition, in this range of temperature, any significant change in the XRD peak profiles of the $Gd_2O_3:Eu^{3+}$ samples were detected.

To better analyze the influence of the precursor thermal decomposition, the $Gd(OH)_3:Eu^{3+}$ and $Gd_2O_3:Eu^{3+}$ powders were submitted to the Rietveld refinement analysis.

The Rietveld refinement is a method in which the profile intensities obtained from step-scanning measurements of the powders allow to estimate an approximate structural model for the real structure [52]. In our work, the Rietveld refinement was performed using the General Structure Analysis (GSAS) program [53]. In these analyses, the parameters like scale factor, background, shift lattice constants, profile half-width parameters (u, v, w), isotropic thermal parameters, strain anisotropy factor, occupancy and atomic functional positions were refined. The background was corrected using a Chebyschev polynomial of the first kind. The peak profile function was modeled using a convolution of the Thompson-Cox-Hastings pseudo-Voigt (pV-TCH) [54] with asymmetry function described by Finger *et al.* [55]. To account for the anisotropy in the half width of the reflections, it was used the model proposed by Stephens [56].

The obtained results from the Rietveld refinement analyses of the crystalline $Gd(OH)_3:Eu^{3+}$ and $Gd_2O_3:Eu^{3+}$

are shown in **Figure 3**. The results showed that the $Gd(OH)_3:Eu^{3+}$ powder crystallize in a hexagonal structure, space group *P63/m* and two clusters per unit cell (Z = 2), while the $Gd_2O_3:Eu^{3+}$ powders crystallize in a cubic structure, space group Ia-3 and sixteen formula units per cell (Z = 16). These results showed a good relation between the observed XRD patterns and the theoretical ones, as shown by a line ($Y_{obs} - Y_{cal}$) at **Figure 3**. Moreover, it was not verified the presence of secondary phases related to the Eu^{3+} ions, probably indicating that these ions were incorporated to the hydroxide and oxide matrixes. The lattice parameters, unit cell volume, unit cell angles and correlation parameters (R_{Bragg}, χ^2 and R_{wp}) obtained by means of the structural refinement data for analyzed powders are listed in **Table 1** and **Table 2**.

From the lattice parameters, unit cell volume and atomic positions obtained from the Rietveld refinement data, it was possible to model a schematic representation of the hexagonal $Gd(OH)_3:Eu^{3+}$ unit cell and space group *P63/m* **Figure 4(A)**, using the Visualization for Electronic and Structural Analysis (VESTA) program version 2.1.6 for Windows [57].

In the literature, the crystalline structure of the $Gd(OH)_3$ has been studied by Chang *et al.* [51]. From this schematic representation is was possible to observed that the hexagonal $Gd(OH)_3$ is composed by $[Gd(OH)_3]^-$

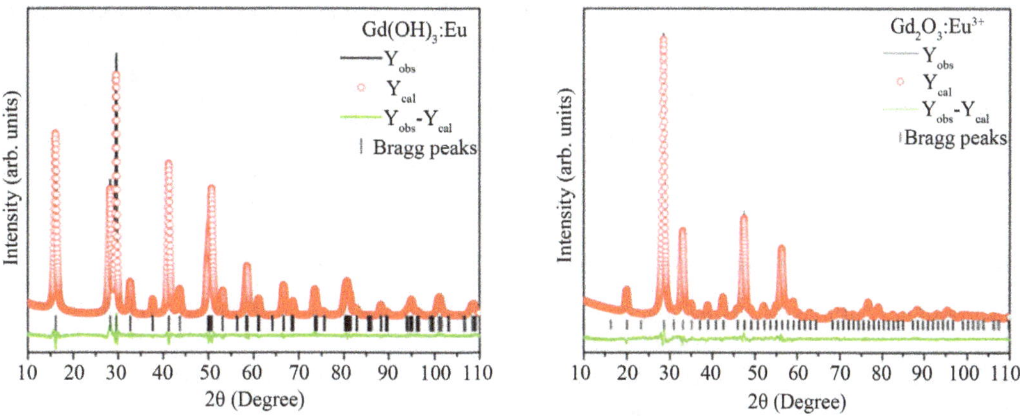

Figure 3. Rietveld refinement data for the $Gd(OH)_3:Eu^{3+}$ and $Gd_2O_3:Eu^{3+}$ powders, respectively.

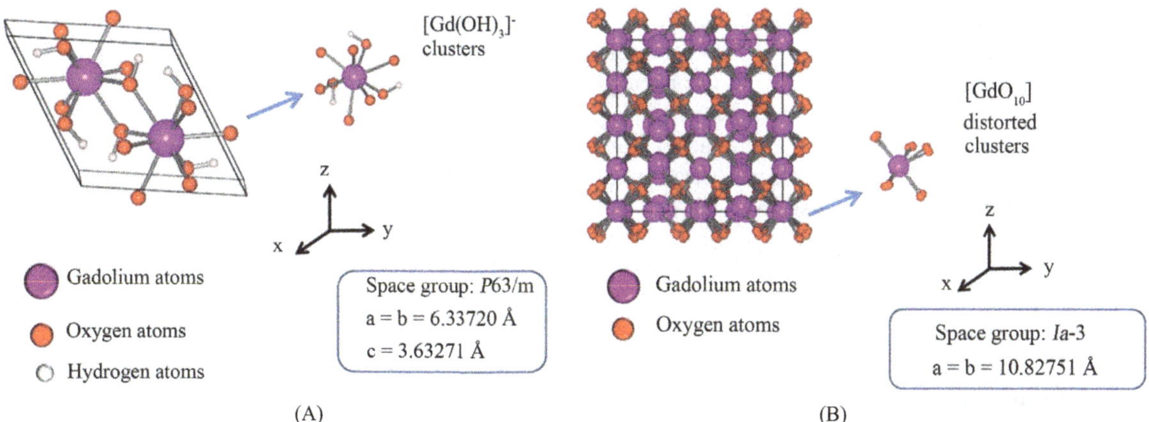

Figure 4. Schematic representation of the $Gd(OH)_3:Eu^{3+}$ (A) and $Gd_2O_3:Eu^{3+}$ (B) unit cell, respectively.

Table 1. Lattice parameters, unit cell volume of $Gd(OH)_3:Eu^{3+}$ obtained by MH processing at 140˚C for 1 min.

	Lattice Parameters		Cell volume (Å³)	Unit cell angle (˚)		R_{Bragg} (%)	χ^2 (%)	R_{wp} (%)	R_p (%)
	a, *b* (Å)	*c* (Å)		$\alpha = \beta$	□				
$Gd(OH)_3:Eu^{3+}$	6.337 (2)	3.632 (7)	126.344 (4)	90	120	2.80	1.98	6.38	5.08
ICSD # 200093	6.329 (2)	3.631 (1)	125.960	90	120	-	-	-	-

Table 2. Lattice parameters, unit cell volume of $Gd_2O_3:Eu^{3+}$ obtained by thermal decomposition of the $Gd(OH)_3:Eu^{3+}$ precursor at 700˚C.

	Lattice parameters (Å)	Cell volume ($Å^3$)	Unit cell angle (˚)	R_{Bragg} (%)	χ^2 (%)	R_{wp} (%)	R_p (%)
	$a = b = c$		$\alpha = \beta = \gamma$				
$Gd_2O_3:Eu^{3+}$	10.82751 (22)	126.344 (4)	90	2.93	1.70	6.46	5.01
ICSD # 94892	10.82311 (20)	126.782	90	-	-	-	-

clusters connected to each other. Moreover, the addition of Eu^{3+} in its crystalline structure promotes the expansion of the $Gd(OH)_3$ unit cell as a consequence of the substitution of some Gd sites by the Eu ion ones.

The unit cell of the $Gd_2O_3:Eu^{3+}$ powder was also simulated through the VESTA software and its schematic representation is illustrated at **Figure 4(B)**. From this schematic representation the $Gd_2O_3:Eu^3$ are composed by the Gd atoms coordinated to ten oxygen atoms forming $[GdO_{10}]$ clusters which are connected to each other and all dispersed to the cubic crystalline structure, which are in good agreement with References [58] and [59].

From **Table 2**, it can be seen any significant changes in lattice parameters (a, b and c) of Gd_2O_3 system as function of the heat treatment. When the $Gd(OH)_3:Eu^{3+}$ material is subjected to a thermal treatment in a microwave oven sintering, there is an interaction of the Gd, O and H atoms with the microwave radiation in the crystalline structure. Consequently, the heating of these systems occurs. This process is able to promote the dehydration of the $Gd(OH)_3:Eu^{3+}$ material giving rise to Gd_2O_3 crystalline structure. Due to low heating rate used (5˚C /min), this interaction occurs slowly and the quantity of surface defects is reduced as well as particle growth is promoted [60].

3.3. Fourier Transmission Infrared Absorption Spectroscopy (FT-IR)

Figure 5 shows the FT-IR transmission spectra of the $Gd(OH)_3:Eu^{3+}$ precursor and of the $Gd_2O_3:Eu^{3+}$ nanocrystals obtained heated at 500˚C, 550˚C, 600˚C, 650˚C and 700˚C for 5 min.

The FT-IR of the $Gd(OH)_3:Eu^{3+}$ precursor presents a sharp absorption band at 3610 cm^{-1} which is characteristic of the Gd-OH matrix. The bands at around 3460 and 1627 cm^{-1} are due to the OH stretching (υ) and OH deformation vibrations (δ), respectively. The broad at absorptions around 2333 cm^{-1} is assigned to the existence of CO_2. The absorption bands at around 1480 and 400 cm^{-1} are ascribed to the CO asymmetric vibration (υ_{as}).

The sharp peak observed for the $Gd(OH)_3:Eu^{3+}$ precursors is ascribed to the OH stretching vibration, and indicates the absence of hydrogen bonds between the hydroxyls group [61]. Thus, it can be supposed that the $[Gd(OH)_3]^-$ clusters in the $Gd(OH)_3$ are connected to each other, in accordance to the Rietveld refinement analysis. In the IR absorption spectra of the $Gd_2O_3:Eu^{3+}$ powders annealed from 500˚C to 700˚C, it was observed the appearance of some absorption at around 525 and 830 cm^{-1} characteristic of Gd-O vibrations, confirming the formation of an oxide matrix after the heat treatment process [62] [63]. Also, the broad absorption bands observed at 3460 cm^{-1} in these samples are only corresponded to the water absorbed in its superficies. All these results are also in agreement with the Rietveld refinement analysis.

The intensity of the absorption bands corresponding to the OH and CO groups are strong dependent of the annealing temperature, indicating that the powder prepared in air atmosphere have strong absorption to water and CO_2 and it can be potentially applied as gas sensors in monitoring gases such as water and CO_2.

3.4. Field Emission Scanning Electron Microscopy (FE-SEM)

Figure 6 shows the FE-SEM images of the $Gd(OH)_3:Eu^{3+}$ precursor (A) and $Gd_2O_3:Eu^{3+}$ samples heated at 500˚C (B), 550˚C (C), 600˚C (D), 650˚C (E) and 700˚C (F), respectively. The FE-SEM images showed that the $Gd_2O_3:Eu^{3+}$ and $Gd(OH)_3:Eu^{3+}$ powders are composed by aggregated and polydispersed particles structured as nanorods-like morphology. Moreover, the size and thickness of the $Gd_2O_3:Eu^{3+}$ powders (**Figures 6(B)-(F)**) vary as function of the annealing temperature. It was also noticed that as the temperature increases the particles have a tendency to agglomerate.

FE-SEM images were also employed to evaluate the average particle size distribution (width) of the $Gd(OH)_3:Eu^{3+}$ and $Gd_2O_3:Eu^{3+}$ nanostructures. During this measurement was considered around 100 nanostructures and the best fit for this system was adjusted as a lognormal function, which is described by the following equation:

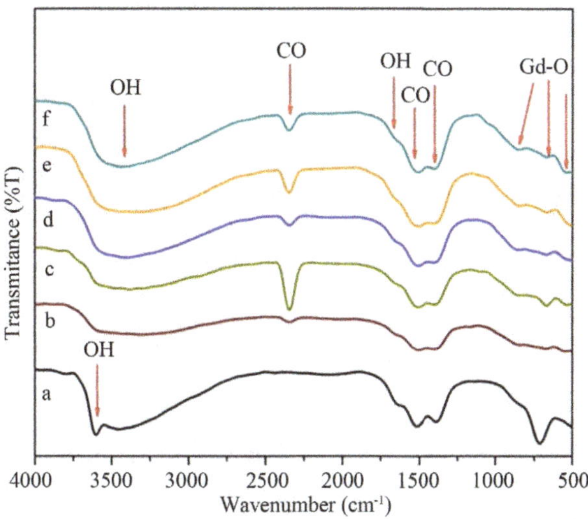

Figure 5. FE-SEM images of $Gd(OH)_3:Eu^{3+}$ precursor (A) and $Gd_2O_3:Eu^{3+}$ samples heat at 500°C (B), 550°C (C), 600°C (D), 650 °C (E) and 700°C (F), respectively.

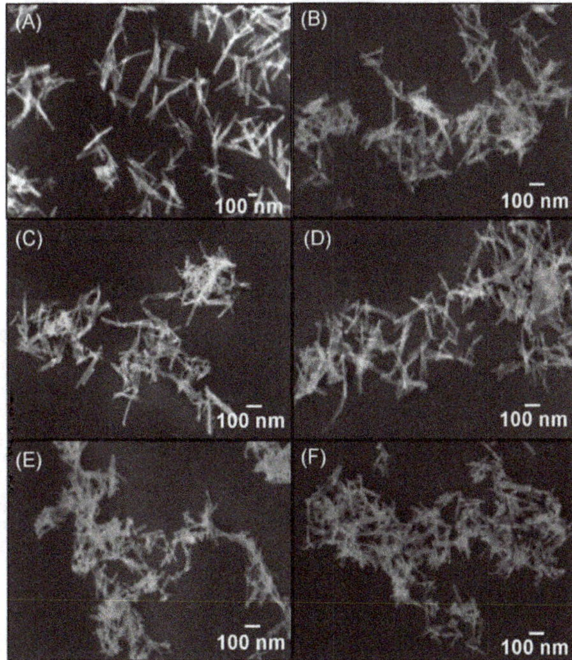

Figure 6. Average thickness distribution of $Gd(OH)_3:Eu^{3+}$ precursor (A) and $Gd_2O_3:Eu^{3+}$ samples heat at 500°C (B), 550°C (C), 600°C (D), 650°C (E) and 700°C (F), respectively.

$$Y = Y_0 + \frac{A}{\sqrt{2\pi}WX} e^{-\frac{\left[\ln\frac{x}{xc}\right]^2}{2w^2}} \tag{1}$$

where y_0 is the first value in y-axis, A is the amplitude, w is the width, π is a constant, x_c is the center value of the distribution curve in x-axis.

The obtained results shown in **Figure 7** presented an assymmetrical distribution on the logarithmic scale of average particle size. In this case, it was noted that almost all particles presented an average width between 8 and 20 nm.

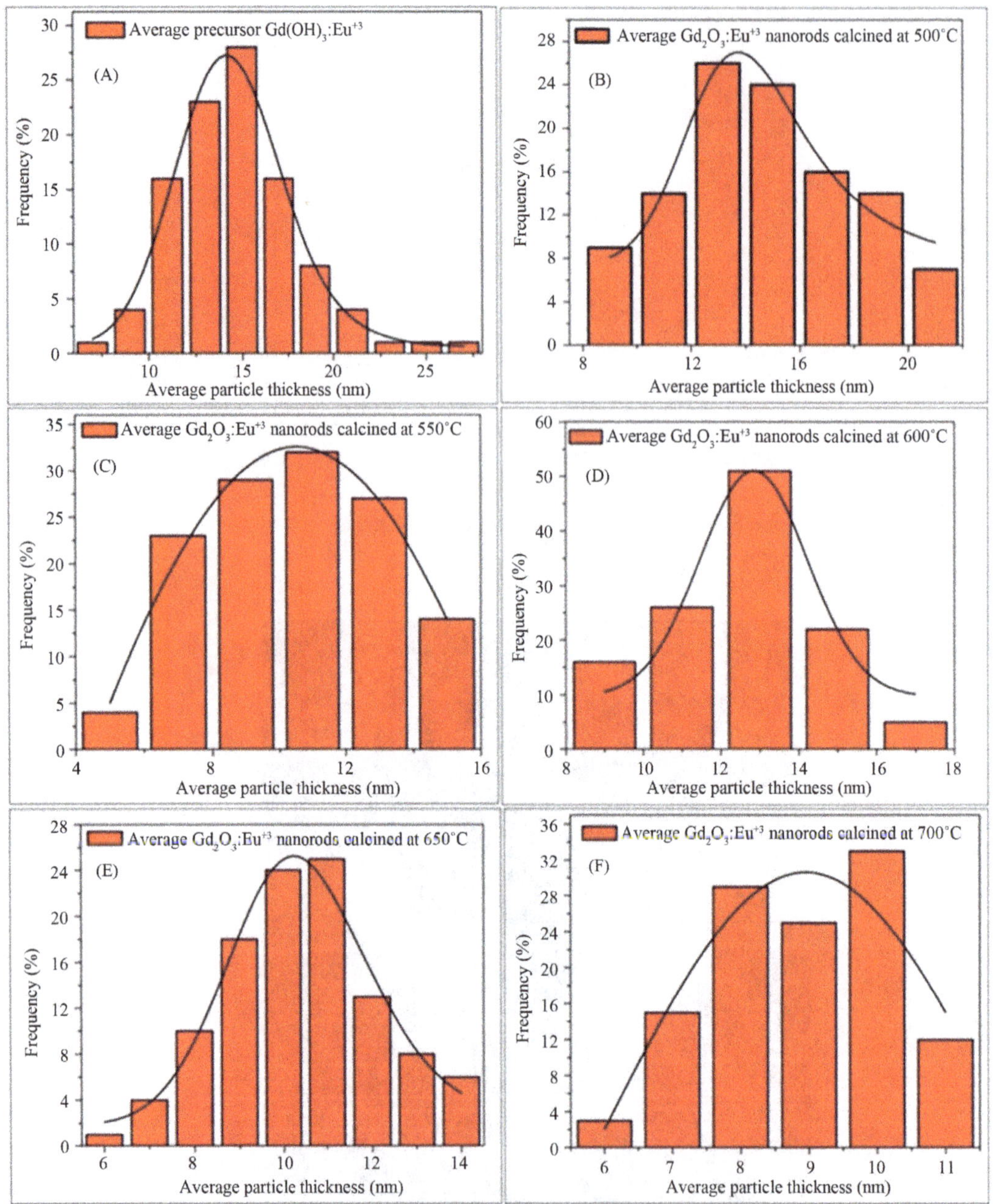

Figure 7. FT-IR spectra of the Gd(OH)$_3$:Eu^{3+} precursor (A) and of the Gd$_2$O$_3$:Eu^{3+} calcined at: 500°C (B); 550°C (C); 600°C (D); 650 °C (E) and 700°C (F).

3.5. Photoluminescence (PL) Emission Measurements

Figure 8(A) illustrates the PL spectra of the Gd(OH)$_3$:Eu^{3+} precursor and Gd$_2$O$_3$:Eu^{3+} samples heat treated at 500°C, 550°C, 600°C, 650°C and 700°C, respectively and (B) the PL spectrum of the non-doped Gd$_2$O$_3$ prepared by the microwave assisted hydrothermal method. All the measurements were recorded at room temperature.

To a better understanding of the PL properties and its dependence on the structural order-disorder in the Gd$_2$O$_3$ lattice, the PL emission spectra of the Gd(OH)$_3$:Eu^{3+} and Gd$_2$O$_3$:Eu^{3+} powders were performed at room temperature, using an excitation of a krypton laser source at 350.7 nm. **Figure 8(A)** also shows a broad emission band from 400 to 600 nm. This band can be ascribed to the emission of the Gd$_2$O$_3$ matrix as it was confirmed by the PL emission spectrum of Gd$_2$O$_3$ powder where a broad band with maximum situated at 449 nm were observed (**Figure 8(B)**). Moreover, in range of 600 and 700 nm, it was possible to noticed the intra-configurational

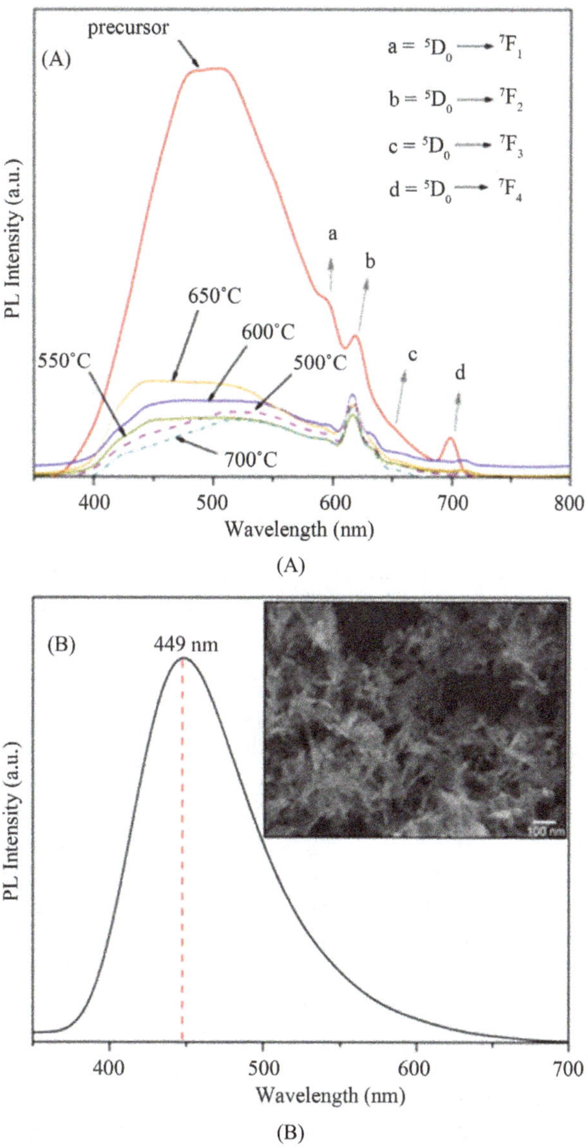

Figure 8. PL emission spectra of the $Gd(OH)_3:Eu^{3+}$ and $Gd_2O_3:Eu^{3+}$ powders heated in different temperatures (A) and pure Gd_2O_3 heat treated at 600°C. Insert: FE-SEM image of pure Gd_2O_3 powder.

$4f_6$ transitions of the Eu^{3+} ions specifically the $^7F_0 \rightarrow {}^5D_J$ (J = 1, 2, 3 and 4) transitions at 590, 615, 633 and 710 nm, respectively, for all analyzed samples.

Figure 9 shows the excitation spectra of $Gd_2O_3:Eu^{3+}$ powders annealed at 500°C, 550°C, 600°C, 650°C and 700°C. The excitation spectra were recorded monitoring the emission wavelength at 612 nm. It can be clearly seen that the excitation spectra consist of a main intense broad band with a maximum at 263 nm attributed to the the $O^{2-} \rightarrow Eu^{3+}$ energy transfer state [64]-[67] and the internal Gd^{3+} $S^8 \rightarrow {}^6I$ and $^8S \rightarrow {}^6P$ transitions situated at 274 and 311 nm, respectively. These transitions are possible to be detected as a consequence of the $Gd^{3+} \rightarrow Eu^{3+}$ energy transfer [65] [66] in the Gd_2O_3 matrix (CTB). Above 330 nm, it was noted the Eu^{3+} $^4f_6$ intra-configurational transitions from the ground state $^7F_0$ to the excited states $^5G_6$ at 362 nm, $^5H_4$ at 380 nm, $^5L_6$ at 393 nm, $^5D_2$ at 464 nm and $^5D_1$ at 532 nm, respectively (**Figure 9**). In this case, we observed that the Eu^{3+} ion can be excited via matrix in a wide range of wavelength and the most intense absorption band is correspondent to the $S^8 \rightarrow {}^6I$ transition of the Gd^{2+} ions at 263 nm. Moreover, as the temperature increases the relative intensity of the bands corresponding to its transitions also increases [68].

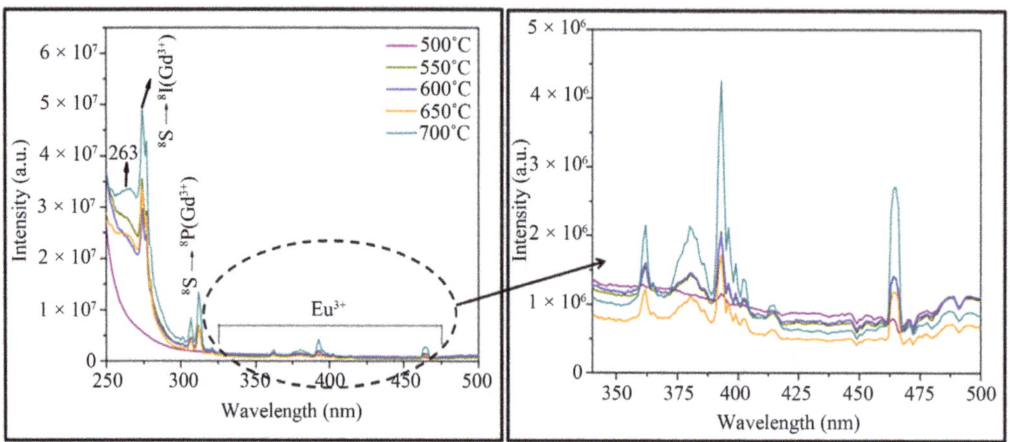

Figure 9. Excitation spectra of $Gd_2O_3:Eu^{3+}$ samples heated at 500°C, 550°C, 600°C, 650°C and 700°C, λ_{em} = 612 nm.

The emission spectra of $Gd_2O_3:Eu^{3+}$ for the samples annealed at 500°C, 550°C, 600°C, 650°C and 700°C are shown in **Figure 10**. These spectra were obtained setting the excitation wavelength into the energy transfer band CTB of Eu^{3+} at 263 nm.

The emission spectra of the Eu^{3+} ion present narrow bans ascribed to the $^5D_0 \rightarrow {}^7F_{0,1,2,3 \text{ and } 4}$ transitions at around 578, 589, 614, 652 and 699 nm, respectively. The most intense band is related to $^5D_0 \rightarrow {}^7F_2$ transition with maximum situate at 612 nm. The hypersensitive transition $^5D_0 \rightarrow {}^7F_2$ is dependent on the local Eu^{3+} environment due to its electric dipole character, while the intensity of $^5D_0 \rightarrow {}^7F_2$, a magnetic dipole transition is almost independent of the Eu^{3+} surroundings. Thus, the ratio $^5D_0 \rightarrow {}^7F_2/{}^5D_0 \rightarrow {}^7F_1$ emission intensity gave us valuable information about the environmental changes around the rare earth ions and can be used as a dimension of the degree of distortion from the inversion symmetry of the Eu^{3+} site in the lattice [69].

The ratio $^5D_0 \rightarrow {}^7F_2/{}^5D_0 \rightarrow {}^7F_1$ values obtained for the $Gd_2O_3:Eu^{3+}$ powders annealed at 500°C, 550°C, 600°C, 650°C and 700°C is of 18.0, 17.1, 15.8, 11.95 and 18.3, respectively. The decrease in the ratio $^5D_0 \rightarrow {}^7F_2/{}^5D_0 \rightarrow {}^7F_1$ values as the temperature increases from 500°C to 650°C is a indicative that the Eu^{3+} ions is occupying higher symmetry sites. In this case, we believe that some Eu^{3+} is already connected to the –OH linkages from the $Gd(OH)_3$ precursor. However, there is an increase of the ratio $^5D_0 \rightarrow {}^7F_2/{}^5D_0 \rightarrow {}^7F_1$ value at 700°C, indicating that now the Eu^{3+} ions is occupying lower symmetric sites. The temperature favored the higher symmetric coordination sites, since there is Eu^{3+} ions linked to the O-Gd linkages in the Gd_2O_3 matrix. Some of these oxygen atoms are in the first coordination sphere of Eu^{3+} ions giving rise to $[EuO_8]^\bullet$ clusters in this powder.

All these results indicates that probably the emission in the $Gd_2O_3:Eu^{3+}$ powders annealed from 500°C to 700°C comes from the energy transfer from the Gd-O linkages to the $[EuO_8]^\bullet$ clusters in the crystalline structure. Moreover, the increase of the relative intensity of the Eu^{3+} $^5D_0 \rightarrow {}^7F_{0,1,2,3 \text{ and } 4}$ transitions is strongly related to the formation of these complex clusters in the Gd_2O_3 matrix. The photoluminescence decay curves and lifetime of the Eu^{3+} $^5D_0 \rightarrow {}^7F_2$ transition with emission and excitation wavelengths set at 612 nm and 263 nm, respectively. In this work, the lifetime value of the $Gd_2O_3:Eu^{3+}$ samples as a function of the annealing temperatures are shown in **Figure 11**. All these curves can be fitted into a single exponential function as $I = I_0 \exp(-t/\tau)$ were (τ is the lifetime of the rare earth ion).

According to these data, it was observed that the lifetime of Eu^{3+} increases as the annealing temperature increases. This behavior is probably due to the increase of the energy transfer from the Gd-O linkages to the $[EuO_8]^\bullet$ clusters. All these results are in accordance to the emission and excitation measurements.

4. Conclusion

In summary, the obtained results showed that the $Gd(OH)_3:Eu^{3+}$ (precursor) was synthesized by the microwave assisted hydrothermal method in a short period of time (30 minutes). After heated treated from 500°C to 700°C,

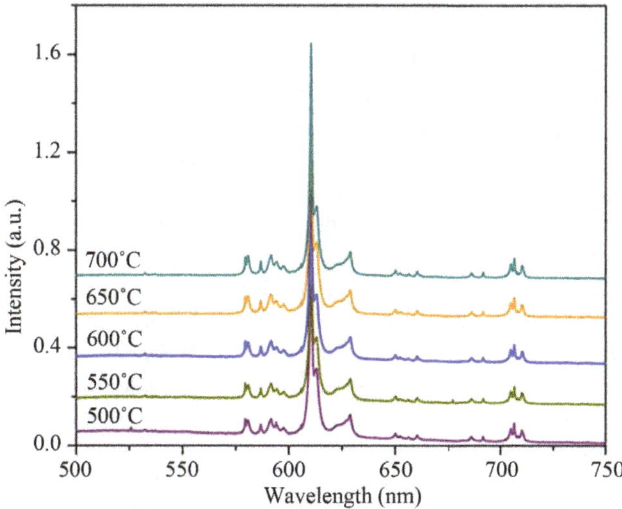

Figure 10. Emission spectra of $Gd_2O_3:Eu^{3+}$ samples calcined at 500°C, 550°C, 600°C, 650°C and 700°C, λ_{ex} = 263 nm.

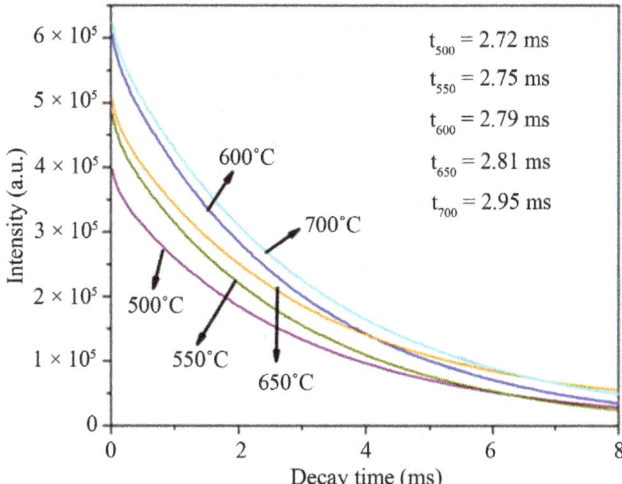

Figure 11. Decay curves and lifetime of the $^5D_0 \rightarrow {}^7F_2$ transition characteristic of the Eu^{3+} of the $Gd_2O_3:Eu^{3+}$ nanorods heat treated at 500°C, 550°C, 600°C, 650°C and 700°C (λ_{ex} = 612 nm and λ_{em} = 612 nm).

the XRD patterns and Rietveld refinement and FT-IR analyses indicated the formation of $Gd_2O_3:Eu^{3+}$ powders which crystallizes in a cubic structure of crystalline Gd_2O_3 and space group Ia-3. No secondary phases related to the Eu^{3+} ions were detected indicating that these ions were incorporated to the hydroxide and oxide matrixes in the analyzed powders. FE-SEM images indicated that the $Gd(OH)_3:Eu^{3+}$ precursor and $Gd_2O_3:Eu^{3+}$ powders are composed by several aggregated particles with nanorods-like morphology, which sizes are in the range of 8 and 20 nm. Eu^{3+} emission and excitation spectra pointed out that the emission in the $Gd_2O_3:Eu^{3+}$ powders comes from the energy transfer from the Gd-O and $[EuO_8]^{\bullet}$ clusters in the crystalline structure. Moreover, these are in accordance to the lifetime values, which presented an increase as the temperature increases. This method is very simple and effective, and can be extended to synthesize some other rare earth and metal oxide nanorods.

Acknowledgements

The authors acknowledge the financial support of the Brazilian research financing institutions: CNPq (INCTMN), CAPES and FAPESP (CEPID). A special thanks for Maria Fernanda Cgnin de Abreu.

References

[1] Iijima, S. (1991) Synthesis of Carbon Nanotubes. *Nature*, **354**, 56-58. http://dx.doi.org/10.1038/354056a0

[2] Ajayan, P.M. (1999) Nanotubes from Carbon. *Chemical Reviews*, **99**, 1787-1800. http://dx.doi.org/10.1021/cr970102g

[3] Hu, J.T., Odom, T.W. and Lieber, C.M. (1999) Chemistry and Physics in one Dimension: Synthesis and Properties of Nanowires and Nanotubes. *Accounts of Chemical Research*, **32**, 435-445. http://dx.doi.org/10.1021/ar9700365

[4] Xia, Y.N., Yang, P.D., Sun, Y.G., Wu, Y.Y., Mayers, B., Gates, B., Yin, Y.D., Kim, F. and Yan, H.Q. (2003) One-Dimensional Nanostructures: Synthesis, Characterization, and Applications. *Advanced Materials*, **15**, 353-389. http://dx.doi.org/10.1002/adma.200390087

[5] Rao, C.N.R., Deepak, F.L., Gundiah, G. and Govindaraj, A. (2003) Inorganic Nanowires. *Progress in Solid State Chemistry*, **31**, 5-147. http://dx.doi.org/10.1016/j.progsolidstchem.2003.08.001

[6] Huang, M.H., Mao, S., Feick, H., Yan, H.Q., Wu, Y.Y., Kind, H., Weber, E., Russo, R., Yang, P.D. (2001) Room-Temperature Ultraviolet Nanowire Nanolasers. *Science*, **292**, 1897-1899. http://dx.doi.org/10.1126/science.1060367

[7] Pan, Z.W., Dai, Z.R. and Wang, Z.L. (2001) Nanobelts of Semiconducting Oxides. *Science*, **291**, 1947-1949. http://dx.doi.org/10.1126/science.1058120

[8] Shi, W.S., Peng, H., Wang, N., Li, C.P., Xu, L., Lee, C.S., Kalish, R. and Lee, S.T. (2001) Free-Standing Single Crystal Silicon Nanoribbons. *Journal of the American Chemical Society*, **123**, 11095-11096. http://dx.doi.org/10.1021/ja0162966

[9] Hu, J., Odom, T.W. and Lieber, C.M. (1999) Chemistry and Physics in One Dimension: Synthesis and Properties of Nanowires and Nanotubes. *Accounts of Chemical Research*, **32**, 435-445. http://dx.doi.org/10.1021/ar9700365

[10] Kazes, M., Lewis, D.Y., Ebenstein, Y., Mokari, T. and Banin, U. (2002) Lasing from Semiconductor Quantum Rods in a Cylindrical Microcavity. *Advanced Materials*, **14**, 317-321. http://dx.doi.org/10.1002/1521-4095(20020219)14:4<317::AID-ADMA317>3.0.CO;2-U

[11] Lee, K.-H., Bae, Y.-J. and Byeon, S.-H. (2008) pH Dependent Hydrothermal Synthesis and Photoluminescence of Gd_2O_3:Eu Nanostructures. *Bulletin of the Korean Chemical Society*, **29**, 2161-2168. http://dx.doi.org/10.5012/bkcs.2008.29.11.2161

[12] Ropp, R.C. (1993) The Chemistry of Artificial Lighting Devices: Lamps, Phosphors, and Cathode Ray Tubes. Elsevier, New York.

[13] Blasse, G. and Grabmaier, B.C. (1994) Luminescent Materials. Springer, New York. http://dx.doi.org/10.1007/978-3-642-79017-1

[14] Wan, J., Wang, Z., Chen, X., Mu, L. and Qian, Y. (2005) Shape-Tailored Photoluminescent Intensity of Red Phosphor $Y_2O_3:Eu^{3+}$. *Journal of Crystal Growth*, **284**, 538-543. http://dx.doi.org/10.1016/j.jcrysgro.2005.07.040

[15] Xu, G.X. and Xiao, J.M. (1985) New Frontiers Rare Earth Science and Application. Academic Press, New York.

[16] Cuif, J.P., Rohart, E., Macaudiere, P., Bauregard, C., Suda, E., Pacaud, B., Imanaka, N., Masui, T. and Tamura, S. (2004) Binary Rare Earth Oxides. Kluwer Academic Publishers, Dordrecht.

[17] Bae, Y.J., Lee, K.H. and Byeon, S.H. (2009) Synthesis and Eu^{3+} Concentration-Dependent Photoluminescence of $Gd_{2-x}Eu_xO_3$ Nanowires. *Journal of Luminescence*, **129**, 81-85. http://dx.doi.org/10.1016/j.jlumin.2008.08.004

[18] Beaurepaire, E., Buissette, V., Sauviat, M.P., Mercuri, A., Martin, J.L., Lahlil, K., Aume, D., Huignard, A., Gacoin, T., Boilot, J.P. and Alexandrou, A. (2004) Functionalized Fluorescent Oxide Nanoparticles: Artificial Toxins for Sodium Channel Targeting and Imaging at the Single-Molecule Level. *Nano Letters*, **4**, 2079-2083. http://dx.doi.org/10.1021/nl049105g

[19] Louis, C., Bazzi, R., Marquette, C.A., Bridot, J.L., Roux, S., Ledoux, G., Mercier, B., Blum, L., Perriat, P. and Tillement, O. (2005) Nanosized Hybrid Particles with Double Luminescence for Biological Labeling. *Chemistry of Materials*, **17**, 1673-1682. http://dx.doi.org/10.1021/cm0480162

[20] Nichkova, M., Dosev, D., Gee, S.J., Hammock, B.D. and Kennedy, I.M. (2005) Quantum Dots as Reporters in Multiplexed Immunoassays for Biomarkers of Exposure to Agrochemicals. *Analytical Letters*, **40**, 1423-1433.

[21] Goldys, E.M., Tomsia, K.D., Jinjun, S., Dosev, D., Kennedy, I.M., Yatsunenko, S. and Godlewski, M. (2006) Optical Characterization of Eu-Doped and Undoped Gd_2O_3 Nanoparticles Synthesized by the Hydrogen Flame Pyrolysis Method. *Journal of the American Chemical Society*, **128**, 14498-14505. http://dx.doi.org/10.1021/ja0621602

[22] Zhou, Y., Lin, J. and Wang, S. (2003) Energy Transfer and Up-Conversion Luminescence Properties of Y_2O_3:Sm and Gd_2O_3:Sm Phosphors. *Journal of Solid State Chemistry*, **171**, 391-395. http://dx.doi.org/10.1016/S0022-4596(02)00219-0

[23] Rossner, W. and Grabmaier, B.C. (1991) Phosphors for X-Ray Detectors in Computed Tomography. *Journal of Luminescence*, **48-49**, 29-36. http://dx.doi.org/10.1016/0022-2313(91)90072-4

[24] Guo, C., Chen, T., Luan, L., Zhang, W. and Huang, D. (2008) Luminescent Properties of $R_2(MoO_4)_3:Eu^{3+}$ (R = La, Y, Gd) Phosphors Prepared by Sol-Gel Process. *Journal of Physics and Chemistry of Solids*, **69**, 1905-1911. http://dx.doi.org/10.1016/j.jpcs.2008.01.021

[25] Pereira, P.F.S., de Moura, A.P., Nogueira, I.C., Lima, M.V.S., Longo, E., de Sousa Filho, P.C., Serra, O.A., Nassar, E.J. and Rosa, I.L.V. (2012) Study of the Annealing Temperature Effect on the Structural and Luminescent Properties of $SrWO_4$:Eu Phosphors Prepared by a Non-Hydrolytic Sol-Gel Process. *Journal of Alloys and Compounds*, **526**, 11-21. http://dx.doi.org/10.1016/j.jallcom.2012.02.083

[26] Rosa, I.L.V., Oliveira, L.H., Suzuki, C.K., Varela, J.A., Leite, E.R. and Longo, E. (2008) SiO_2-GeO_2 Soot Perform as a Core for Eu_2O_3 Nanocoating: Synthesis and Photophysical Study. *Journal of Fluorescence*, **18**, 541-545. http://dx.doi.org/10.1007/s10895-007-0297-7

[27] Morais, E.A., Scalvi, L.V.A., Tabata, A., De Oliveira, J.B.B. and Ribeiro, S.J.L. (2008) Photoluminescence of Eu^{3+} Ion in SnO_2 Obtained by Sol-Gel. *Journal of Materials Science*, **43**, 345-349. http://dx.doi.org/10.1007/s10853-007-1610-1

[28] Marques, A.P.A., Tanaka, M.T.S., Longo, E., Leite, E.R. and Rosa, I.L.V. (2011) The Role of the Eu^{3+} Concentration on the $SrMoO_4$:Eu Phosphor Properties: Synthesis, Characterization and Photophysical Studies. *Journal of Fluorescence*, **21**, 893-899. http://dx.doi.org/10.1007/s10895-010-0604-6

[29] Yan, M.F., Huo, T.C.D. and Ling, H.C.J. (1987) Preparation of $Y_3Al_5O_{12}$-Based Phosphor Powders. *Journal of the Electrochemical Society*, **134**, 493-498. http://dx.doi.org/10.1149/1.2100487

[30] Shea, L.E., McKittrick, J., Lopez, O.A. and Sluzky, E. (1996) Synthesis of Red-Emitting, Small Particle Size Luminescent Oxides Using an Optimized Combustion Process. *Journal of the American Ceramic Society*, **79**, 3257-3265. http://dx.doi.org/10.1111/j.1151-2916.1996.tb08103.x

[31] Ravichandran, D., Roy, R., White, W.B. and Erdei, S. (1997) Synthesis and Characterization of Sol-Gel Derived Hexa-Aluminate Phosphor. *Journal of Materials Research*, **12**, 819-824. http://dx.doi.org/10.1557/JMR.1997.0119

[32] Erdei, S., Roy, R., Harshe, G., Juwhari, S., Agrawal, H.D., Ainger, F.W. and White, W.B. (1995) The Effect of Powder Preparation Processes on the Luminescent Properties of Yttrium Oxide Based Phosphor Materials. *Materials Research Bulletin*, **30**, 745-753. http://dx.doi.org/10.1016/0025-5408(95)00052-6

[33] Santos, M.L., Lima, R.C., Riccardi, C.S., Tranquilin, R.L., Bueno, P.R., Varela, J.A. and Longo, E. (2008) Preparation and Characterization of Ceria Nanospheres by Microwave-Hydrothermal Method. *Materials Letters*, **62**, 4509-4511. http://dx.doi.org/10.1016/j.matlet.2008.08.011

[34] Lima, R.C., Macario, L.R., Espinosa, J.W.M., Longo, V.M., Erlo, R., Marana, N.L., Sambrano, J.R., Santos, M.L.D., Moura, A.P., Pizani, P.S., Andres, J., Longo, E. and Varela, J.A. (2008) Toward an Understanding of Intermediate- and Short-Range Defects in ZnO Single Crystals. A Combined Experimental and Theoretical Study. *The Journal of Physical Chemistry A*, **112**, 8970-8978. http://dx.doi.org/10.1021/jp8022474

[35] Moura, A.P., Cavalcante, L.S., Sczancoski, J.C., Stroppa, D.G., Paris, E.C., Ramirez, A.J., Varela, J.A. and Longo, E. (2010) Structure and Growth Mechanism of CuO Plates Obtained by Microwave-Hydrothermal without Surfactants. *Advanced Powder Technology*, **21**, 197-202. http://dx.doi.org/10.1016/j.apt.2009.11.007

[36] de Moura, A.P., Lima, R.C., Moreira, M.L., Volanti, D.P., Espinosa, J.W.M., Orlandi, M.O., Pizani, P.S., Varela, J.A. and Longo, E. (2010) ZnO Architectures Synthesized by a Microwave-Assisted Hydrothermal Method and Their Photoluminescence Properties. *Solid State Ionics*, **181**, 775-780. http://dx.doi.org/10.1016/j.ssi.2010.03.013

[37] Motta, F.V., Lima, R.C., Marques, A.P.A., Li, M.S., Leite, E.R., Varela, J.A. and Longo, E. (2010) Indium Hydroxide Nanocubes and Microcubes Obtained by Microwave-Assisted Hydrothermal Method. *Journal of Alloys and Compounds*, **497**, L25-L28. http://dx.doi.org/10.1016/j.jallcom.2010.03.069

[38] de Moura, A.P., Lima, R.C., Paris, E.C., Li, M.S., Varela, J.A. and Longo, E. (2011) Formation of β-Nickel Hydroxide Plate-Like Structures under Mild Conditions and Their Optical Properties. *Journal of Solid State Chemistry*, **184**, 2818-2823.

[39] Bohr, H. and Bohr, J. (2000) Microwave-Enhanced Folding and Denaturation of Globular Proteins. *Physical Review E*, **61**, 4310-4314. http://dx.doi.org/10.1103/PhysRevE.61.4310

[40] Blanco, C. and Auerbach, S.M. (2002) Microwave-Driven Zeolite-Guest Systems Show Athermal Effects from Nonequilibrium Molecular Dynamics. *Journal of the American Chemical Society*, **124**, 6250-6251. http://dx.doi.org/10.1021/ja017839e

[41] Favretto, L., Nugent, W.A. and Licini, G. (2002) Highly Regioselective Microwave-Assisted Synthesis of Enantiopure C_3-Symmetric Trialkanolamines. *Tetrahedron Letters*, **43**, 2581-2584. http://dx.doi.org/10.1016/S0040-4039(02)00306-4

[42] Hoz, A.D.L., Diaz-Ortiz, A. and Moreno, A. (2004) Selectivity in Organic Synthesis under Microwave Irradiation. *Current Organic Chemistry*, **8**, 903-918. http://dx.doi.org/10.2174/1385272043370429

[43] Bren, M., Janežič, D. and Bren, U. (2010) Microwave Catalysis Revisited: An Analytical Solution. *The Journal of Phy-*

sical Chemistry A, **114**, 4197-4202. http://dx.doi.org/10.1021/jp100374x

[44] Sun, L.D., Yao, J., Liu, C., Liao, C. and Yan, C.H. (2000) Rare Earth Activated Nanosized Oxide Phosphors: Synthesis and Optical Properties. *Journal of Luminescence*, **87-89**, 447-450. http://dx.doi.org/10.1016/S0022-2313(99)00471-8

[45] Kappe, C.O., Stadler, A. and Dallinger, D. (2012) Microwaves in Organic and Medicinal Chemistry. 2nd Edition, Vol. 52, Wiley-VCH, Weinheim. http://dx.doi.org/10.1002/9783527647828

[46] Obermayer, D., Gutmann, B. and Kappe, C.O. (2009) Microwave Chemistry in Silicon Carbide Reaction Vials: Separating Thermal from Nonthermal Effect. *Angewandte Chemie International Edition*, **48**, 8321-8324. http://dx.doi.org/10.1002/anie.200904185

[47] Yao, B.D. and Wang, N. (2001) Carbon Nanotube Arrays Prepared by MWCVD. *The Journal of Physical Chemistry B*, **105**, 11395-11398. http://dx.doi.org/10.1021/jp011849k

[48] Zhu, Y.J., Wang, W.W., Qi, R.J. and Hu, X.L. (2004) Microwave-Assisted Synthesis of Single-Crystalline Tellurium Nanorods and Nanowires in Ionic Liquids. *Angewandte Chemie International Edition*, **43**, 1410-1414.

[49] Tompsett, G.A., Conner, W.C. and Yngvesson, K.S. (2006) Microwave Synthesis of Nanoporous Materials. *ChemPhysChem*, **7**, 296-319. http://dx.doi.org/10.1002/cphc.200500449

[50] de Moura, A.P., de Oliveira, L.H., Paris, E.C., Li, M.S., Andrés, J., Varela, J.A., Longo, E. and Rosa, I.L.V. (2011) Photolumiscent Properties of Nanorods and Nanoplates Y_2O_3:Eu^{3+}. *Journal of Fluorescence*, **21**, 1431-1438. http://dx.doi.org/10.1007/s10895-010-0827-6

[51] Chang, C. and Mao, D. (2007) Thermal Dehydration Kinetics of a Rare Earth Hydroxide, $Gd(OH)_3$. *International Journal of Chemical Kinetics*, **39**, 75-81. http://dx.doi.org/10.1002/kin.20221

[52] Rietveld, H.M. (1969) A Profile Refinement Method for Nuclear and Magnetic Structures. *Journal of Applied Crystallography*, **2**, 65-71. http://dx.doi.org/10.1107/S0021889869006558

[53] Larson, C.A. and Von Dreele, R.B. (2001) The Regents of the University of California, Copyright 1985-2000, Los Alamos National Laboratory, Los Alamos, EUA.

[54] Thompson, P., Cox, D.E. and Hastings, J.B. (1987) Rietveld Refinement of Debye-Scherrer Synchrontron X-Ray Data from Al_2O_3. *Journal of Applied Crystallography*, **20**, 79-83. http://dx.doi.org/10.1107/S0021889887087090

[55] Finger, L.W., Cox, D.E. and Jephcoat, A.P. (1994) A Correction for Powder Diffraction Peak Asymmetry Due to Axial Divergence. *Journal of Applied Crystallography*, **27**, 892-900. http://dx.doi.org/10.1107/S0021889894004218

[56] Stephens, P.W. (1999) Phenomenological Model of Anisotropic Peak Broadening in Powder Diffraction. *Journal of Applied Crystallography*, **32**, 281-289. http://dx.doi.org/10.1107/S0021889898006001

[57] Momma, K. and Izumi, F. (2008) VESTA: A Three-Dimensional Visualization System for Electronic and Structural Analysis. *Journal of Applied Crystallography*, **41**, 653-658. http://dx.doi.org/10.1107/S0021889808012016

[58] Buijs, M., Meyerink, A. and Blasse, G. (1987) Energy Transfer between Eu^{3+} Ions in a Lattice with Two Different Crystallographic Sites: Y_2O_3:Eu^{3+}, Gd_2O_3:Eu^{3+} and Eu_2O_3. *Journal of Luminescence*, **37**, 9-20. http://dx.doi.org/10.1016/0022-2313(87)90177-3

[59] Kevorkov, A.M., Karyagin, V.F., Munchaev, A.I., Uyukin, E.M., Bolotina, N.B., Chernaya, T.S., Bagdasarov, K.S. and Simonov, V.I. (1995) Y_2O_3 Single Crystals: Growth, Structure and Photoinduced Effects. *Crystallography Reports*, **40**, 23.

[60] Godinho, M., Ribeiro, C., Longo, E. and Leite, E.R. (2008) Influence of Microwave Heating on the Growth of Gadolinium-Doped Cerium Oxide Nanorods. *Crystal Growth Design*, **8**, 384-386. http://dx.doi.org/10.1021/cg700872b

[61] Baraldi, P. and Davolio, G. (1989) An Electrochemical and Spectral Study of the Nickel Oxide Electrode. *Materials Chemistry and Physics*, **21**, 143-154. http://dx.doi.org/10.1016/0254-0584(89)90109-0

[62] Guo, H., Yang, X., Xiao, T., Zhang, W., Lou, L. and Mugnier, J. (2004) Structure and Optical Properties of Sol-Gel Derived Gd_2O_3 Waveguide Films. *Applied Surface Science*, **230**, 215-221. http://dx.doi.org/10.1016/j.apsusc.2004.02.032

[63] Jayasimhadri, M., Ratnam, B.V., Jang, K., Lee, H.S., Chen, B., Yi, S.S., Jeong, J.H. and Moorthy, L.R. (2011) Combustion Synthesis and Luminescent Properties of Nano and Submicrometer-Size Gd_2O_3:Dy^{3+} Phosphors for White LEDs. *International Journal of Applied Ceramic Technology*, **8**, 709-717. http://dx.doi.org/10.1111/j.1744-7402.2010.02499.x

[64] Liu, G., Hong, G., Wang, J. and Dong, X. (2007) Hydrothermal Synthesis of Spherical and Hollow Gd_2O_3:Eu^{3+} Phosphors. *Journal of Alloys and Compounds*, **432**, 200-204. http://dx.doi.org/10.1016/j.jallcom.2006.05.127

[65] Liu, G., Hong, G., Dong, X. and Wang, J. (2008) Preparation and Characterization of Gd_2O_3:Eu^{3+} Luminescence Nanotubes. *Journal of Alloys and Compounds*, **466**, 512-516. http://dx.doi.org/10.1016/j.jallcom.2007.11.108

[66] Schmechel, R., Kennedy, M., von Seggerm, H., Winkler, H., Kolbe, M., Fischer, R.A., *et al.* (2001) Luminescence

Properties of Nanocrystalline $Y_2O_3:Eu^{3+}$ in Different Host Materials. *Journal of Applied Physics*, **89**, 1679-1686. http://dx.doi.org/10.1063/1.1333033

[67] Pang, M.L., Lin, J., Fu, J., Xing, R.B., Luo, C.X. and Han, Y.C. (2003) Preparation, Patterning and Luminescent Properties of Nanocrystalline Gd_2O_3:A (A = Eu^{3+}, Dy^{3+}, Sm^{3+}, Er^{3+}) Phosphor Films via Pechini Sol-Gel Soft Lithography. *Optical Materials*, **23**, 547-558. http://dx.doi.org/10.1016/S0925-3467(03)00020-X

[68] Teotonio, E.E.S., Felinto, M.C.F.C., Brito, H.F., Malta, O.L., Najjar, A.C.R. and Strek, W. (2004) Synthesis, Crystalline Structure and Photoluminescence Investigations of the New Trivalent Rare Earth Complexes (Sm^{3+}, Eu^{3+} and Tb^{3+}) Containing 2-Thiophenecarboxylate as Sensitizer. *Inorganica Chimica Acta*, **357**, 451-460. http://dx.doi.org/10.1016/j.ica.2003.08.009

[69] Rosa, I.L.V., Maciel, A.P., Longo, E., Leite, E.R. and Varela, J.A. (2006) Synthesis and Photoluminescence Study of $La_{1.8}Eu_{0.2}O_3$ Coating on Nanometric α-Al_2O_3. *Materials Research Bulletin*, **41**, 1791-1797. http://dx.doi.org/10.1016/j.materresbull.2006.03.026

2

Enhancing Production Efficiency of Oil and Natural Gas Pipes Using Microwave Technology

Wissam M. Alobaidi[1*], Entidhar A. Alkuam[2], Eric Sandgren[1], Hussain M. Al-Rizzo[1]

[1]Systems Engineering Department, Donaghey College of Engineering & Information Technology, University of Arkansas at Little Rock, Little Rock, Arkansas, USA

[2]Department of Physics and Astronomy, College of Arts, Letters, and Sciences, University of Arkansas at Little Rock, Little Rock, Arkansas, USA

Email: [*]wmalobaidi@ualr.edu

Abstract

The research reported in this paper aims at developing means of Non Destructive testing (NDT) to increase the line efficiency of pipe production in oil and natural gas pipe manufacturing plants using the Standard Allowed Minutes (SAM) method. Existing line production stations encounter difficulties in maintaining the recommended testing speed of smaller diameter pipe, due to limitations in the Visual Inspection (VI) station. We propose to implement one additional technique which will prevent the decline of line efficiency in a pipe production factory. The range of diameters identified as a problem in this research is from 254 mm to 762 mm. Microwave techniques are expected to improve the line efficiency by increasing the production of the plant. This happens as a consequence of maintaining the production rates of the identified pipe diameters, so that they equal the production output of the larger pipe diameters. We analyze the velocity traveled by the pipe through Radiographic Testing (RT) according to the VI output (production). The RT velocity is decreased for the diameters identified above, in order to maintain quality control and cover the shortcoming of the VI. The number of pipes produced is computed during shift hours of the factory and pipe lengths of the forming department are determined. We compare the output (production) of a series of NDT line stations with and without the microwave technique for the first of the three pipe cases considered in this study, classified as perfect pipe (PP), repair pipe (RP) and scrap pipe (SP). The velocity of RT stations analyzed in the paper ranges from 50 mm/s for larger diameter pipe, and decline to 16.667 mm/s for the identified diameters. The analytical calculations of line output (production) and line efficiency demonstrate the solution of this velocity problem after the microwave technique is introduced. It demonstrates that an economical and precise methodology to extend the production capability of the pipe plant has been determined.

[*]Corresponding author.

Keywords

Microwave Technology, Non Destructive Testing (NDT), Spiral Pipe Process (SPP), Pipe Inspection, Standard Allowed Minutes (SAM)

1. Introduction

Oil pipes with diameters up to 500 mm are used as the standard method of transporting petroleum and derivatives or natural gas for great distances overland. Corrosion of the inner pipe wall often occurs at the bottom of the pipe in the field because of the presence of water in the pipelines [1]. This corrosion leads to removal of material, creating a reduction in the wall of the pipe, called Pipe Wall Reduction (PWR) [2] [3]. Defects develop even during the manufacturing process, but these are often caught in factory due to standard testing guidelines. Such discontinuities can be assessed and repaired per the standards and codes, before the pipe is shipped to the field. During its time in service, any pipeline will develop discontinuities that are of a greater extent than those show up in manufacturing. Once again, there are standards that can be used to determine whether repair or replacement is the proper action on a case by case basis [4] [5].

Seeking a solution for determination of whether a field-defective pipeline is still suitable for service has inspired a great deal of research. A number of analytical approaches have been tested for detection and sizing of such defects in situ [6].

NDT techniques have been developed during the past few decades to determine the presence and extent of such discontinuities. Our previous survey paper reviewing the application of several of these NDT methods looked at digital X-ray, ultrasonic, visual inspection, radiographic testing and so on [5]. Detection of surface defects requires a large amount of time and labor for each of these techniques. For larger diameter pipes the techniques can be used effectively, but smaller diameter pipelines prevent entry of human inspectors to perform visual inspection, for example [5].

Previous studies have demonstrated that microwave NDT techniques can reliably detect surface defects in pipes. The technique uses a microwave emitter/detector that introduces a microwave signal into the waveguide (the pipe itself) and detects the reflected waves [3] [7]. Normal pipe walls generate a constant resonant frequency (no loss in energy) in the reflected waves. PWR creates variances in the waveforms produced, allowing calculation of the location and extent of PWR defects. Inspecting the pipe inner walls can be done without entering the pipe, and without moving the detector during the test, while detection proceeds in real time. This has obvious advantages in the case of smaller diameter pipeline [3].

After the advantages of microwave NDT techniques become widely known, we foresee the inclusion of microwave testing stations among the series of NDT testing stations in pipe manufacturing facilities.

Standard Allowed Minutes (SAM) was used to establish a general plan for the target (line output and line efficiency standard) and was developed by analysis of the time required to produce the product [8] [9].

Many of the techniques listed in our survey paper work are well with larger diameter pipes, but are difficult or impossible to use on smaller diameter pipes [5].

We present an analytical study of inspection routines including Visual Inspection (VI) and Radiographic Testing (RT) with and without the microwave technique (which will be placed between VI and RT) for smaller diameter pipe, and then study the case of the PP with the SAM for the two routines.

It is worth mentioning that this paper essentially aims to enhance the production and line efficiency of oil and gas pipes by using microwave technology. The observational study employs microwave technology to carry out the testing of the inner surface of the pipe to meet the VI requirement. The commonly used VI station can carry out the test for all diameters except the group ranging in size from 254 mm to 762 mm, which cannot be inspected from the inside with VI.

This limitation requires slowing the velocity of the test pipe in the RT station, as mentioned in our previous study, from 50 mm/s to 16.667 mm/s [5]. Placing microwave technology between the two stations has three advantages. First, the velocity of the RT station up to the customer requirement is got. Second, to inspect the pipe from inside for those smaller diameters listed above. Consequently, the pipe can be released to the RT station without interruption, resulting in improved line efficiency and line output of the product. Third, the high quality of the product during the test is ensured.

This research will confirm the times needed for examination of the pipe after its formation for the diameter range listed above. Then we will compute the line production values to yield line efficiency for each station, by analyzing the necessary decrease in speed of the wagon that carries the pipe to conduct the ultrasonic test for the weld before the radiography test. Calculations will be performed for various lengths of pipe: 24,384 mm, 21,336 mm, 18,288 mm and 15,240 mm respectively. Line production and line efficiency will be computed for various shift lengths: 8 hrs, 9 hrs, 10 hrs, 11 hrs and 12 hrs. After adding microwave technology the computations will be repeated for the new setup. Results for the previous sequencing of stations after VI will be compared to the new sequence including the microwave technique which evaluates the pipe from inside.

2. Sequential Stations of Non-Destructive Examination in Spiral Pipe Plants

- **Coil Ultrasonic Testing (COIL UT):** This station is responsible for detecting discontinuities in the materials of the pipe body before it is spirally formed. The station uses straight beam ultrasonic signals to test the roll stock, as shown in **Figure 1**. COIL UT allows identification of discontinuities with enough precision to allow the ones that exceed the standards to be cut away before the pipe is formed. In most cases the manufacturer specifications require a clean run of at least 40 feet up to 80 feet, for the roll stock to be usable [10] [11].
- **Visual Inspection (VI):** After the formation of the pipe and the application of spiral welds internally and externally, the pipe is sent to the VI station to evaluate the pipe body and its welds for outer diameter (OD) and inner diameter (ID) [5] [12].
- **Radiography Testing (RT):** Inspection in this station is confined to the spiral weld only, from the front end to the tail end of the pipe. This station contains two examination methods, as shown in **Figure 1**. First is the Ultrasonic machine to give an indication of the manpower needed. The examination process follows the weld by spiral rotational movement of the pipe, to ensure the probe is placed reliably to provide indication of any discontinuity for the ID and OD (two separate groups of probes are used to test the ID and OD of the weld simultaneously). The second is the RT machine which is used when UT indicates that follow-up inspection is necessary [5] [13].
- **Hydraulic System (HS):** This is a system to inject water inside the pipe to create the amount of internal pressure called for by the requirements and standards code. Pipes are checked for any leakage [14].
- **Seam Ultrasonic Testing (SEAM UT):** This machine is limited to the main section of the weld. Probes are used to test the ID and OD of the weld, beginning 200 mm from the front end, and ending at 200 mm from

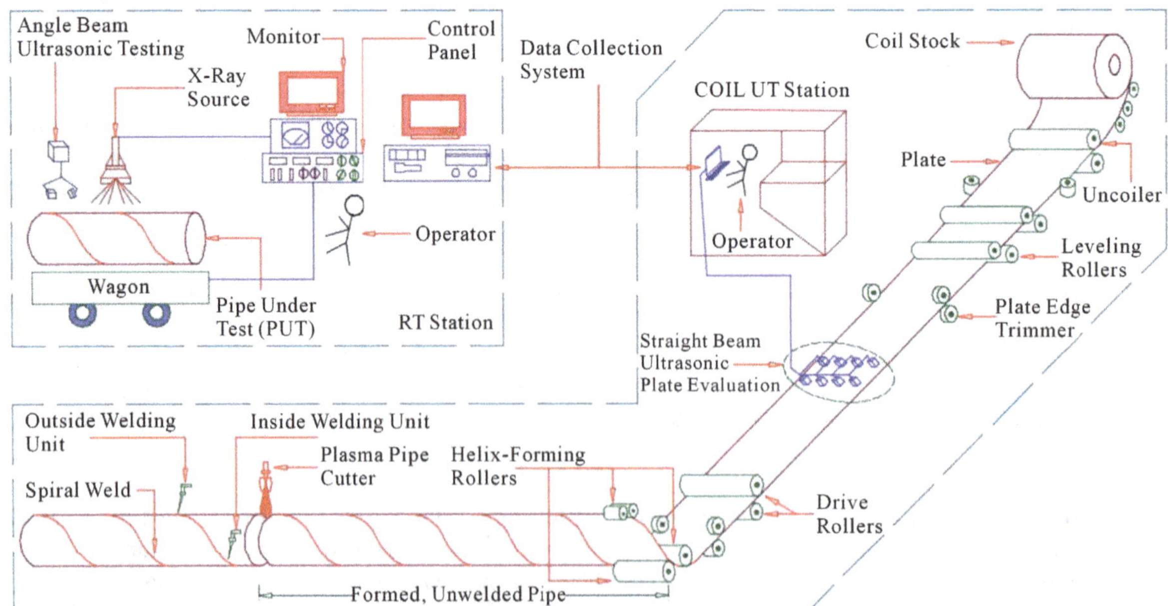

Figure 1. Illustration of oil and natural gas pipe manufacturing process.

the tail end (to protect the probes from damage due to contact with the flanges of the pipe). The pipe is moved helically for scanning. This test paints a red line on any section of the weld that shows a discontinuity. The red lines indicate to the NDT person which areas must be tested at the next station [11] [15].

- **Manual Ultrasonic Testing (MUT):** This testing station includes two checks of the pipe. First is the straight beam probe. This is used to check the body of the pipe circumferentially from the outside within a 254 mm linear distance from both front end and tail end (to cover the pipe body in the areas not covered by SEAM UT). Second is the angle beam probe. This checks the spiral weld at both front end and tail end, within a 254 mm linear distance. The angle beam probe is then used to check any indication from the SEAM UT (red painted areas) of a discontinuity along the spiral weld [11] [16].
- **Digital X-Ray (D-XR):** This last station is for verifying the places that have been pointed out by MUT. After this test, pipes with defects needing repair (RP) are sent to the repair station. Scrap pipe (SP) go to the scrap area. Perfect pipe (PP) goes to the beveling station, then to final inspection [5] [17].
- **Microwave Inspection:** High-frequency Weld (HFW) or Electrical Resistance Weld (ERW) small diameter steel pipe is sometimes tested with microwave signals. This kind of pipe is useful in situations with liquids under reduced pressure, such as the transfer of water, oil, and in equipment manufacturing [18] [19]. All of the research done on microwave as an NDT technique was done with smooth inner wall pipe of this type [3] [7]. We propose to incorporate the microwave technique to test spiral weld pipe to reduce the test times spent in the stations following visual inspection (where the pipe cannot be entered by manpower for direct visual inspection, due to small inner diameter). **Figure 2** shows the schematic of the microwave process.

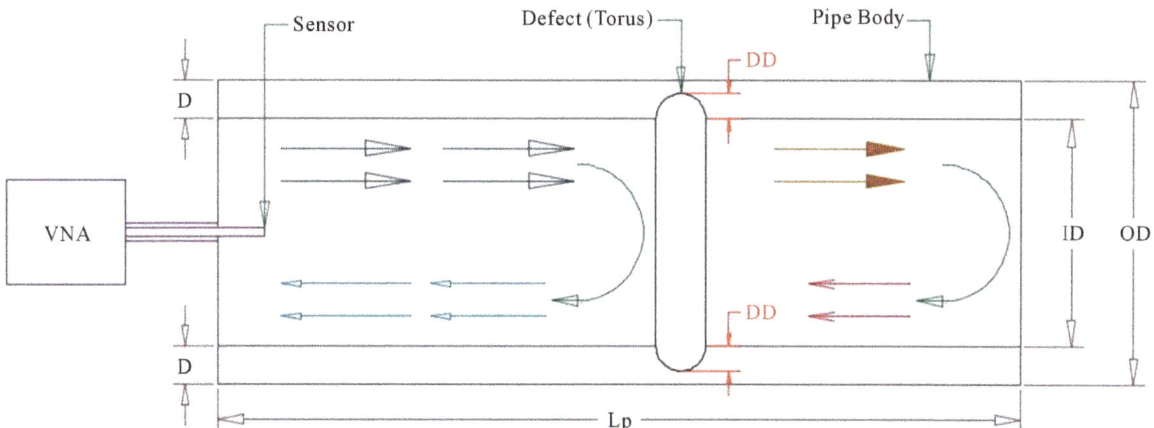

Figure 2. Illustration of a typical setup for microwave propagation for NDT of small diameter pipe. VNA is a vector network analyzer, L_p is the length of the pipe, D is the material thickness and DD is depth of the defect.

3. Theoretical Analysis

3.1. Confirming the Time Inspection of the VI and RT Stations for PP Case

In manufacturing plants is always considered the perfect product for setting up the output line charts daily, weekly and monthly [20]. To customize the standard timing, when pipes containing defects are found, the material is divided into three cases: Perfect Product (PP), Repair Product (RP) and Scrap Product (SP). Furthermore, the correct timing of the product is always based on the product free of defects. This concept is extensively used in various industries [5]. As shown in **Figure 3**, the path overview of the series of NDT stations for evaluating pipes in manufacturing and testing commences with visual inspection and ends at the production area. For this evaluation, we apply the concept of SAM to pipe manufacturing [21] [22]. It is worth mentioning that, while the time the PP spends in the NDT area is not exactly a SAM, it is considered as an exact time to conduct the partial test and full test as shown in **Figure 4**, in order to inspect the pipe in the VI station. The pipes are split into two groups based on diameter: the first group (G1) consists of pipes ranging from 254 to 762 mm in diameter, and the second group (G2) ranging from 889 mm to 1270 mm in diameter. Generally, these two groups will control the increase and decrease in the line output for each station. Thereby, line efficiency (LE) is different for G1 and G2.

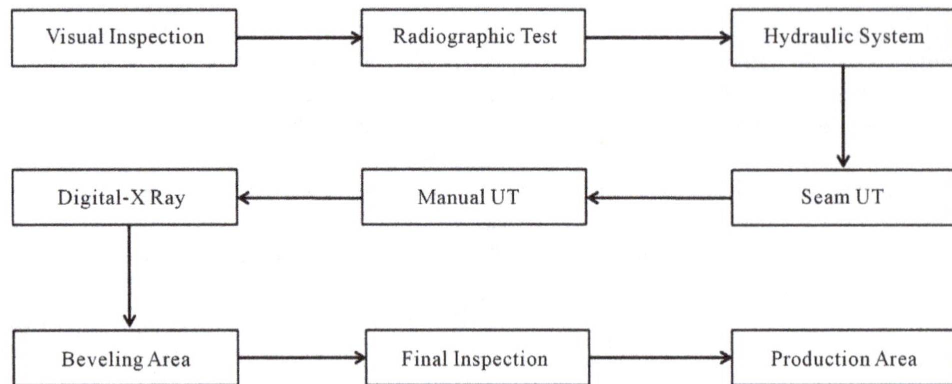

Figure 3. Path overview of NDT stations, used in the evaluation.

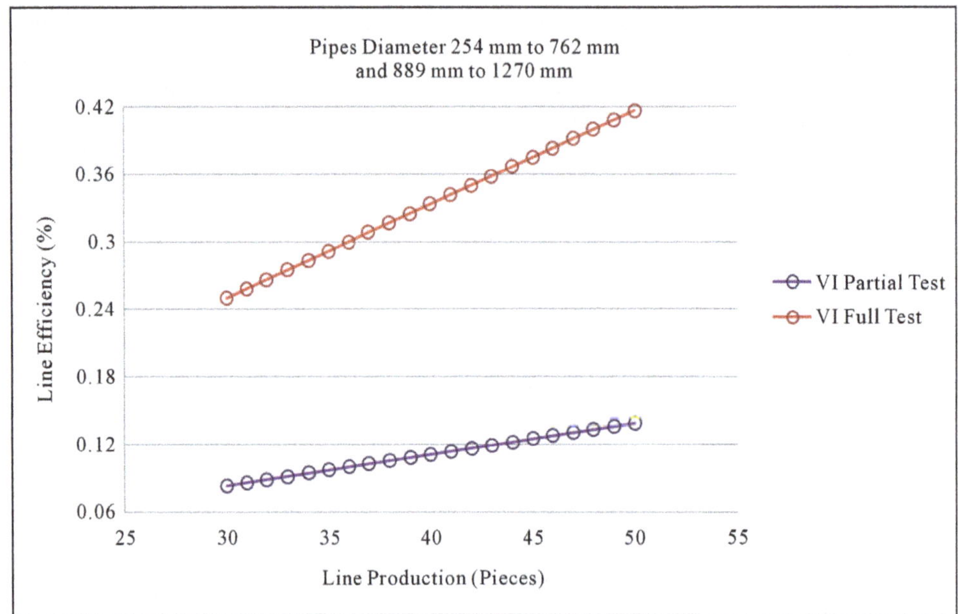

Figure 4. Evaluated times for VI station as per the G1 & G2, with pipe length 24,384 mm, number of operator 3 and shift work 8 hrs.

The three equations below are excerpted from [21].

$$\text{Line Efficiency} = \frac{\text{Total Minutes Produced (TMP)}}{\text{Total Minutes Attended (TMA)}}\% \tag{1}$$

Exact time to inspect outside diameter (ETOD) = 4 min, as well as exact time to inspect inside diameter (ETID) = 11 min. This term is total exact time (TET) = 15 min for each pipe. Moreover, the line output (LO) of this station ranges from 30 to 50 pieces/shift, per the working conditions for the shift period of 8 hrs. So, G1 matches to the ETOD and G2 to the TET.

$$\text{TMP} = \text{LO} * \text{TET} \tag{2}$$

Because of the difficulty the operator has in testing and going through the G1 pipes from the inside. At this point only the process called partial test is appropriate for G1. Therefore, the full test relates to G2, because of the ease of inspecting the pipe from the inside as well as outside. Once the full and partial testing modes of the VI station for PP are quantified, in accordance to what has been mentioned above, the time for the full test is set up to be higher than that for the partial test for pipe length 24,384 mm.

Further, the LEF depends on the TMP, and is changed through Equation (1), depending on the constancy of TMA, which is described below.

$$TMA = NOP * WH * 60 \quad (3)$$

where NOP is number of operators and WH is working hours, both per shift.

Inspection time of the PP through the VI station will decrease because only the VI partial test is done on the product (G1) to insure the quality of the weld only, not for the body; but the partial VI test increases the time spent in RT. The PP case of various length and diameters of pipe in G2 were successfully inspected and evaluated at the recommended velocity of 50 mm/s when the RT station comes after the VI station [5]. Whereas, in the case of PP from G1, VI followed by microwave technique (MWT) could be used to decrease the inspection time in the RT station.

The velocity of the wagon utilized to move pipe through the testing stations controls the time spent in each station. The TET is relative to the length of the pipe. For PP we get the minimum test processing (time interval) for any station. The velocity can be mathematically computed through displacement function to the velocity and PP time interval. Descending velocity occurs for G1 as shown in **Figure 5**, which leads to increased time interval. The velocities of the G1 PP inspection decrease, because quality control for the pipe weld can then be given as the indication for decreasing the LO and LE.

When using G1 to calibrate the RT wagon velocity, the recommended velocity decreases as demonstrated in **Figure 6**, for different proposed shift periods and different pipe lengths.

3.2. Time and Speed Analysis for Inspect PP with in VI, MW and RT

The microwave (MW) station has been proposed to work with G1 pipe. It is expected that the MW station located after VI will increase both the LO and the LE of the PP to be inspected by the RT station. With the MW station working, the partial test for a PP in **Figure 4** is not needed. As shown in **Figure 7**, PP ranging from 30 to 50 pieces for each station (VI, MWT & RT) per shift were suggested for different cases as a LO, and improvements in the LE were obtained automatically. The TETs in the MW station are 2.5 min, 3 min, 3.5 min and 4 min with the tests applicable to the PP lengths 15,240 mm, 18,288 mm, 21,336 mm and 24,384 mm respectively. These lengths were evaluated with MWT and each length was run with different shift hours in order to analyze the MWT efficiency at the recommended velocity (50 mm/s) for the RT station.

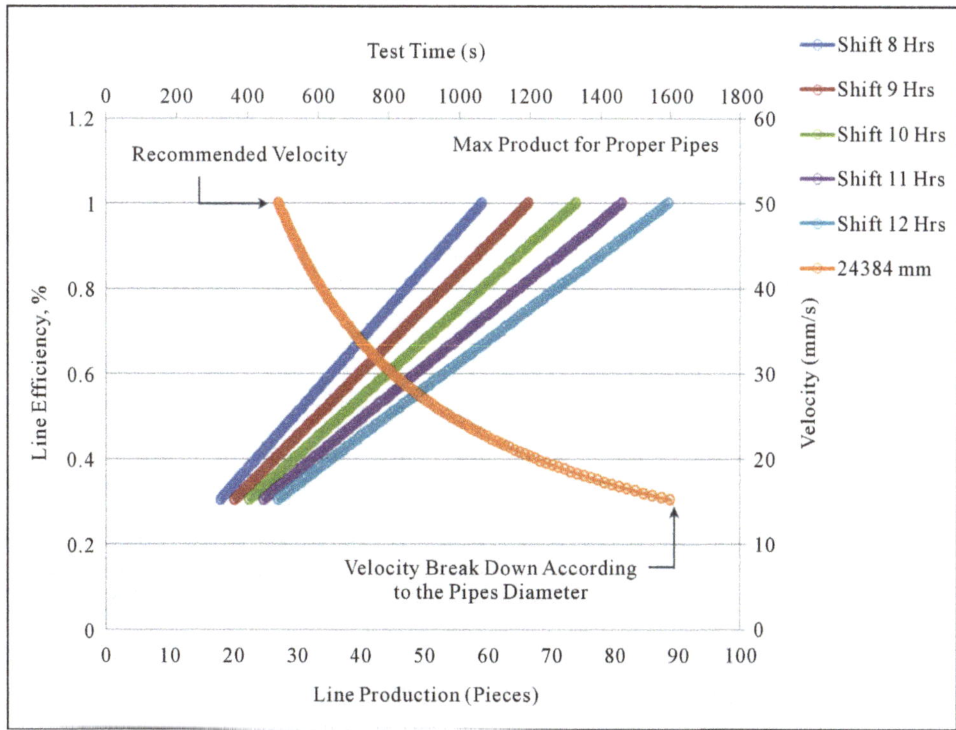

Figure 5. Inspection times for RT station as per the G1, with long pipe 24,384 mm, NOP = 1 and various shift hours.

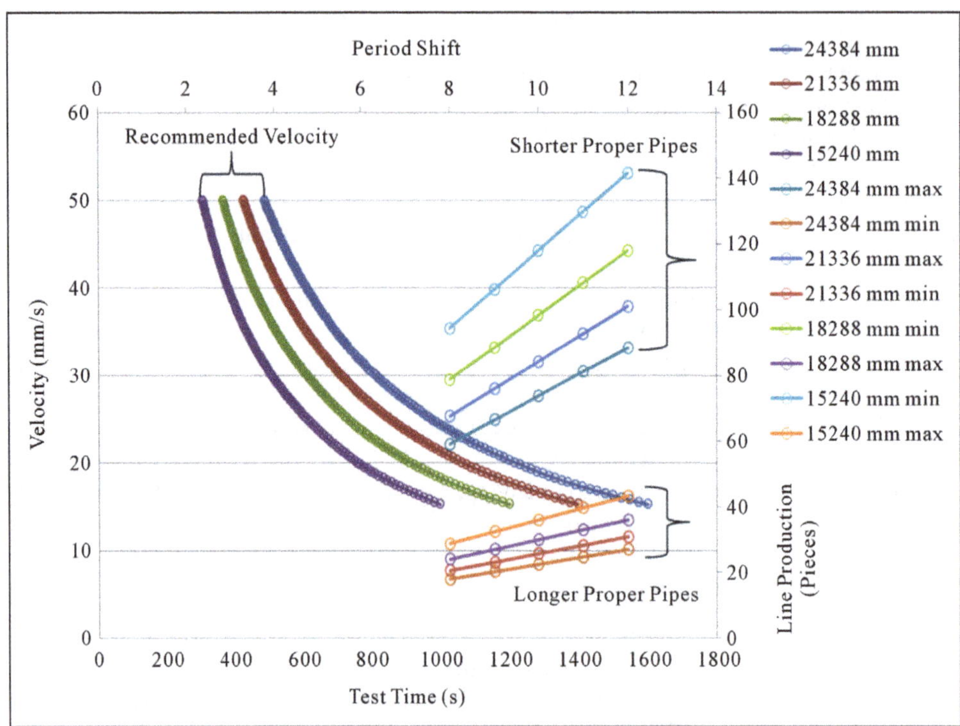

Figure 6. Calibrate velocity as per the G1, with long pipes 15,240 mm, 18,288 mm, 21,336 mm and 24,384 mm, NOP = 1 and different period shift hours.

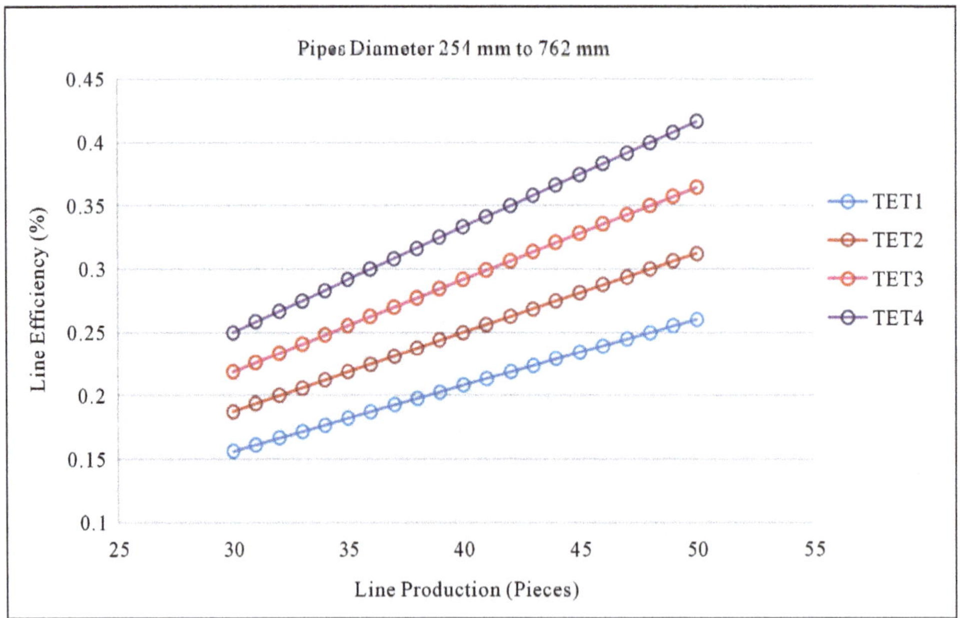

Figure 7. Line efficiency result of microwave station as per the G1, with various long pipe, and NOP = 1.

The computed results of the maximum LO for each pipe length and shift hours 8, 9, 10, 11 and 12 hrs are shown in **Figure 8**. It should be noted that we do not consider random events or calibration errors, or necessary repairs to broken testing machinery that occur at work and increase the time interval in the RT station beyond the standard time allowed. For our analysis the recommended velocity of PP processing is assumed to be constant.

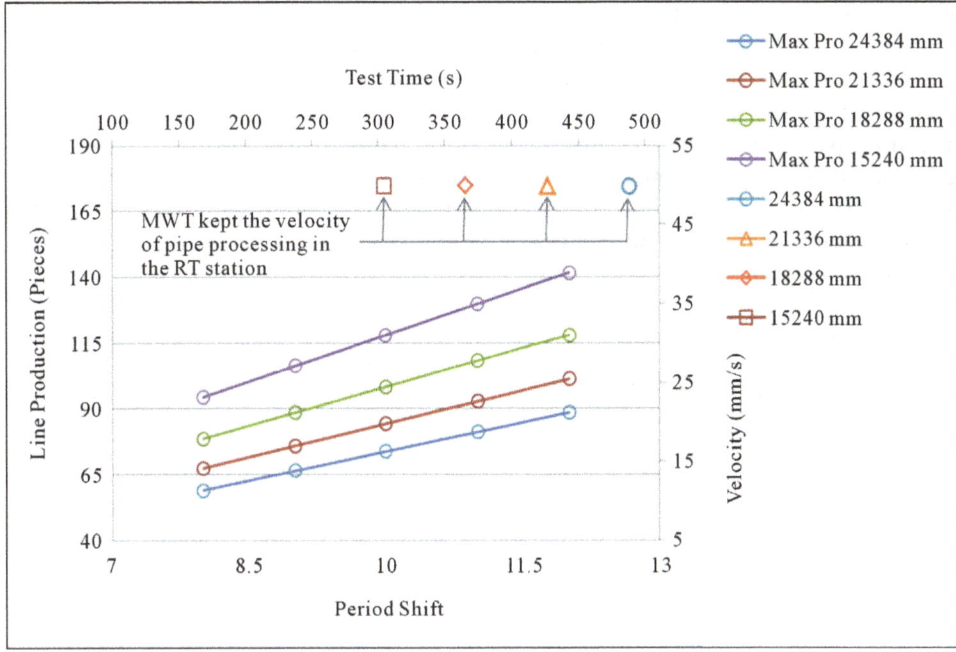

Figure 8. Maximum LO results after microwave station add it to inspect pipes G1, with different length pipe and shift hours.

3.3. Assess the Ratio of Work Done through Proposed MW Technology

To evaluate the defects qualitatively and quantitatively using VI for G1, the microwave technique (MWT) was used as the input factor of the LO and LE theoretical analysis, in order to complete the inspection for the defect locations quantitatively. SAM is factored into the proposed method. During the analysis, the exact times to inspect the OD and ID by the inspector were factored into LE calculations. Pipes are inspected in VI and according to the exact time, the ratio of work done was calculated, to be 26.667% approximately work done by VI for G1. Hence, the rest of the work ratio is 73.334% approximately done by MWT. Finally, the percentage of work completed by the microwave (MW) station is a high, and this is confirmed in **Figure 7** above. MWT kept the velocity of the wagon which carries the pipe during the testing process in the RT station at the recommended velocity according to the standards and codes, as mentioned in [5]. For G2, work done ratios are 100% for VI and RT individually.

4. Results and Discussion

The new generation of line output and efficiency has been managed in view of the complexity of the nondestructive evaluation sequence integrating the latest developments of microwave technology-based inspection systems. MW can be used to control the declining pipe process velocity through the RT station, and to overcome the deficiencies in the aforementioned VI method for G1 pipes.

Figure 4 shows the values of LE for VI station per the characteristics of standardized evaluation of the pipes in this station. The values are divided into two parts, as shown in the graph, partial test and full test according to the diameter groups, G1 & G2. In G1, the VI has a physical limitation to test the pipes related to pipeline specifications, such as dimension, pipe formation, pipe material and purpose of the pipe. But those inspections must be made because the specifications reflect standard codes. These codes specify to what level pipes are examined in terms of acceptance of sizes of defects (whether a particular size of defect size is acceptable or not) that exist in the body of the pipe or in welded areas. Naturally, the testing standards for oil and gas pipeline vary from, for example a chemical pipeline, during the manufacturing phase. Let us suggest without arguing that the efficiency and production are based on pipe inspection time. And by analyzing the SAM, the LE is controlled by TMP which changes according to the actual time expended to test the pipe. In G2, the pipes are inspected both outside and inside, and thus satisfied the full test. But the LE declines when we compute the LE for G1 which is represented only by the partial test. The reason is TMP includes only a partial test for VI which means less time spent to inspect only the outside diameter, the time varies and TMA is constant, as we mentioned above. This is because it depends on the number of operators which is constant for each station according to the requirements, and the length of the work shift is constant according to the production time.

Because of the partial test in the VI station, the RT station must reduce the velocity of the wagon that carries the pipe during the test from 50 mm/s to 16.667 mm/s, as shown in **Figure 5**. LO for pipe length 24,384 mm falls from 60 to 18 pieces and from 89 to 27 pieces per shift worked 8 hrs and 12 hrs, respectively. Similarly, **Figure 6** shows that reducing the inspection velocity to keep up with quality control requirements causes the LO to decline as shown in **Table 1**.

Table 1. Summary of the reduced velocity per quality control during different shift periods.

Pipe Length (mm)	Maximum Production (Pieces)	Minimum Production (Pieces)	Period Shift (hr)
24,384	89	27	12
	60	18	8
21,336	102	31	12
	68	21	8
18,288	119	36	12
	79	24	8
15,240	142	44	12
	95	29	8

Figure 6 and the table show that the shorter the length of the pipe, the more pieces is produced within a shift.

Figure 7 shows the increase in LE as per LO associated with using the microwave station, where improved TET were achieved. More specifically, the curves in **Figure 7** demonstrate that the TET's (2.5 min, 3 min, 3.5 min and 4 min) are directly correlated to the length of pipe (15,240 mm, 18,288 mm, 21,336 mm and 24,384 mm) respectively.

Finally, **Figure 8** shows the measured maximum LO related to the various shift lengths (8 hrs, 9 hrs, 10 hrs, 11 hrs and 12 hrs) obtained for four PP lengths as mentioned above. Velocity for the RT wagon improved to 50 mm/s for all four PP pipe lengths as shown in **Figure 8** with time intervals 487.68 s, 426.72 s, 365.76 s and 304.8 s, starting from the longer to shorter PP's. Thus, the final result is that pipe velocity in the RT station was maintained at the recommended velocity for all pipe lengths.

5. Conclusions

To analyze the line output and line efficiency of a pipe manufacturing plant at arbitrary and random production capacity due to a partial VI test, we propose to add a microwave station for a particular pipe diameter for the analysis of the PP.

Microwave technology can be situated in diverse production system line configurations in manufacturing plants per the stakeholder and customer needs. In our study, microwave kept the recommended velocity of the RT inspection process synchronized for the PP case, handled a 73.334% approximately as a ratio of work done for sequence system in the plant, and covered the shortcoming of the VI.

Also MWT kept the pipes at a high level of quality assurance. Because if it is compared to the previous status, before and after using the microwave technique, significant changes may be observed by tracking curves as mentioned above. Even with reduction in velocities of the wagon carrying the pipe out to be examined during the RT station. As mentioned in section 2, it is noted that the inspection is only for the weld areas. Consequently, if there is any defect in the inner wall of the pipe body, it will not be detected. Categorically, the microwave is shown to be a promising technology since it allows the RT station to keep the velocities as recommended. It can detect defects in both the weld and the body of the pipes easily after calibration and without human effort, thus preserving the quality of the product is expected to enhance production efficiency for oil and natural gas pipe production.

References

[1] Okamoto, J., Adamowski, J.C., Tsuzuki, M.S.G., Buiochi, F. and Camerini, C.S. (1999) Autonomous System for Oil Pipelines Inspection. *Mechatronics*, **9**, 731-743. http://dx.doi.org/10.1016/S0957-4158(99)00031-8

[2] Contreras, A., Hernández, S.L., Orozco-Cruz, R. and Galvan-Martínez, R. (2012) Mechanical and Environmental Effects on Stress Corrosion Cracking of Low Carbon Pipeline Steel in a Soil Solution. *Materials & Design*, **35**, 281-289. http://dx.doi.org/10.1016/j.matdes.2011.09.011

[3] Ju, Y. (2007) Remote Measurement of the Pipe Thickness Reduction by Microwaves. *ASME* 2007 *Pressure Vessels and Piping Division Conference*, Texas, 22-26 July 2007, 177-179. http://dx.doi.org/10.1115/pvp2007-26565

[4] Cosham, A. and Hopkins, P. (2004) The Effect of Dents in Pipelines-Guidance in the Pipeline Defect Assessment Manual. *International Journal of Pressure Vessels and Piping*, **81**, 127-139. http://dx.doi.org/10.1016/j.ijpvp.2003.11.004

[5] Alobaidi, W., Sandgren, E. and Al-Rizzo, H. (2015) A Survey on Benchmark Defects Encountered in the Oil Pipe Industries. *International Journal of Scientific and Engineering Research*, **6**, 844-853.

[6] Ahammed, M. and Melchers, R.E. (1996) Reliability Estimation of Pressurised Pipelines Subject to Localised Corrosion Defects. *International Journal of Pressure Vessels and Piping*, **69**, 267-272. http://dx.doi.org/10.1016/0308-0161(96)00009-9

[7] Abbasi, K., Ito, S. and Hashizume, H. (2008) Prove the Ability of Microwave Nondestructive Method Combined with Signal Processing to Determine the Position of a Circumferential Crack in Pipes. *International Journal of Applied Electromagnetics and Mechanics*, **28**, 429-439.

[8] Sarkar, P. (2015) OCS Online Clothing Study. What is the Meaning of SAM in Garment Industry? http://www.onlineclothingstudy.com/2013/12/what-is-meaning-of-sam-in-garment.html

[9] Education Year (2015) Calculate SAM of a Garment. http://educationyears.blogspot.com/2014/05/calculate-sam-of-garment.html

[10] Blue Star, Professional Electronics & Industrial Systems Division (2015) Material Testing—Non Destructive, Automation Project Capabilities (Ultrasonic Testing System). http://electronics.bluestarindia.com/MatTestNonDest/Auto Proj.htm

[11] Alobaidi, W.M., Alkuam, E.A., Al-Rizzo, H.M. and Sandgren, E. (2015) Applications of Ultrasonic Techniques in Oil and Gas Pipeline Industries: A Review. *American Journal of Operations Research*, **5**, 274-287. http://dx.doi.org/10.4236/ajor.2015.54021

[12] AMERICAN, THE RIGHT WAY (2015) AMERICAN Steel Pipe Manufacturing Process. http://www.american-usa.com/resources/american-steel-pipe-resources/american-steel-pipe-manufacturing-process

[13] YXLON, Technology with Passion (2015) Inspection of Pipe Segments with Longitudinal and Spiral Welds at Pipe Mills. http://www.yxlon.com/Applications/Welds/Pipe-mills

[14] HUNGER, Hydraulik (2015) Hydraulic Equipment for the Mechanical Engineering. http://hunger-hydraulik.de/Mechanical-engineering.html

[15] Deutsch, W., Gessinger, M. and Joswig, M. (2012) ECHOGRAPH Ultrasonic Testing of Helical Submerged Arc-Welded (HSAW) Pipes. 18*th World Conference on Nondestructive Testing*, Durban, 16-20 April 2012, 1-12.

[16] GE Measurement & Control (2015) Krautkramer USN 60/60L, Portable Ultrasonic Flaw Detectors. http://www.gemeasurement.com/sites/gemc.dev/files/usn_60_60l_brochure.pdf

[17] YXLON, Technology with Passion (2015) Y. Pipe Solutions—X-Ray Systems for Pipe Inspection. http://www.yxlon.com/Products/Pipe-inspection-systems

[18] U. S. Steel Tubular Products (2015) High Frequency Electric Weld (ERW) Line Pipe and Standard Products. http://usstubular.com/standard-and-line-steel-pipe/high-frequency-electric-weld-(erw)-line-pipe-and-s

[19] U.S. Department of Transportation, Pipeline & Hazardous Materials Safety Administration (2015) Fact Sheet: Pipe Manufacturing Process. https://primis.phmsa.dot.gov/comm/FactSheets/FSPipeManufacturingProcess.htm

[20] Ujam, A.J., Ekere, P.O. and Chime, T.O. (2013) Performance Evaluation of a Gas Turbine Power Plant by the Application of Compressor Off-Line and On-Line Water Washing Techniques. (A Case Study of 450MW Sapele Power Station in Delta State, Nigeria). *IOSR Journal of Engineering*, **3**, 29-41. http://dx.doi.org/10.9790/3021-031112941

[21] Sarkar, P., OCS Online Clothing Study (2015) How to Calculate Efficiency of a Production Batch or Line? http://www.onlineclothingstudy.com/2011/09/4-how-to-calculate-efficiency-of.html

[22] Song, B.L., Wong, W.K., Fan, J.T. and Chan, S.F. (2006) A Recursive Operator Allocation Approach for Assembly Line-Balancing Optimization Problem with the Consideration of Operator Efficiency. *Computers & Industrial Engineering*, **51**, 585-608. http://dx.doi.org/10.1016/j.cie.2006.05.002

3

Active and Capacitive Conductance of the Diode in a Strong Microwave Field

Muhammadjon Gulomkodirovich Dadamirzaev

Namangan Engineering Pedagogical Institute, Namangan, Uzbekistan
Email: dadamirzaev70@umail.uz

Abstract

It is shown that the mean value of the capacitive current arising in the p-n-junction in a microwave field is zero, and the average value of the active current independently of the current value is different from zero and is equal to the current generated by the diode.

Keywords

Hot Electrons, The Microwave Field, The Active and Capacitive Conductivity, Fault Current, p-n-Junction

1. Introduction

Under the influence of the electromagnetic wave the average energy of the charge carriers increases in the p-n-junction. As a result, through the potential barrier, current of hot electrons and holes flows. On the other hand, due to changing of the height of the barrier at the contact, alternating current generated by the electric field of the wave will be rectified. Due to the fact that the directions of the currents of hot carriers and rectified currents are the same, so defining the true mechanism of generation of electromotive force (EMF) on diode is an important issue. The interaction of an electromagnetic field with a semiconductor heats the free electrons and holes, and the electrons directly interact with phonons, which lead to heating of the semiconductor [1] [2].

In [3] [4] the effect of heating of the lattice on the kinetics of the establishment of the thermoelectric power of hot carriers in p-n-junction is theoretically studied. It is shown that the lattice heating leads to an additional third stage of the establishment of the thermoelectric power and thermoelectric hot carriers during the relaxation time determined by the thermal conductivity and heat capacity of the sample [4]. It is also shown that the third stage is slower than the previous two stages, which are established by A. I. Veynger and S. M. Sargsyan [5].

Effect of warming up of the charge carriers in the operation of semiconductor rectifying structures was first considered by G. M. Avakyantsom [6]. In the future, this task was carried out many studies [1] [7] [8].

In the p-n-junction with the hot carriers thermoelectric power for the different currents was studied and CVC was theoretically obtained in thin p-n-junction. In the paper of Guliamov G. and Shamirzaev S. H. [9] CVC of p-n-junction was considered, taking into account the heating of the crystal lattice and carriers. Work [10] is devoted to theoretical research of action of effect of Frenkel on the recombination centers p-n-junction, located in a microwave field. Later, the mechanisms of occurrence of EMF and current in the p-n-junction in strong microwave fields were explained in detail [4] [11]. Hitherto in the literature for calculating the current arising in the p-n-junction in the microwave fields the active current is not taken into account relative to the capacitive current, considering it negligible. On the external circuit comes out the average values of these currents. However, the average values of these currents in the literature are not evaluated.

The aim of this work is to study and compare the active and capacitive conductivity arising in the p-n-junction in a strong microwave field.

2. Theoretical Calculations of the Active and Capacitive Conductivity of the Diode in a Strong Microwave Field

To calculate the EMF and currents generated in the p-n junction the differential conductance in the form of active and reactive components I_a and I_r can be written in the following form [12]:

$$I(t) = I_a(t) + iI_r(t). \tag{1}$$

Here,

$$I_a = \frac{ep_n\mu_p}{L_p}\exp\left(\frac{eU_0}{kT}\right), \tag{2}$$

$$I_r = \frac{ep_n\mu_p}{L_p}\frac{\omega\tau_p}{2}\exp\left(\frac{eU_0}{kT}\right), \tag{3}$$

here, e is the charge of an electron, T is the lattice temperature, k is the Boltzmann constant, p_n—concentration and L_p—diffusion length of holes into the n region, τ_p and μ_p—respectively the lifetime and mobility of holes, U_0—applied voltage, ω—cyclic frequency.

Consider the ratio of the second term to the first in Equation (1):

$$\frac{I_r}{I_a} = \frac{\frac{ep_n\mu_p}{L_p}\frac{\omega\tau_p}{2}\exp\left(\frac{eU_0}{kT}\right)}{\frac{ep_n\mu_p}{L_p}\exp\left(\frac{eU_0}{kT}\right)} = \frac{\omega\tau_p}{2} = \frac{2\pi\nu\tau_p}{2}. \tag{4}$$

We estimate Equation (4) in the microwave range. For this we use the typical parameters of the samples used in the experiments A. I. Veyngera and others [13] [14], where $\nu = 10^{10}$ Hz, $\tau_p = 10^{-5}$ s. Then the ratio of the first term to the second, we have $\sim 3 \times 10^5$. This shows that the reactive current by several orders of magnitude higher active current conduction.

Based on these facts we can conclude that in the experiments capacitance diode completely bypasses its active differential resistance R_g. Indeed, at frequencies of the order of $\nu = 10^{10}$ Hz capacitive diode current will be much larger than the active current through the p-n junction. In this case, the active conductivity of the p-n-junction is heavily shunted by the capacitive conduction of the diode and the sample is almost completely loses its rectifying property. In experimental studies [13] [14] based on these considerations, believed that the generated current to the diode in a microwave field only occurs due to heating of charge carriers, and the convection current neglected.

On the other hand the total current density is made up of convection currents and displacement currents:

$$\boldsymbol{I}(t) = \boldsymbol{I}_c(t) + \frac{\partial(\varepsilon \boldsymbol{E})}{\partial t} = -e\int_{\vartheta} \upsilon f(\upsilon)\mathrm{d}^3\upsilon + \frac{\partial(\varepsilon \boldsymbol{E})}{\partial t}. \tag{5}$$

Here, the first term is an active, and a second capacitive current, $I_c(t)$—current density of convection associated with the motion of free charges, $f(\upsilon)$—the distribution function of the velocity of charge carriers; ε—dielectric permeability of crystal lattice, E—intensity electric field.

We estimate the current convection and the bias current in the p-n-junction, located in a microwave field. Convection current can be written as:

$$J_c = \frac{U(t)}{R_g}. \tag{6}$$

Here $U(t)$—can be estimated by the formula [15]:

$$P = \frac{U^2(t)}{2R_g}, \tag{7}$$

where P—Power, $R_g = \frac{\partial U(t)}{\partial J}$—the differential resistance of the p-n-junction. If we assume that the field of the p-n-junction $|E| = \frac{U(t)}{d}$ (where d—width of the space charge region of p-n-junction) the formula for the total current through the diode can be written as follows:

$$I(t) = \frac{U(t)}{R_g} + \frac{\varepsilon\varepsilon_0 S}{d}\frac{\partial U(t)}{\partial t}. \tag{8}$$

Here $C = \frac{\varepsilon\varepsilon_0 S}{d}$ the geometric capacitance of p-n-junction.

Leaving the diode current when exposed to microwave field is determined by the average current during the period of the wave. We average the total current in the microwave field:

$$I = \frac{1}{T}\int_{t_1}^{t_1+T} I(t)\mathrm{d}t_1 = \frac{1}{T}\int_{t_1}^{t_1+T}\left(I_c(t) + \frac{\partial(\varepsilon E)}{\partial t}\right)\mathrm{d}t. \tag{9}$$

We calculate the average value of the bias current:

$$\frac{1}{T}\int_{t_1}^{t_1+T}\frac{\partial(\varepsilon E)}{\partial t}\mathrm{d}t = \varepsilon E(t_1+T) - \varepsilon E(t_1) = \varepsilon\left(E(t_1+T) - E(t_1)\right). \tag{10}$$

The electric field is a periodic function of time $E(t_1+T) = E(t_1)$, resulting in the average value of the bias current is always zero. Then the average value of the current generated by the diode is located in a microwave field will be equal to:

$$\bar{I} = \frac{1}{T}\int_{t_1}^{t_1+T} I_c(t)\mathrm{d}t. \tag{11}$$

This current is due to the active current, a capacitive current will not affect the average value of the total current emitted from the diode. This is a simple but important conclusion to explain averaged currents arising on different diodes when exposed to the electromagnetic field. It should be noted that the expression of BAX p-n-junction (11) corresponds to any diode with an asymmetric p-n-junction. For example, contact metal-semiconductor, Schottky diode, a tunnel diode, metal-insulator-metal and others can relate to these diodes.

3. Conclusion

Based on the analysis of the results obtained it can be concluded that no matter how large the capacitive current is, its average value is zero. Because the electric field is a periodic function of time with period T. No matter how small the active current is, the average value is not zero, and the average current generated by diode is determined precisely by this current. In p-n junction, the active current is closed with the recombination current and total average current is determined by recombination of electrons and holes.

Acknowledgements

The author would like to thank G. Gulyamov, G. Dadamirzayev and N. Yu. Sharibayev for numerous discussions.

References

[1] Buss, A.G. and Gurevich, Y.G. (1975) Hot Electrons, and Strong Electromagnetic Field in the Plasma of Semiconductors and Gas Discharge. Nauka, Moscow, 389 p.

[2] Bass, F.G., Bochkov, B.S. and Gurevich, Y.G. (1984) The Electrons and Phonons in Bounded Semiconductors. Nauka, Moscow, 288 p.

[3] Sah, C.T., Noyce, R.H. and Shockly, W. (1957) Carrier Generation and Recombination in p-n-Junctions and p-n-Junctions Characteristics. *Proceedings of the IRE*, **45**, 1228-1243. http://dx.doi.org/10.1109/JRPROC.1957.278528

[4] Guliamov, G., Dadamirzaev, M.G. and Boydedaev, S.R. (2000) Kinetics of the Establishment of the Thermoelectric Power of Hot Carriers in the p-n-Junction, Taking into Account Heating. *Semiconductors*, **34**, 266-269.

[5] Veynger, A.I. and Sargsiyan, M.P. (1980) The Kinetics of the Thermopower Arising in the p-n-Junction with Heating Carrier. *Semiconductors*, **14**, 2020-2027.

[6] Avakyants, G.M. (1954) On the Theory of Contact Phenomena When. *Journal of Experimental and Theoretical Physics*, **27**, 333-338.

[7] Veynger, A.I., Paritssky, L.G., Akopyan, E.A. and Dadamirzaev, G. (1975) Thermoelectric Power of Hot Carriers in the p-n-Junction. *Semiconductors*, **9**, 216-224.

[8] Bonch-Bruyevich, V.L., Zvyagin, I.P. and Mironov, A.F. (1972) Blast Electrical Instability in Semiconductors. Nauka, Moscow, 414 p.

[9] Gulyamov, G. and Shamirzaev, S.H. (1981) Thermoelectric Power of Hot Carriers in the p-n-Junction, Taking into Account Heating. *Lattice Semiconductors*, **15**, 1858-1861.

[10] Gulyamov, G., Dadamirzaev, M.G., Boydedaev, S.R. and Gulyamov A.G. (2007) Nonideality Factor of CVC of p-n-Junction in a Strong Microwave Field. *International Scientific Conference*: *Nonequilibrium Processes in Semiconductors and Semiconductor Structures*, Tashkent, 1-3 February 2007, 147-148.

[11] Guliamov, G., Dadamirzaev, M.G. and Boydedaev, S.R. (2000) EMF Hot Carriers Due to the Modulation of the Surface Potential in a Strong Microwave Field. *Semiconductors*, **34**, 572-575.

[12] Picos, G.E. (1965) Fundamentals of the Theory of Semiconductor Devices. Nauka, Moscow, 448 p.

[13] Ablyazimova, N.A., Veynger, A.I. and Pitanov, V.S. (1988) The Electrical Properties of Silicon p-n-Junctions in Strong Microwave Fields. *Semiconductors*, **22**, 2001-2007.

[14] Ablyazimova, N.A., Veynger, A.I. and Pitanov, V.S. (1992) The Impact of a Strong Microwave Field in the Photovoltaic Properties of Silicon p-n-Junctions. *Semiconductors*, **26**, 1041-1047.

[15] Usanov, D.A., Skripal, A.V. and Ugryumova, N.V. (1998) The Negative Resistance in Structures Based on the p-n-Junction in a Microwave Field. *Semiconductors*, **32**, 1399-1402. http://dx.doi.org/10.1134/1.1187600

4

Microwave Heating of Liquid Foods

Vittorio Romano*, Rino Apicella

Dipartimento di Ingegneria Industriale, Università degli Studi di Salerno, Fisciano, Italy
Email: *vromano@unisa.it

Abstract

A mathematical model has been formulated to describe the heat transfer in liquid foods flowing in circular ducts, subjected to microwave irradiations. Three types of liquids with different rheological behavior are considered: skim milk (Newtonian), apple sauce and tomato sauce as non-Newtonian fluids. Each one can flow with different velocities but always in laminar way. The temperature profiles have been obtained solving the transient momentum and heat equations by numerical resolution using the Finite Element Method. The generation term due to the microwave heating has been evaluated according to Lambert's law. Dielectric properties are considered to be temperature dependent.

Keywords

Continuous Microwave Heating, Lambert's Law, Non-Newtonian Fluids, FEM

1. Introduction

Microwave heating has been utilized since the 1940s [1] in different fields such as polymer and ceramics industries [2] [3] and medicine [4] [5]. However, food processing is the largest consumer of microwave energy, that can be employed for cooking, thawing, tempering, drying, freeze drying, pasteurization, sterilization, baking, heating and re-heating. In microwave heating, electromagnetic field polarizes the molecules of dielectric materials and creates dipole moments that cause these molecules to rotate. The resulting molecular friction causes heat generation in the body. Due to intrinsic heat generation capability, microwave heating can provide prompt rise of temperature within the low thermal conductive products, especially in food items. However, this technique is blamed for uneven heating since the food product processed with microwaves shows alternate hot and cold spots. This is primarily caused by the non-uniform distribution of microwave energy in the foodstuff, due to factors such as dielectric loss, penetration depth, thickness, shape and size of the sample.

Microwave heating of solid foods has been largely investigated in the last years. The energy equation is writ-

*Corresponding author.

ten as a conductive heat transfer with a generation term. The latter has been modeled by many authors, using two different approaches to evaluate the effects of the microwave distribution: by solving the Maxwell's equations [2] [6] [7] or by applying the Lambert's law [8] [9]. The Lambert's law is a simple power formulation that was believed to simulate temperature profiles for Cartesian geometries and for cylindrical geometries with high radius. Ayappa *et al.* [2] deduced the minimum value of the characteristic sample dimension to successfully apply the law for slabs; Oliveira and Franca [7] found that this value was higher for cylinders than for slabs. Finally, Romano *et al.* [10] correctly applied Lambert's law considering the geometry of the sample, thus finding the same field of application for both the shapes and a power concentration along the central axis for a cylindrical domain, according with the experimental observations [9].

As regards fluids, there are fewer studies and the most are about batch processes [11] [12]. The references in continuous processes instead are represented by Salvi *et al.* [13] and J. Zhu *et al.* [15] whose mathematical model considers the generation term according to Maxwell's equations. In this paper, Lambert's law has been applied, that is much less expensive in terms of time of calculus. The aim is to verify whether similar results are obtained or not.

2. Mathematical Model

2.1. Physical System

The physical system represented in **Figure 1** is a cylindrical horizontal tube with a length $L = 0.3$ m and a radius $R = 0.02$ m in which different liquid foods are heated by microwave irradiance in the radial direction.

A laminar flow at various velocities is realized by changing the difference of pressure between the inlet and the outlet sections. Only the axial component of the velocity is different from zero and it is a function of the variable r: $v_z = v_z(r)$. The temperature, even though the microwave penetration is only radial, is also function of the axial direction z by the effect of the flow field: $T = T(r, z)$.

2.2. Transport Equations

In order to find the temperature profile, the following mathematical model has been constructed.

It consists of the following three differential equations in cylindrical coordinate system with their boundary conditions [14]

$$\frac{\partial v_z}{\partial z} = 0 \qquad \text{(Equation of continuity)}$$

$$\rho\frac{\partial v_z}{\partial t} = -\frac{1}{r}\frac{\partial}{\partial r}\left(r\tau_{rz}\right) + \frac{P_{(z=0)} - P_{(z=L)}}{L} \qquad \text{(Equation of motion)}$$

As initial condition, the fluid is stationary.

$$@\ t = 0,\ v_z = 0\ \ \forall r, z \tag{1}$$

As boundary conditions, no slipping at the wall has been assumed

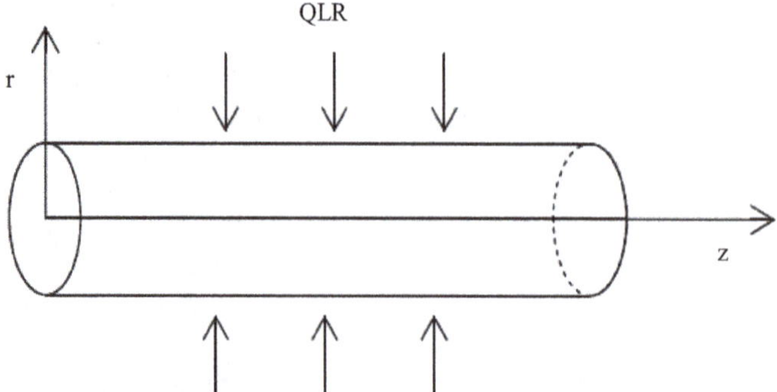

Figure 1. Representation of the cylindrical duct.

$$@\ r = R, v_z = 0\ \ \forall z, t \tag{2}$$

and symmetry on the axis has been considered.

$$@\ r = 0, \tau_{rz} = 0\ \ \forall z, t \tag{3}$$

$$\rho c_p \left(\frac{\partial T}{\partial t} + v_z \frac{\partial T}{\partial z} \right) = k \left[\frac{1}{r} \frac{\partial}{\partial r} \left(r \frac{\partial T}{\partial r} \right) + \frac{\partial^2 T}{\partial z^2} \right] + QLR \qquad \text{(Energy equation)}$$

As initial condition, the temperature of the fluid is assumed to be uniform

$$@\ t = 0, T = T_0\ \ \forall r, z \tag{4}$$

As boundary conditions for the radial direction, symmetry respect to the axis and heat convective flux at the wall have been considered.

$$@\ r = 0, \frac{\partial T}{\partial r} = 0\ \ \forall z, t \tag{5}$$

$$@\ r = R, -k \frac{\partial T}{\partial r} = h_c \left(T - T_{air} \right)\ \forall z, t \tag{6}$$

while for the z direction uniform temperature in the inlet section and Danckwerts condition in the outlet one have been imposed.

$$@\ z = 0, T = T_{in}\ \ \forall r, t \tag{7}$$

$$@\ z = L, \frac{\partial T}{\partial z} = 0\ \ \forall r, t \tag{8}$$

Initial temperature T_0 and input temperature T_{in} are both equal to environmental temperature T_{air}.

Heat generation due to microwaves has been modeled according to the Lambert's law along the radial direction of a cylindrical sample [10]:

$$QLR = 2\alpha \left(\frac{R}{r} \right) Q_1 \left[\mathrm{e}^{-2\alpha(R-r)} + \mathrm{e}^{-2\alpha(R+r)} \right]$$

where

$$Q_1 = \frac{Q_0}{2\pi R (R + L)}$$

$$\alpha = \frac{2\pi}{\lambda_0} \sqrt{\frac{\varepsilon' \left\{ \left[1 + \left(\frac{\varepsilon''}{\varepsilon'} \right)^2 \right]^{\frac{1}{2}} - 1 \right\}}{2}}$$

$\frac{\varepsilon''}{\varepsilon'}$ is the loss tangent, tanδ.

The attenuation factor for each fluid has been considered as a function of temperature, calculated by interpolation starting from graphic relationships for dielectric constant and loss tangent versus temperature in a range 10°C - 90°C [15].

Such a system, with the equations and boundary conditions written before, results to be axial-symmetric.

3. Materials and Methods

To solve the previous partial differential equations, a Finite Elements Method (FEM) has been used. To practically implement this solution, COMSOL Multiphysics® has been utilized with the following mesh features: 2049 mesh points, 3840 triangular elements, 256 boundary elements and 4 vertex elements.

Figure 2 shows an example of mesh that the software uses to discretize the domain.

Three fluid foods have been considered: skim milk, with a Newtonian behavior (constant viscosity), apple sauce and tomato sauce as non Newtonian fluids, modeled with a power law having different fluid consistency coefficient and flow behavior index. All the physical properties of the three fluids are resumed in **Table 1**.

As ε' and ε'' are temperature functions, average values have been obtained by integrating in the entire domain and in the time (range 0 - 50 s). They are fundamental for microwave heating, because they determine respectively the energy absorbed and the fraction converted in heat power.

4. Results

It is possible to make a qualitative analysis of the results observing the following temperature maps reported in **Figures 3-5**. They have been obtained with an incident power Q_1 = 20,000 W·m^{-2}, for each fluid, for two different velocities (2 mm/s and 4 mm/s) and for two different instants of time (30 s and 50 s). They show the temperature profiles in a *rz* plane section of the real physical system on the left and in a 3D plot on the right. The higher heating is obtained for longer times and lower velocities. The nature of liquid also have a great influence: skim milk reaches higher temperature with a less uniform profile.

The graphs in **Figures 6-8** provide a more significant analysis since they show the temperature profiles along the radial direction, obtained with the same incident power of 20,000 W·m^{-2} and appointing the outlet section instead of the instant of time. In this way, each fluid element has the residence time due to its velocity, which is related to its position.

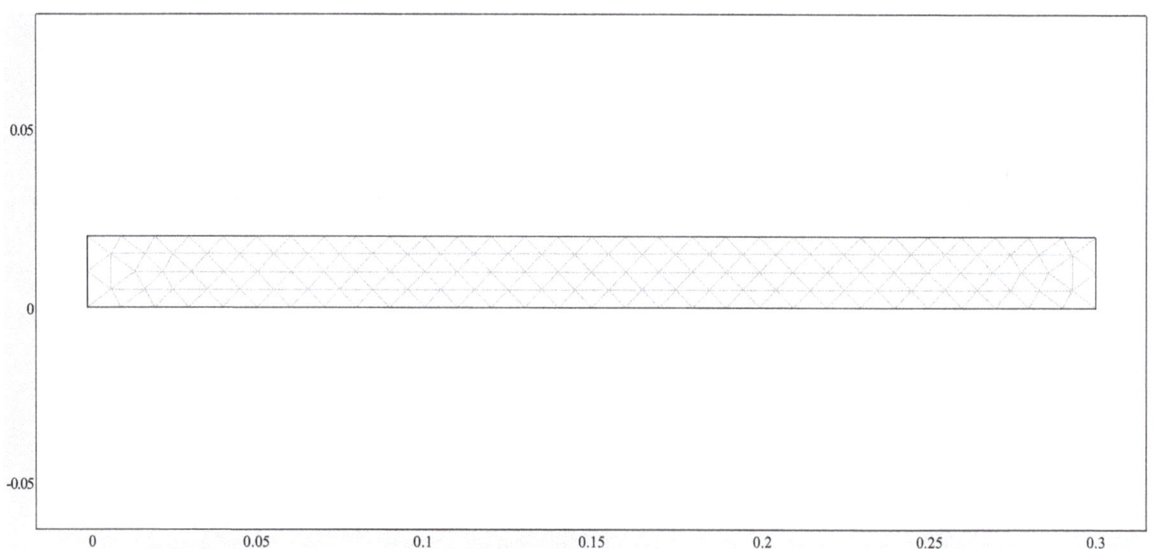

Figure 2. Example of a generic mesh.

Table 1. Physical, dielectric and transport properties (Zhu *et al.*, 2007).

	Skim milk	Apple sauce	Tomato sauce
Density, ρ [kg·m^{-3}]	1047.7	1104.9	1036.9
Specific heat, c_p [J·kg^{-1}·K^{-1}]	3943.7	3703.3	4000.0
Thermal conductivity, k [W·m^{-1}·K^{-1}]	0.5678	0.5350	0.5774
Viscosity, μ [Kg·m^{-1}·s^{-1}]	0.0059		
Fluid consistency coefficient		32.734	3.9124
Flow behavior index		0.197	0.097
Dielectric constant, ε'	66.31	68.97	74.27
Dielectric loss, ε''	13.26	5.30	46.42

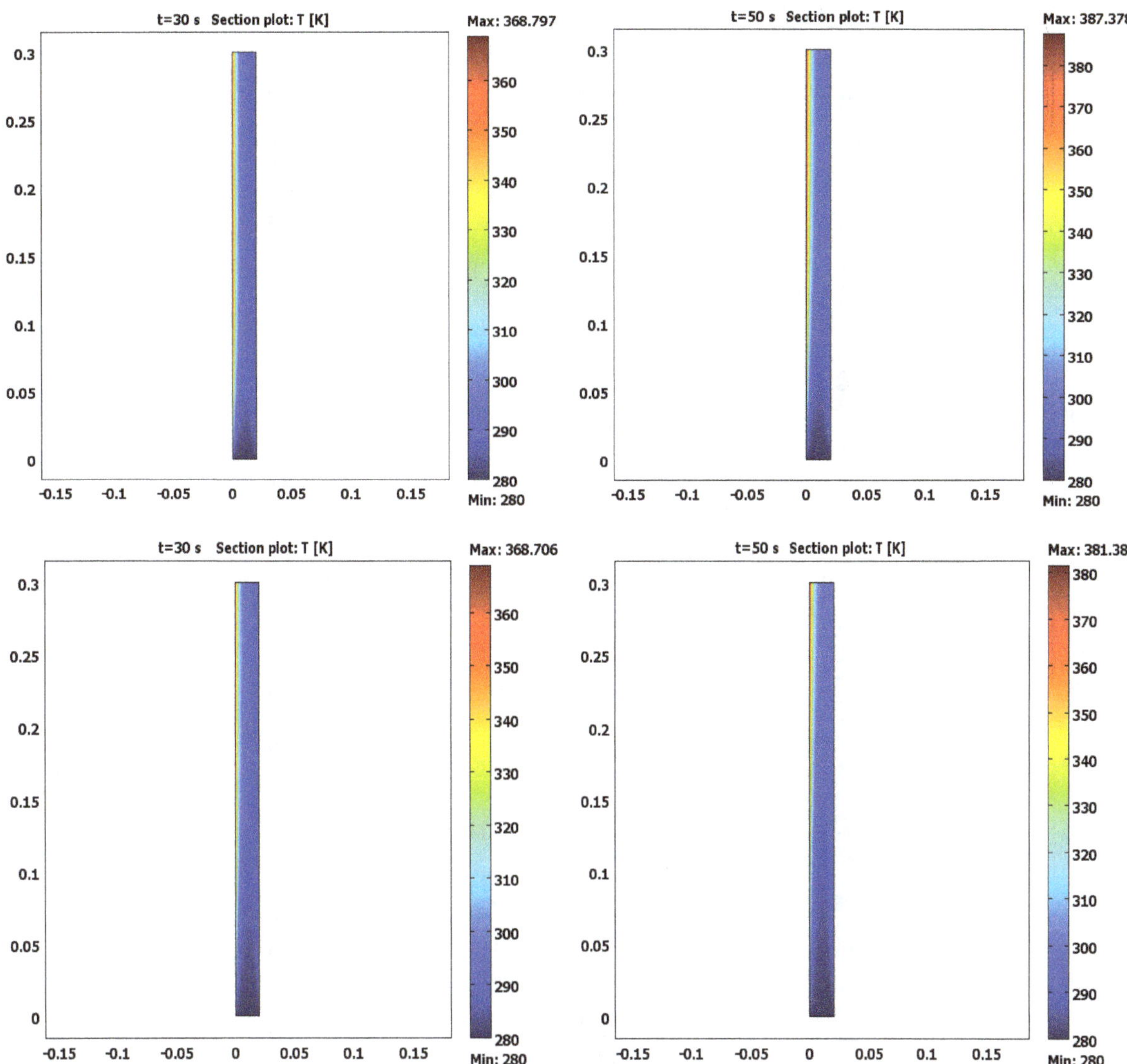

Figure 3. Temperature maps for skim milk, with v_m = 2 mm/s (on the top) and v_m = 4 mm/s (on the bottom).

In these graphs all the fluids show a minimum (cold spot) located at about half the radium. This is the sum of two effects: Lambert's law that predicts a different heating along the radial direction and the flow field that produces different residence times inside the tube. In particular, near the axis (r = 0) the velocity is higher and so the residence times are lower, but according to the Lambert's law, the power density is maximum by effect of the term R/r; conversely, near the wall (r = R) the velocity is lower and the times of exposition to the microwaves are higher, but the Lambert's power density undergoes the effect of the exponential decay. These are opposite effects and their combination produces the previous profiles, with higher temperatures on the axis and on the wall.

The kind of fluid plays an important role, both for the rheological behavior and the dielectric properties, while the physical properties are quite the same for all of them. On one hand, pseudo plastics having a flatter velocity profile than Newtonian fluids, can't fully balance the effect of Lambert's law. On the other hand, the dielectric properties, in particular dielectric loss, determinant for the absorbed heat, are different fluid by fluid. Such dielectric properties cause a lower difference in the absorbed heat between axis and wall in the case of tomato sauce, as it can be noticed in **Table 2**.

As the ε' is quite the same for all the fluids, they absorb about the same quantity of energy from microwaves, but tomato sauce transforms the higher fraction of this energy into heat ($Q_{converted}$) by a higher value of ε'' and so it shows a higher average temperature. These results are resumed in **Table 3**.

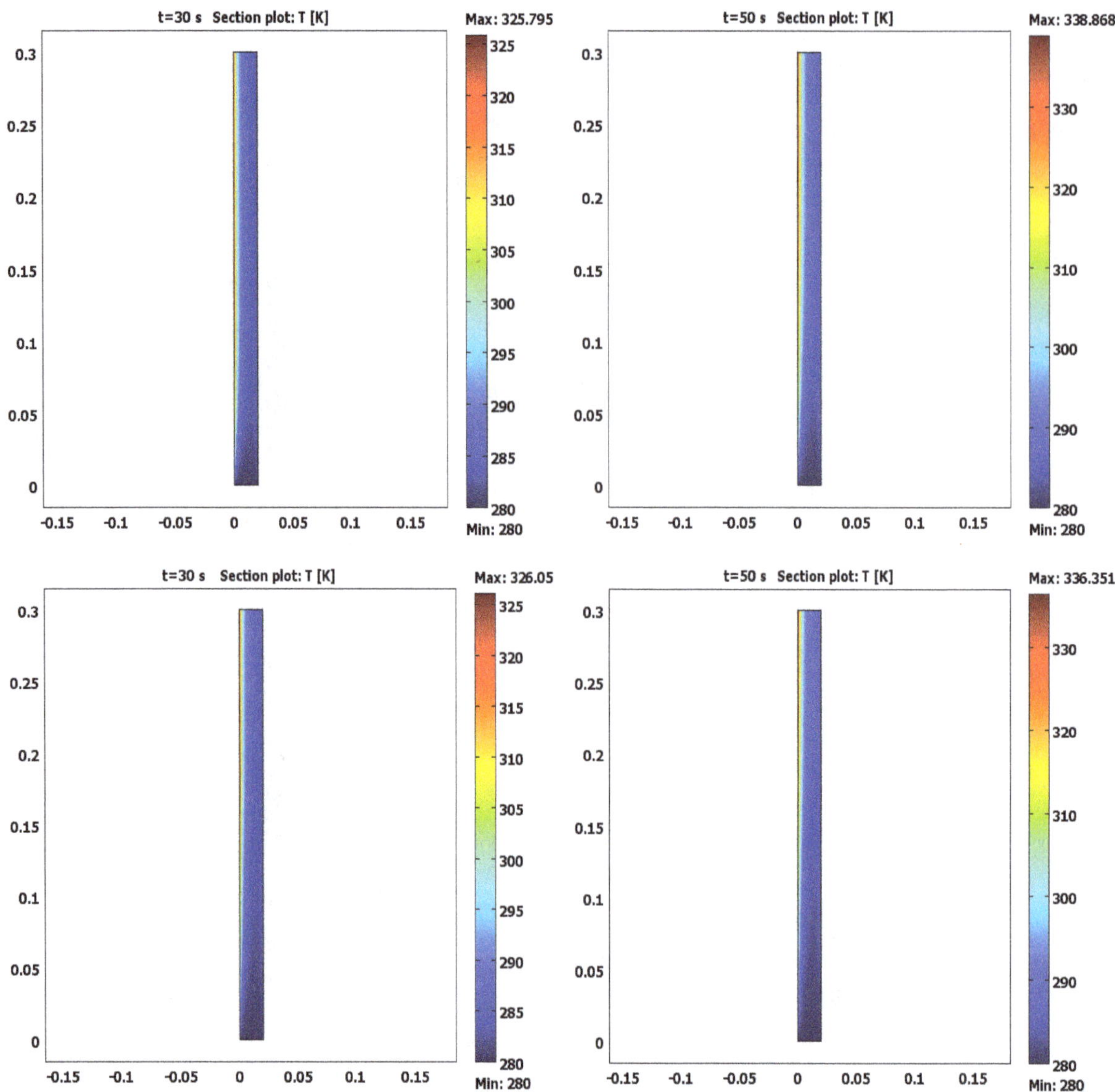

Figure 4. Temperature maps for apple sauce, with $v_m = 2$ mm/s (on the top) and $v_m = 4$ mm/s (on the bottom).

5. Conclusions

In this paper, microwave heating of three liquid foods moving in a cylindrical duct with a laminar flow has been analyzed.

The multiphysics mathematical model considers the momentum and energy transport at unsteady state.

The analysis has been achieved with the Finite Element Method solving the mathematical model with Comsol®3.5.

Contrasting effects of Lambert's law for microwave heating in case of a cylindrical geometry and distribution of residence times of the fluids in the duct have been taken into account.

Different factors play a role in case of microwave heating of a liquid. In particular, rheological and dielectric properties of the fluids have been considered and two different operational conditions have been obtained by varying the average velocity of the fluids.

Results show that absorbed power has always a maximum on the axis, caused by the ratio R/r appearing in the radial contribution to the microwave heat source for a cylindrical geometry; however, near the wall of the tube a high quantity of absorbed heat has been found, in this case due to the long times of residence of the fluid inside the tube.

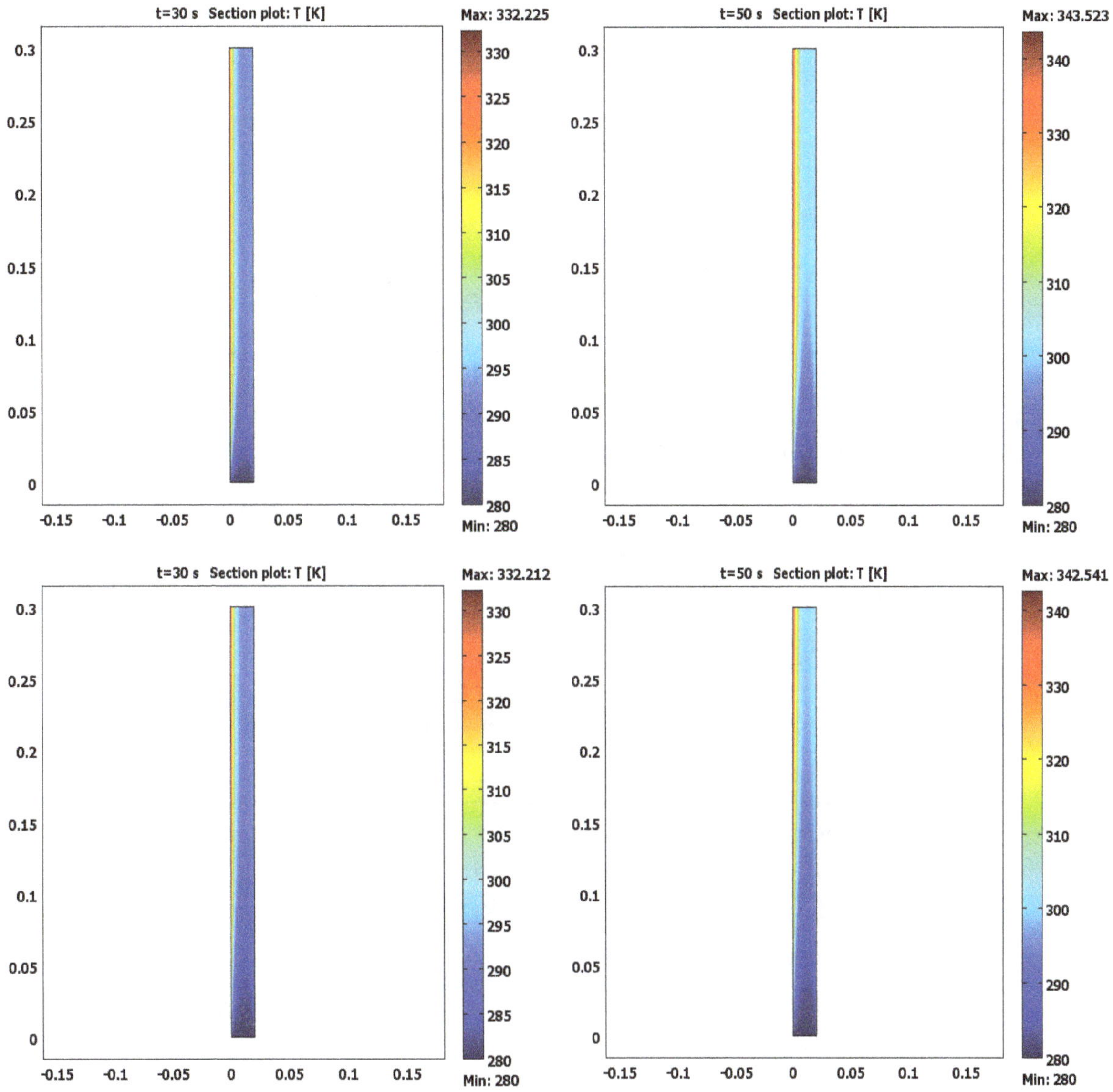

Figure 5. Temperature maps for tomato sauce, with $v_m = 2$ mm/s (on the top) and $v_m = 4$ mm/s (on the bottom).

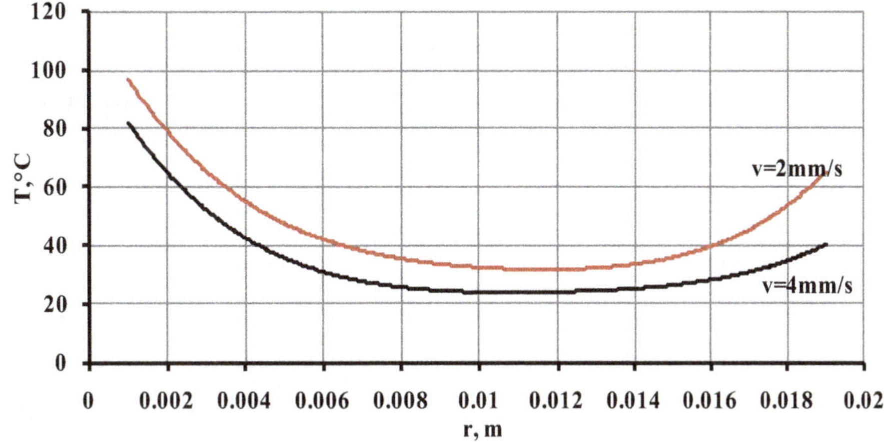

Figure 6. Temperature profiles for skim milk, $Q_1 = 20{,}000$ W/m^2.

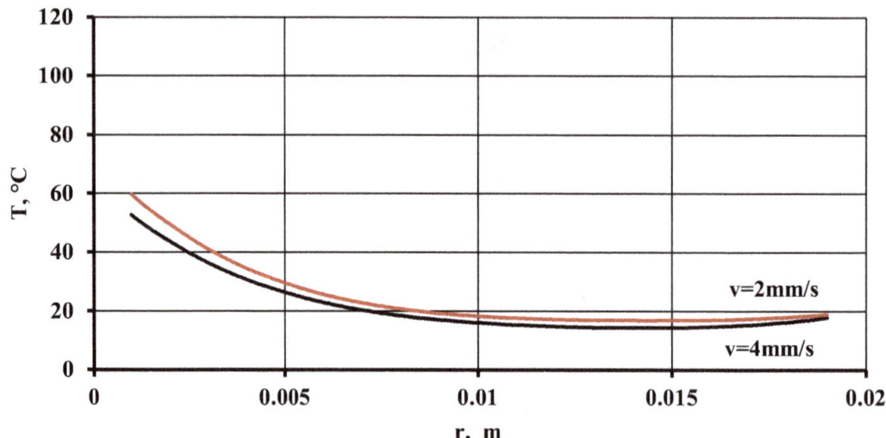

Figure 7. Temperature profiles for apple sauce, $Q_1 = 20{,}000$ W/m^2.

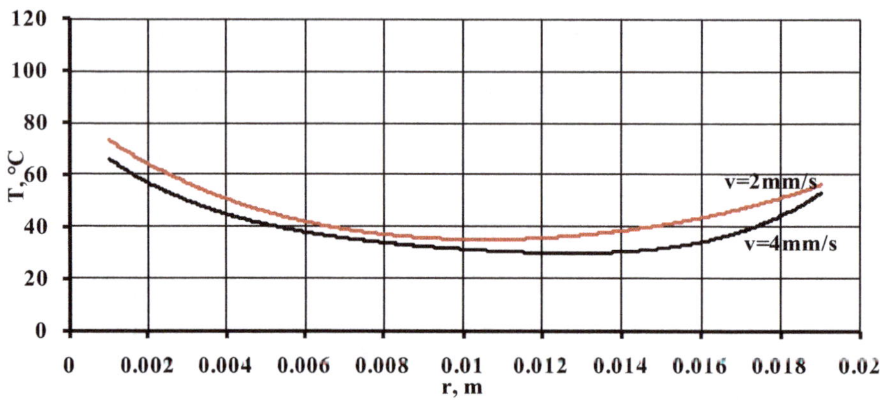

Figure 8. Temperature profiles for tomato sauce, $Q_1 = 20{,}000$ W/m^2.

Table 2. Power density at axis and wall.

	Axis QLR 10^{-6} [W·m^{-3}]	Wall QLR 10^{-6} [W·m^{-3}]
Skim milk	14.71	1.187
Apple sauce	7.063	0.362
Tomato sauce	7.790	2.614

Table 3. Average over space and time of power density and temperature.

	Average QLR 10^{-6} [W·m^{-3}]	Average T [K]
Skim milk	1.491	287
Apple sauce	1.108	285
Tomato sauce	2.887	294

The final result is a double pick of absorbed heat on the axis and on the wall; between the two opposite effects the term R/r overcomes the one due to the residence time.

In any case, the absorbed powers are almost equal for the three fluids because the dielectric constants ε' are of the same order of magnitude, whereas the fraction of this energy which is dissipated into heat is different in the three cases as the dielectric losses ε'' are different.

The tomato sauce gives the higher average temperature and also the more uniform temperature distribution.

We have compared these results with those obtained studying a cylindrical system with anon-symmetric microwaves exposure [15] in which Maxwell equations are used to evaluate the electromagnetic field. The temperature profiles are analogous even if obtained with different computational times: 50 hours with a 3.0 GHz processor using Maxwell equations, few minutes with a 2.1 GHz processor adopting Lambert's law.

In both the cases, the temperature is higher on the axis, it decreases until about half the radius and it increases again near the wall. Furthermore, the lower difference between the hot and the cold spot in the temperature profile can be found in the tomato sauce case.

References

[1] Mermelstein, N.H. (1997) How Food Technology Covered Microwaves over the Years. *Food Technology*, **51**, 82-84.

[2] Ayappa, K.G., Davis, H.T., Davis, E.A. and Gordon, J. (1991) Analysis of Microwave Heating of Materials with Temperature Dependent Properties. *AIChE Journal*, **37**, 313-322. http://dx.doi.org/10.1002/aic.690370302

[3] Chatterjee, A., Basak, T. and Ayappa, K.G. (1998) Analysis of Microwave Sintering of Ceramics. *AIChE Journal*, **44**, 10.

[4] O'Brien, K.T. and Mekkaoui, A.M. (1993) Numerical Simulation of the Thermal Fields Occurring in the Treatment of Malignant Tumors by Local Hyperthermia. *Journal of Biomechanical Engineering*, **115**, 247-253. http://dx.doi.org/10.1115/1.2895482

[5] Paulsen, K.D., Lynch, D.R. and Strohbehn, J.W. (1998) Three-Dimensional Finite, Boundary, and Hybrid Element Solutions of the Maxwell Equations for Lossy Dielectric Media. *IEEE Transactions on Microwave Theory and Techniques*, **36**, 682-693.

[6] Ayappa, K.G., Davis, H.T., Davis, E.A. and Gordon, J. (1992) Two Dimensional Finite Elements Analysis of Microwave Heating. *AIChE Journal*, **38**, 1577-1592. http://dx.doi.org/10.1002/aic.690381009

[7] Oliveira, M.E.C. and Franca, A.S. (2002) Microwave Heating of Foodstuffs. *Journal of Food Engineering*, **53**, 347-359. http://dx.doi.org/10.1016/S0260-8774(01)00176-5

[8] Lin, Y.E., Anantheswaran, R.C. and Puri, V.M. (1995) Finite Element Analysis of Microwave Heating of Solid Foods. *Journal of Food Engineering*, **25**, 85-112. http://dx.doi.org/10.1016/0260-8774(94)00008-W

[9] Zhou, L., Puri, V.M., Anantheswaran, R.C. and Yeh, G. (1995) Finite Element Modeling of Heat and Mass Transfer in Food Materials during Microwave Heating-Model Development and Validation. *Journal of Food Engineering*, **25**, 509-529. http://dx.doi.org/10.1016/0260-8774(94)00032-5

[10] Romano, V.R., Marra, F. and Tammaro, U. (2005) Modelling of Microwave Heating of Foodstuff: Study on the Influence of Sample Dimensions with a FEM Approach. *Journal of Food Engineering*, **71**, 233-241. http://dx.doi.org/10.1016/j.jfoodeng.2004.11.036

[11] Ratanadecho, P., Aoki, K. and Akahori, M. (2002) A Numerical and Experimental Investigation of the Modeling of Microwave Heating for Liquid Layers Using a Rectangular Wave Guide (Effects of Natural Convection and Dielectric Properties). *Applied Mathematical Modelling*, **26**, 449-472. http://dx.doi.org/10.1016/S0307-904X(01)00046-4

[12] Zhang, Q., Jackson, T.H. and Ungan, A. (2000) Numerical Modeling of Microwave Induced Natural Convection. *Journal of Heat and Mass Transfer*, **43**, 2141-2154. http://dx.doi.org/10.1016/S0017-9310(99)00281-1

[13] Salvi, D., Boldor, D., Aita, G.M. and Sabliov, C.M. (2011) COMSOL Multiphysics Model for Continuous Flow Microwave Heating of Liquids. *Journal of Food Engineering*, **104**, 422-429. http://dx.doi.org/10.1016/j.jfoodeng.2011.01.005

[14] Bird, R.B., Stewart, W.E. and Lightfoot, E.N. (2002) Transport Phenomena. 2nd Edition, John Wiley and Sons, Inc.

[15] Zhu, J., Kuznetsov, A.V. and Sandeep, K.P. (2007) Mathematical Modeling of Continuous Flow Microwave Heating of Liquids (Effects of Dielectric Properties and Design Parameters). *International Journal of Thermal Sciences*, **46**, 328-341. http://dx.doi.org/10.1016/j.ijthermalsci.2006.06.005

Nomenclature

L	tube length	[m]
P	pressure	[Pa]
Q_0	incident microwave power at the sample surface	[W]
Q_1	incident microwave power per unit surface	$[W \cdot m^{-2}]$
QLR	heat generation due to microwave	$[W \cdot m^{-3}]$
R	tube radius	[m]
T	T temperature	[K]
c_p	specificheat	$[J \cdot Kg^{-1} \cdot K^{-1}]$
h_c	convective heat transfer coefficient	$[W \cdot m^{-2} \cdot k^{-1}]$
r	radialdirection	[m]
t	time	[s]
v_m	averageaxialfluidvelocity	$[m \cdot s^{-1}]$
v_z	axialfluidvelocity	$[m \cdot s^{-1}]$
z	axialdirection	[m]
α	attenuationfactor	$[cm^{-1}]$
ε'	dielectric constant	[–]
ε''	dielectric loss	[-]
ρ	density	$[kg \cdot m^{-3}]$
τ_{rz}	flux of z-momentum in the positive r direction-	[Pa]

5

Influence of Deformation on CVC p-n-Junction in a Strong Microwave Field

Muhammadjon Gulomkodirovich Dadamirzayev

Namangan Engineering Pedagogical Institute, Namangan, Uzbekistan
Email: gulyamov1949@mail.ru

Abstract

This paper investigates the current-voltage characteristics (CVC) strain of p-n-junction in a strong microwave (MW) field and shows that the deformation increases the current generated in the p-n-junction. We analyze the current-voltage characteristics of p-n-junction in which three-dimensional space (I,U,ε) gives more complete information than the two-dimensional.

Keywords

Microwave Electromagnetic Field, Deformation Effects in Semiconductors, The Concentration of Minority Carriers, The Current-Voltage Characteristic of the p-n-Junction, The Temperature of the Electrons and Holes

1. Introduction

The first experimental study of the heating of the charge carriers by the strong electric field in inhomogeneous semiconductors began with thermopower measurements of hot charge carriers occurring at the p-n-junction. [1] [2] studied the current-voltage characteristics (CVC) and investigated the variation of photoelectric characteristics of silicon p-n-junctions in strong microwave fields. And in [3] [4], the results of theoretical and experimental studies of the effect of occurrence of negative differential resistance of the diode structures based on p-n-junction theoretically described experimentally observed effect of occurrence of negative differential resistance mode and switch to the tunnel diode when exposed to high levels of microwave power. EMF of hot carriers U_{oc}, generated at unsymmetrical p-n-junction in the microwave electromagnetic field, despite the fact that the electron temperature much higher than that of the holes is determined by the hot holes [5]. In [6] theoretically studied the effect of lattice heating on the kinetics of the establishment of the thermopower of hot carriers in the p-n-junction. It is shown that the heating of the lattice leads to an additional third stage of establishing thermoelectric current and thermopower of hot carriers with a relaxation time determined by the thermal conductivity and heat capacity of the sample. In [7] studied the effect of distortion of the heating wave on the recombination currents and electromotive force generated at the p-n-junction in a strong microwave field. It is shown that high-frequency perturbation of the surface potential and the height of the p-n-junction in the mode of the short circuit current decreases the effective barrier height, and idling to anomalously large values of the EMF. In addition, the effect of deformation on the electrical properties of semiconductor devices was presented in many papers [8]-[10], where it showed the action of the pressure change-voltage characteristic of p-n-junctions, Schottky diodes, transistors and other semiconductor devices. The study of these phenomena is of interest from two points of view. Firstly, changes in the characteristics of semiconductor devices can be used for indicating pressure or other mechanical quantities in pressure sensors, audio receivers and similar devices. Secondly, these phenomena are of interest in terms of the definition of the physical processes causing the reliability of semiconductor devices, as the deformation occurring in the unit, for whatever reasons, lead to disturbances in his work. However, the above studies did not investigate the influence of deformation on the CVC p-n-junction in a strong microwave field.

According to the concept of elementary excitations of a semiconductor is the electron-hole and phonon gas. Electromagnetic wave (microwave, light) excites an electron-hole and phonon system. In the pulsed mode of microwave waves or light mainly heats the electron-hole system. Phonon system remains unexcited. When the impact of variable deformation excited phonon system, and the electron-hole system is rebuilt to the changes of the crystal lattice. Thus, the microwave and the deformation from different sides excites semiconductor systems, from the electron-hole and phonon system. Investigation using the microwave [1]-[7] and the variable deformation [8]-[10] allows two different sides examine the semiconductor structure.

Investigation of the effect of deformation on the current and voltage in strong microwave fields of interest, both from a physical point of view, and in the practical application. When the strain changes the energy spectrum, in particular, the band gap can change the saturation currents of p-n-junction. This can greatly affect the currents generated in the p-n-junction of the microwave field. On the other hand, p-n-transitions in strong microwave fields may often be exposed to strong mechanical vibration. It may also affect the current and EMF generated by p-n-junction. You can expect a change CVC p-n-junction with simultaneous exposure to the microwave field and deformation.

However, the previous studies [1]-[10] did not consider the simultaneous influence of mechanical deformation and the microwave field at the CVC p-n-junction.

The aim of this work is to study the effect of deformation on the CVC p-n-junction in a strong microwave field.

2. Calculations of the CVC Strain of p-n-Junction in a Strong Microwave Field

The average value of the total current through the diode under the influence of an electromagnetic wave consists of the electron and hole currents [5].

At high power wave when $T_e \neq T_h \neq T$, CVC is determined by the formula [5]:

$$\begin{aligned}\bar{j} = j_{se}&\left\{\left(\frac{T_e}{T}\right)^{\frac{1}{2}}\exp\left(\frac{e\varphi_0}{kT}-\frac{e(\varphi_0-U)}{kT_e}\right)\int_0^{2\pi}\exp\left(-\frac{eU_B\cos(\omega t)}{kT_e}\right)\frac{\mathrm{d}(\omega t)}{2\pi}-1\right\}\\ &+j_{sh}\left\{\left(\frac{T_h}{T}\right)^{\frac{1}{2}}\exp\left(\frac{e\varphi_0}{kT}-\frac{e(\varphi_0-U)}{kT_h}\right)\int_0^{2\pi}\exp\left(-\frac{eU_B\cos(\omega t)}{kT_h}\right)\frac{\mathrm{d}(\omega t)}{2\pi}-1\right\}\end{aligned} \tag{1}$$

where, $j_{se}=\frac{eD_e n_p}{L_e}$, $j_{sh}=\frac{eD_h p_n}{L_h}$—saturation currents of electrons and holes; φ_0—height of the potential barrier in the absence of an electromagnetic wave; $\varphi=\varphi_0-U$; U—the resulting voltage across the diode; $U_B=-\int_0^d E_B\mathrm{d}x$—AC voltage of the incident wave, created by the barrier diode; T is the temperature of the lattice; k—Boltzmann constant; T_e and T_h—temperature of electrons and holes; E_B—electric field of the wave; e is the charge of the electron; D_e and D_h—diffusion coefficients of electrons and holes, L_e and L_h—

their diffusion length; n_p and p_n—the concentration of minority carriers.

If the value of the diffusion coefficient, diffusion length, mobility and concentration of electrons and holes for a silicon p-n-junction to take as in [5] [11], and given the dependence of the intrinsic concentration of charge carriers from the lattice temperature and the band gap of silicon [11]:

$$n_i^2 = 1.5\times10^{33}\times T^3\times \mathrm{e}^{-\frac{\Delta\varepsilon}{kT}} \tag{2}$$

$$\Delta\varepsilon = 1.27\text{эВ} \tag{3}$$

then the formula (1) can be written as follows:

$$\begin{aligned}\bar{j} = {}& \sqrt{\frac{ekT_e}{\tau_e}\times T^{3.4}\times10^9}\times\frac{3\times10^{33}\times \mathrm{e}^{-\frac{\varepsilon_g}{kT}}}{p_p} \\ &\times\left\{\left(\frac{T_e}{T}\right)^{\frac{1}{2}}\exp\left(\frac{e\varphi_0}{kT}-\frac{e(\varphi_0-U)}{kT_e}\right)\int_0^{2\pi}\exp\left(-\frac{eU_B\cos(\omega t)}{kT_e}\right)\frac{\mathrm{d}(\omega t)}{2\pi}-1\right\} \\ &+\sqrt{\frac{ekT_h}{\tau_h}\times T^{3.7}\times2.5\times10^8}\times\frac{1.5\times10^{33}\times e^{-\frac{\varepsilon_g}{kT}}}{n_n} \\ &\times\left\{\left(\frac{T_h}{T}\right)^{\frac{1}{2}}\exp\left(\frac{e\varphi_0}{kT}-\frac{e(\varphi_0-U)}{kT_h}\right)\int_0^{2\pi}\exp\left(-\frac{eU_B\cos(\omega t)}{kT_h}\right)\frac{\mathrm{d}(\omega t)}{2\pi}-1\right\}\end{aligned} \tag{4}$$

If in this equation will take into account permanent deformation that:

$$\varepsilon_g(\varepsilon,T) = \varepsilon_g(0)+\Delta\varepsilon \tag{5}$$

Then the formula (4) has the following form:

$$\begin{aligned}\bar{j} = {}& \sqrt{\frac{ekT_e}{\tau_e}\times T^{3.4}\times10^9}\times\frac{3\times10^{33}\times \mathrm{e}^{-\frac{\varepsilon_g(0)+\Delta\varepsilon}{kT}}}{p_p} \\ &\times\left\{\left(\frac{T_e}{T}\right)^{\frac{1}{2}}\exp\left(\frac{e\varphi_0}{kT}-\frac{e(\varphi_0-U)}{kT_e}\right)\int_0^{2\pi}\exp\left(-\frac{eU_B\cos(\omega t)}{kT_e}\right)\frac{\mathrm{d}(\omega t)}{2\pi}-1\right\} \\ &+\sqrt{\frac{ekT_h}{\tau_h}\times T^{3.7}\times2.5\times10^8}\times\frac{1.5\times10^{33}\times \mathrm{e}^{-\frac{\varepsilon_g(0)+\Delta\varepsilon}{kT}}}{n_n} \\ &\times\left\{\left(\frac{T_h}{T}\right)^{\frac{1}{2}}\exp\left(\frac{e\varphi_0}{kT}-\frac{e(\varphi_0-U)}{kT_h}\right)\int_0^{2\pi}\exp\left(-\frac{eU_B\cos(\omega t)}{kT_h}\right)\frac{\mathrm{d}(\omega t)}{2\pi}-1\right\}\end{aligned} \tag{6}$$

Using formula (6), first consider the CVC p-n-junction for different strains (**Figure 1**). From these results it follows that with increasing strain the current p-n-junction increases.

Now consider using the formula (6) CVC p-n-junction (**Figure 2**): 1) without distortion and the microwave field (when $\varepsilon = 0$; $T = 300$ K; $T_e = 300$ K; $T_h = 300$ K; $U_B = 0.2$ B; $\varphi_0 = 0.5$ B; $\tau_e = 10^{-9}$ c; $\tau_h = 10^{-7}$ c; $\varepsilon_g = 1.21$эВ); 2) under strain without the microwave field (when $\varepsilon = 10^{-3}$; $T = 300$ K; $T_e = 300$ K; $T_h = 300$ K; $U_B = 0.2$ B; $\varphi_0 = 0.5$ B; $\tau_e = 10^{-9}$ c; $\tau_h = 10^{-7}$ c; $\varepsilon_g = 1.21$эВ); 3) without distortion at microwave exposure (when $\varepsilon = 0$; $T = 300$ K; $T_e = 500$ K; $T_h = 400$ K; $U_B = 0.2$ B; $\varphi_0 = 0.5$ B; $\tau_e = 10^{-9}$ c; $\tau_h = 10^{-7}$ c; $\varepsilon_g = 1.21$эВ); 4) upon deformation and microwave exposure (where $\varepsilon = 10^{-3}$; $T = 300$ K; $T_e = 500$ K; $T_h = 400$ K; $U_B = 0.2$ B; $\varphi_0 = 0.5$ B; $\tau_e = 10^{-9}$ c; $\tau_h = 10^{-7}$ c; $\varepsilon_g = 1.21$эВ).

Figure 2 shows that the strain increases currents through the p-n-junction in the absence of the microwave field (**Figure 2**, curves 1, 2) and under strong heating (**Figure 2**, curves 3, 4) of electrons and holes.

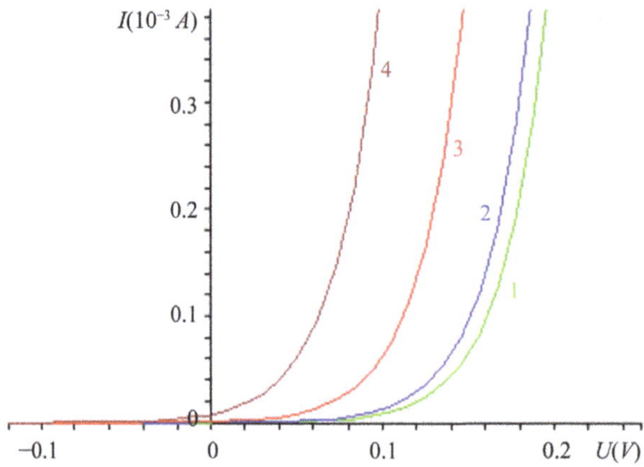

Figure 1. CVC p-n-junction under strain: $T_e = 300$ K, $T_h = 300$ K; 1—$\varepsilon = 0$; 2—$\varepsilon = 0.001$; 3—$\varepsilon = 0.005$; 4—$\varepsilon = 0.01$.

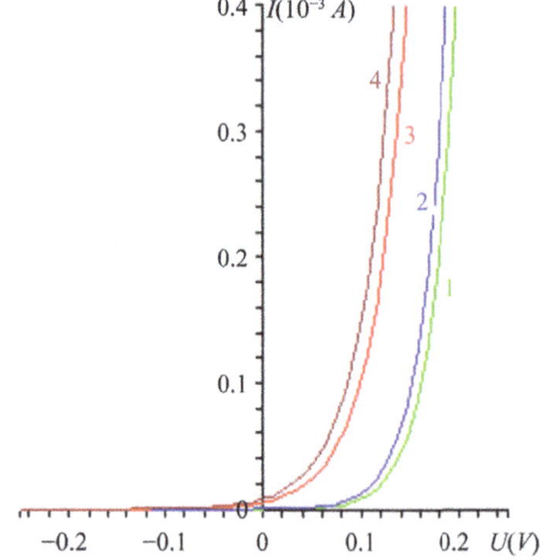

Figure 2. CVC p-n-junction: 1—without distortion and the microwave field; 2—under strain without the microwave field; 3—without distortion at microwave exposure; 4—upon deformation and microwave exposure.

Figure 3 shows the IV characteristics of p-n-junction in three-dimensional space I, U, ε: without microwave (when $T_e = 300$ K; $T_h = 300$ K field); when exposed to microwave (when $T_e = 500$ K; $T_h = 400$ K; $T_e = 600$ K; $T_h = 500$ K) field.

We showed that as the microwave is free and deformation increases currents of hot carriers in the p-n-junction. This is due to the fact that the microwave effect on the electron-hole gas, and the deformation changes the energy structure of semiconductors. It turned out that even in a strong microwave field deformation increases currents at the p-n-junction. This is due to the fact that the currents of p-n-junction affect both the heating of the electron-hole gas and the change in the deformation of the band gap of the semiconductor.

3. Conclusions

Firstly, we examined voltage-current characteristic strain p-n junction in a strong microwave field. The variation of the band gap semiconductor due to deformation influences the concentration of minority carriers in p-n junction. This leads to increase saturation currents of the diode. The strongly changing of the saturation increases the currents of hot electrons and holes through the p-n junction. It is shown that the deformation increases the currents generated in the p-n junction which is located in the microwave field.

We analyzed current-voltage characteristics of p-n transition in three-dimensional space (I, U, ε). This gives

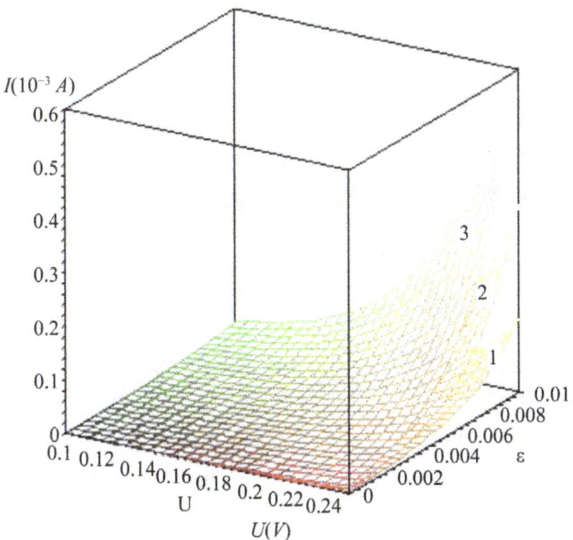

Figure 3. CVC p-n-junction in the three I, U, ε space: 1—no microwave (when T_e = 300 K; T_h = 300 K) field; 2—when exposed to microwave (when T_e = 500 K; T_h = 400 K) field; 3—when exposed to microwave (when T_e = 600 K; T_h = 500 K) field.

more complete information about the mode of deformation of the diode under various values in a strong microwave field.

References

[1] Ablyazimova, N.A. and Veynger, A.I. (1988) *Semiconductors*, **22**, S2001-S2007.

[2] Ablyazimova, N.A. and Veynger, A.I. (1992) *Semiconductors—St. Petersburg*, **26**, S1041-S1047.

[3] Usanov, D.A., Skripal, A.V. and Ugryumova, N.V. (1998) *Semiconductors*, **32**, S1399-S1402.

[4] Usanov, D.A., Skripal, A.V., Ugryumova, N.V., Venig, S.B. and Orlov, V.E. (2000) *Semiconductors*, **34**, S567-S571.

[5] Dadamirzaev, M.G. (2011) *Semiconductors—St. Petersburg*, **45**, 299-302.

[6] Gulyamov, G., Dadamirzaev, M.G. and Boydedaev, S.R. (2000) *Semiconductors—St. Petersburg*, **34**, 266.

[7] Gulyamov, G., Dadamirzaev, M.G. and Boydedaev, S.R. (2000) *Semiconductors—St. Petersburg*, **34**, 572.

[8] Polyakova, A.L. (1979) Deformation of Semiconductors and Semiconductor Devices.

[9] Polyakov, A. (1972) *Acoustical Physics*, **18**, 1-22.

[10] Bahadyrhanov, M.K. and Abdurahimov, A. (1987) *FTP*, **21**, S1710-S1712.

[11] Baransky, P.I., Klochkov, V.P. and Potykevich, I.V. (1975) Semiconductor Electronics. Kyiv., Naukova Dumka, 704.

6

Using Microwave Heating to Completely Recycle Concrete

Heesup Choi[1,2], Myungkwan Lim[1,2*], Hyeonggil Choi[3], Ryoma Kitagaki[3], Takafumi Noguchi[3]

[1]Department of Civil Engineering, Kitami Institute of Technology, Hokkaido, Japan
[2]Graduated School of Engineering, Hankyong National University, Ansung, Korea
[3]Department of Architecture, The University of Tokyo, Tokyo, Japan
Email: [*]limmk79@naver.com

Abstract

The aim of this study was to develop a technique for the complete recycling of concrete based on microwave heating of surface modification coarse aggregate (SMCA) with only inorganic materials such as cement and pozzolanic materials (silica fume, fly ash). The mechanical properties of SMCA, which was produced using original coarse aggregate (OCA) and inorganic admixtures, as well as its separation from the cement matrix and recovery performance were quantitatively assessed. The experimental results showed that micro structural reinforcement of the interfacial transition zone, which is a weak part of concrete, by coating the surface of the OCA with cement and admixtures such as pozzolanic materials can help suppress the occurrence of micro-cracks and improve the mechanical performance of the OCA. Microwave heating was observed to cause micro-cracking and hydrate decomposition. Increasing the void volume and weakening the hydrated cement paste led to the effective recovery of recycled coarse aggregate.

Keywords

Recycling, Surface Modification, Interfacial Transition Zone, Pozzolanic Reaction, Microwave, Recovery

1. Introduction

Concrete, which is used in large quantities in civil engineering and building construction, becomes weak with time; thus, old structures must be demolished and replaced [1] [2]. The handling of old concrete is a major problem

[*]Corresponding author.

for society to adhere to the 3R concept (reduce, reuse, and recycle). The accumulation and storage of concrete in huge piles cannot be a long-term solution because of the reduced natural resources and lack of space. Moreover, decreases in road construction work, which is the main use for recycled concrete, are expected to lead to less demand for sub-base coarse material for roads [3] [4]; in the long term, this calls for measures to expand and diversify the use of recycled concrete waste and to use recycled aggregates for concrete manufacture [5]. Thus, research on recycled aggregates is being conducted from various angles worldwide, and the Japanese Industrial Standards (JIS) has been revised for recycled aggregates [6]. However, there are still problems related to the production of high-quality recycled aggregates such as high energy consumption and the generation of large amounts of fine powder during crushing [7] [8]. On the other hand, using low-quality recycled aggregates can lower the concrete performance, which impedes the spread of recycled aggregate use [9] [10]. Because the aggregate resources that can be newly used are limited, an efficient and reliable mechanism for concrete recycling with low energy consumption is necessary. There are a variety of benefits to recycling concrete rather than dumping or burying it in a landfill [2] [11]:

- Keeping concrete debris out of landfills saves space.
- Using recycled material as gravel reduces the need for gravel mining.
- Recycling 1 ton of cement can save 1360 gallons of water and 900 kg of CO_2.
- Using recycled concrete as the base material for roadways reduces the pollution involved in trucking material.

2. Technical Overview

Concrete recycling via microwave heating is a completely new technique and is shown in **Figure 1** [21] [22]. Admixtures (e.g., pozzolanic materials) improve the chemical bonding and mechanical friction between aggregates in the coating layer of the original coarse aggregate (OCA) surface and cement matrices at the interfacial transition zone (ITZ), which is the weak part of concrete. Thus, recycled coarse aggregate (RCA) can be recovered for concrete structures because the mechanical performance of the concrete is improved, as shown in **Figure 2(a)** and **Figure 2(b)**. This technique involves coating the OCA with iron oxide (Fe_2O_3), which has a high dielectric constant, as a binder and then selectively heating and weakening the aggregate interface with microwaves to manufacture RCA following the dismantling of a structure, as shown in **Figure 3(a)** and **Figure 3(b)** [5]. This technique allows almost complete recycling of the aggregates by recovering high-quality RCA while using a small amount of energy [21] [22]. This technique allows for a trade off [9] between improvement in the concrete strength and aggregate recovery rate. Concrete fabricated with this technique comprises OCA, surface modification coarse paste (SMCP), surface modification coarse aggregate (SMCA), and Fe_2O_3, as shown in **Figure 1** [5].

3. Mechanical Performance of OCA-SMCP-Cement Matrix

3.1. Experiment Overview

In general, because the bond strength between the aggregate and paste is less than the individual tensile strengths of either the aggregate or paste, a crack tends to initiate from the aggregate-paste interface (*i.e.*, ITZ) owing to bleeding in the fresh concrete or a load-induced crack in the hardened concrete [12]. Therefore, the bond strength of aggregate-paste is somewhat directly related with the strength of the concrete, and this shear bond strength is generated by chemical and physical adhesion [13] [14]. In order to review the effectiveness of surface modification at improving the OCA-SMCP (interface)-cement matrix in detail, as shown in **Figure 2**, a shear bond strength experiment was conducted by coating the OCA surface with either cement or an admixture comprising cement and pozzolans. The changes in mechanical properties were assessed.

3.2. Experimental Method

In this experiment, specimens with OCA, SMCP, and a cement matrix structure were fabricated in order to characterize the chemical and physical bonds that form in the interface between the modified aggregate and cement matrices, as shown in **Figure 4**. The compressive and tensile shear bond strengths were measured at interface angles (α, β) of 30˚, 45˚, and 60˚, and the failure load at each angle was compared and analyzed [15] [16]. The OCA specimens were cut from crushed hard sandstone (standard density: 2.66 g/cm^3, water absorption ratio: 0.70%); the compressive and tensile shear specimens had dimensions of 10 cm × 10 cm × 40 cm and 10 cm ×10

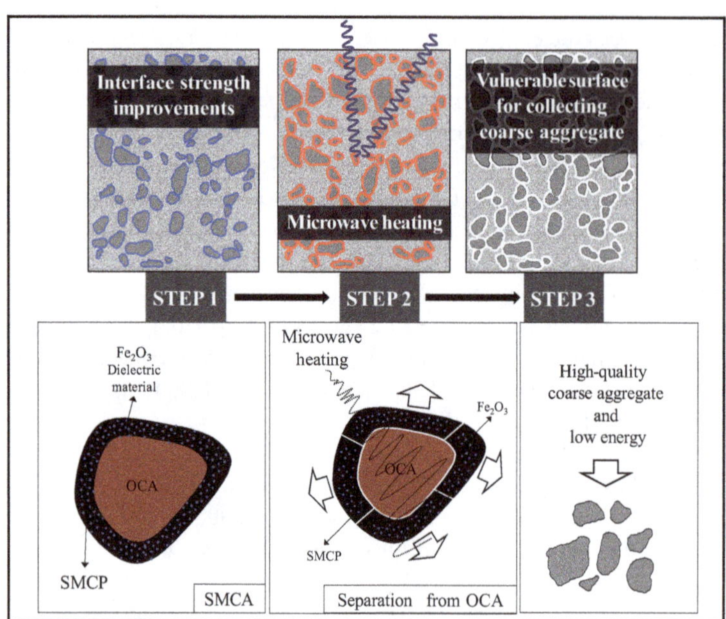

Figure 1. Improvement in concrete strength by modification and recovery by microwave heating [21] [22].

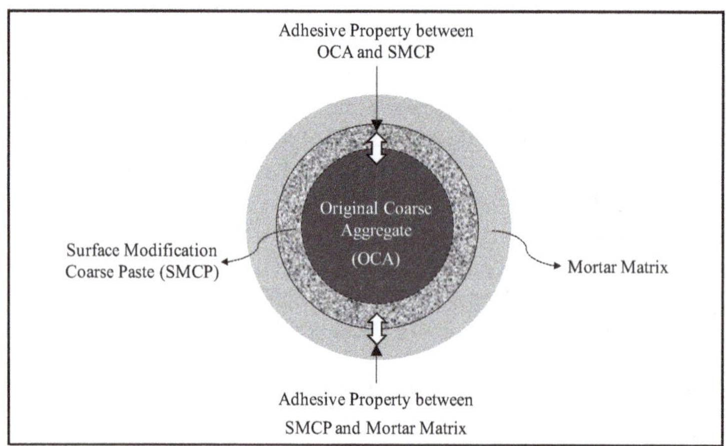

(a)

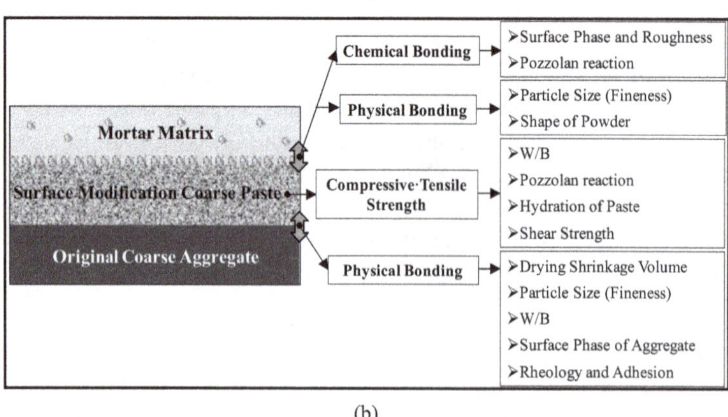

(b)

Figure 2. Mechanism of adhesion between modified aggregate and interface [5]. (a) Concept of surface adhesion of modified aggregate; (b) Control factor of surface adhesion of modified aggregate.

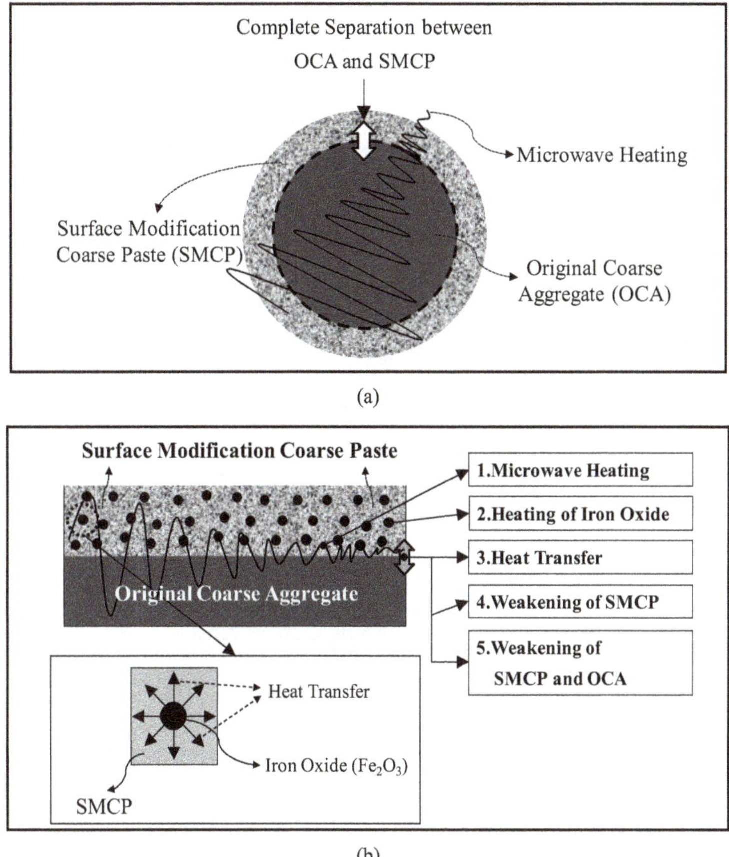

Figure 3. Mechanism of weakening between OCA and SMCP [5]. (a) Concept of surface separation between OCA and SMCP; (b) Control factor of surface separation of modification aggregate.

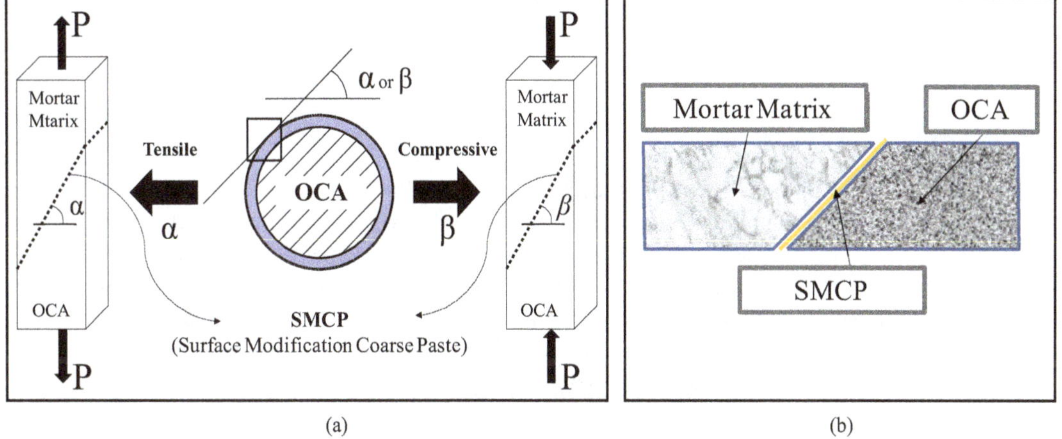

Figure 4. Outline of shear bond testing of SMCA concrete. (a) Concept of shear bond testing; (b) Manufacture of specimen.

cm × 20 cm, respectively (**Table 1**). The SMCP was mixed based on the mixing ratio presented in **Table 2**; for the materials, cement and cement substituted with pozzolanic materials (silica fume, fly ash) were used. The W/C ratio for the mortar was set to 55%; this ratio is generally applied in the field to satisfy the general strength condition given in **Table 3**. In the experiment, the modified aggregate specimens were cut in advance to ensure that the interface angle would be 30°, 45°, and 60°; these were installed in the form before SMCP was coated on

Table 1. Experimental factors and conditions.

Experimental factors		Conditions
OCA		Cutting specimen of crushed hard sandstone (Standard density: 2.66 g/cm^3, Water absorption ratio: 0.70%)
SMCP	W/C	30%
	Modification materials	Cement, Fly ash (F/A), Blast furnace slag (BFS), Silica fume (S/F)
	Replace ratio of modification materials (%)	Cement = 100 Cement + Fly-ash (F/A) = 70:30 Cement + Blast Furnace Slag (BFS) = 70:30 Cement + Silica-fume (S/F) = 90:10
Cement matrix		Normal strength W/C = 55%
Interface angle		30°, 45°, 60°

Table 2. Mix proportions of SMCP.

Type	W/B (%)	W (ml)	Volume (ml/l)				Weight (g/l)				Fe_2O_3 (g/l)
			C	FA	BFS	SF	C	FA	BFS	SF	Fe_2O_3
C	30	487	513	0	0	0	1620	0	0	0	1620
FA30			359	154	0	0	1134	363	0	0	1497
BFS30			359	0	154	0	1134	0	446	0	1580
SF10			461	0	0	51	1458	0	0	113	1571

Note: Density of each material (unit: g/cm^3). C: Cement: 3.16; FA: Fly ash: 2.22; BFS: Blast furnace slag: 2.91; SF: Silica fume: 2.24.

Table 3. Mix proportion of mortar.

Type	W/C (%)	Air (%)	Unit weight (kg/m^3)			
			W	C	Fine aggregate	Admixture
OCA's concrete (O) SMCA's concrete (M)	55	4.5 ± 1.5	175	318	833	1.59 (a*)

Note: O: OCA concrete; M: SMCA concrete; a*: Plasticizer.

the specimen surfaces. The specimens were then air-cured for 28 days in a room at constant temperature and humidity (20°C, 60%RH), and the mortar was placed. In order to prevent failure in the aggregate and cement matrix surface when the specimen was taken out of the form, the specimens were air- and water-cured for 5 and 28 days, respectively, in the room at the same constant temperature and humidity. Then, a load was applied at a rate of 1mm/min to measure the bond failure in the interface. In the tensile shear bond strength test, epoxy resin adhesive was used to attach the specimen to the tensile testing instrument, and the load was applied using a round loop to prevent eccentricity. **Table 4** presents the experimental levels of the SMCA concrete.

3.3. Compressive Shear Bond Strength

Figure 5 shows the results of the compressive shear bond strength test; the SMCA concrete demonstrated an approximately 50% increase in strength compared to the OCA concrete regardless of the interface angle. The compressive shear bond strength tended to decrease as the interface angle increased; this was deemed to be caused by the sliding effect resulting from the increase in the interface angle irrespective of the bonding surface. On the other hand, adding silica fume to the SMCP led to higher compressive shear bond strength because of the micro-filler effect and pozzolanic reaction. A 30% substitution with fly ash resulted in the second-highest compressive shear bond strength following the SMCP containing silica fume. Thus, the pozzolanic reaction caused by the

Table 4. Experimental levels.

Experiment		Modified paste	Compressive shear	Tensile shear
O		N/A		
M	C	A	P	P
	C + FA			
	C + BFS			
	C + SF			

Note: O: OCA; M: Modified coarse aggregate (SMCA concrete), C: Cement; C + FA: Cement + Fly ash; C + BFS: Cement + Blast furnace slag; C + SF: Cement + Silica fume, N/A: Not applicable; A: Application; P: Performed.

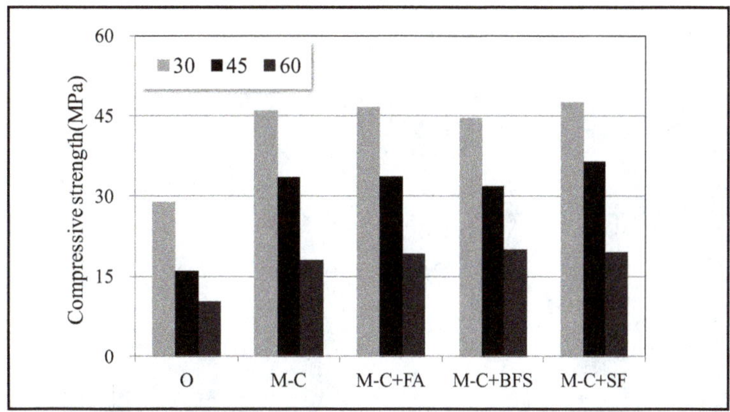

Figure 5. Compressive shear bond strength.

substituted fly ash may improve the strength.

3.4. Tensile Shear Bond Strength

Figure 6 shows the results of the tensile shear bond strength test; all of the modified aggregate specimens had improved shear bond strength compared to the OCA. In addition, the tensile shear bond strength was observed to increase with the interface angle. In other words, a larger bonding surface meant a larger area of the microstructure of ITZ was improved, which in turn improved the shear bond strength. In particular, when a pozzolanic material such as silica fume or fly ash was added to the SMCP, the tensile shear bond strength increased even further; this may be a result of the structural densification caused by the micro-filler effect and pozzolanic reaction. **Figure 7(a)** and **Figure 7(b)** show the fracture surfaces and scanning electron microscope images of the ITZ for the OCA and SMCA concretes. As shown in **Figure 7(a)**, failure occurred in the ITZ between the OCA and cement matrices, which were observed to contain calcium hydroxide (C-H) and ettringite. However, failure of the SMCA concrete occurred in the cement matrices and not in the interface, as shown in **Figure 7(b)**. This may indicate that a denser and stronger ITZ with a high level of calcium silicate hydrate (C-S-H) was strengthened by the surface modification treatment and pozzolanic reaction.

4. Weakening between OCA and SMCP by Microwave Heating

4.1. Experiment Overview and Method

An experiment was carried out to measure the weakening between OCA and SMCP caused by microwave heating (frequency of 2.45 GHz and high-frequency output of 1800 W) at heating times of 0, 60, 120, or 180 sin the SMCA concrete, as shown in **Figure 3**. The microwave heating characteristics and changes in the pores before and after heating were measured by mercury intrusion porosimetry (MIP). The temperature was measured using thermography before and after microwave heating was applied, and the temperature characteristics under each condition were assessed (**Figure 8**). The experimental specimens were manufactured by selecting O and M-C

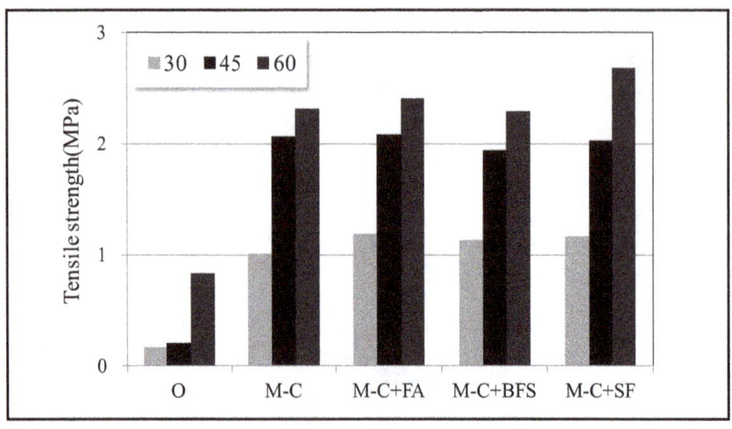

Figure 6. Tensile shear bond strength.

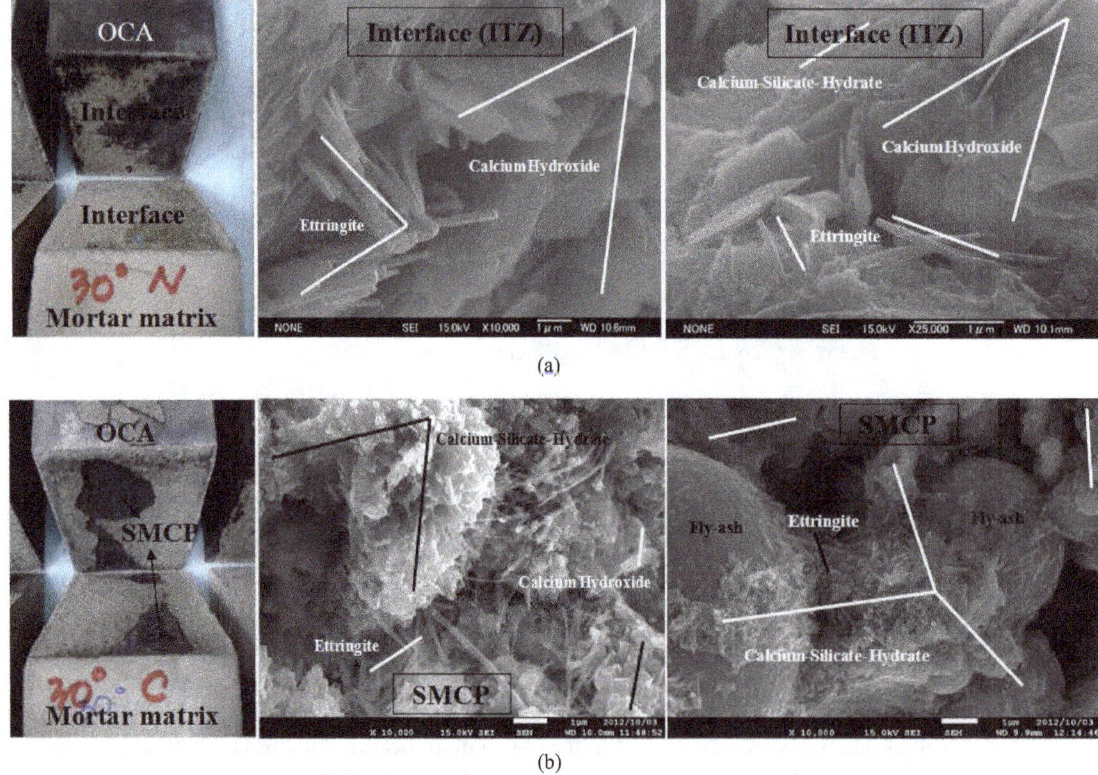

Figure 7. Fracture surfaces and scanning electron microscope images for ITZ of each concrete specimen. (a) OCA concrete; (b) SMCA concrete.

from the pozzolanic materials in Section 3. Also, aggregates with a significant amount of cement paste attached were collected from the concrete for use as test specimens in the MIP experiment, as shown in **Figure 9**.

4.2. Temperature Performance of SMCA by Microwave Heating

Figure 10 shows the experimental results for the OCA concrete. The temperature increased to 110°C after 60 s, 200°C, after 120 s, and 280°C after 180 s. In comparison, the temperature of the SMCA concrete increased to 190°C after 60 s, 280°C after 120 s, and 405°C after 180 s, as shown in **Figure 10**. The SMCA concrete showed a greater temperature increase of approximately 80°C - 130°C compared to the OCA concrete. Thus, microwave heating was determined to have a greater effect on the former. In particular, when the SMCA was heated for 180 s, the increase in temperature was approximately 130°C, higher than that of the OCA. This may be due to the

Figure 8. Microwave heating.

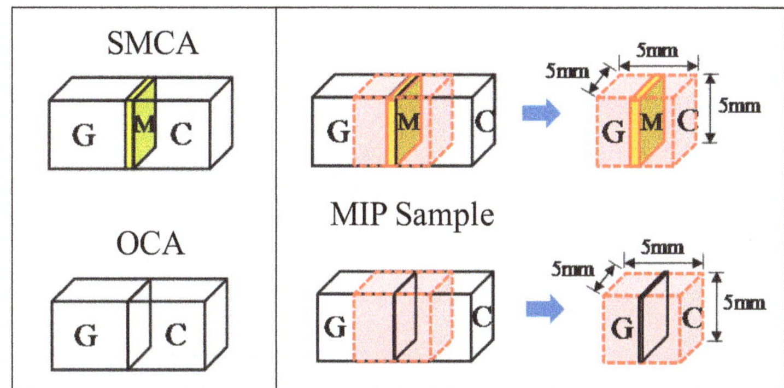

Figure 9. Microwave heating. Note: G: Gravel; M: SMCP; C: Cement matrix.

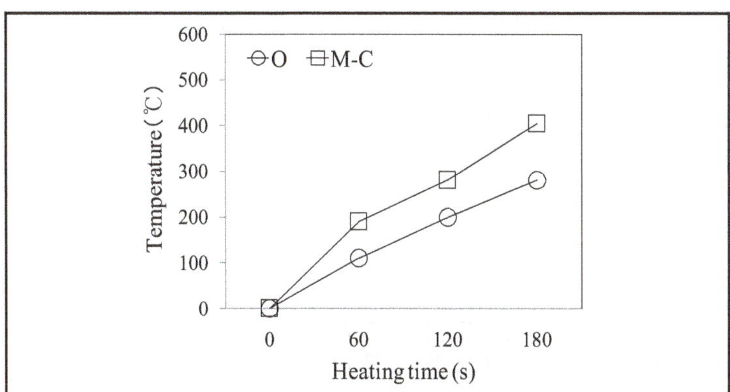

Figure 10. Temperature characteristics of SMCA and OCA with microwave heating.

effective heating of iron oxide [17], which was mixed in the SMCP as a dielectric material, as shown in **Figure 11**.

4.3. Weakening of SMCA Due to Changes in Void Volume

The experimental results showed that the temperatures of the specimens increased because of the thermal conductivity of the iron oxide being heated by the microwaves, as shown in **Figure 12** and **Figure 13**. The total increase in void volume and peak of the pore distribution moved to the part with the largest void volume. In particular, when the heating time was over 180 s, the number of pores with a diameter of 0.05 μm or less decreased, whereas pores with a diameter of 0.05 - 0.1 μm moved toward spores with a diameter of 0.1 μm as the void volume increased (**Figure 13**). As shown in **Figure 14**, when the heating time was over 180 s, the void volume

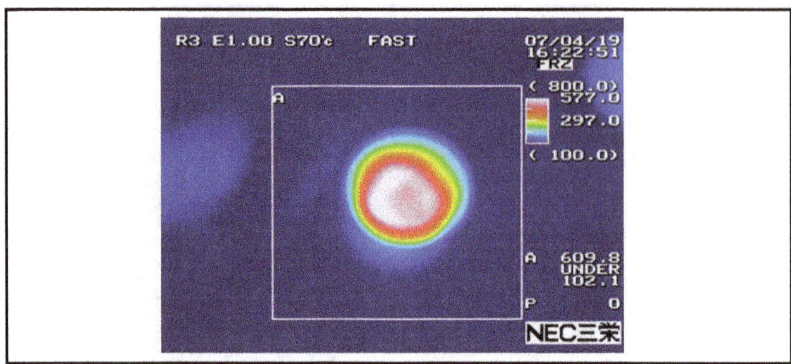

Figure 11. Heating of iron oxide by microwave heating (577°C after 30 s).

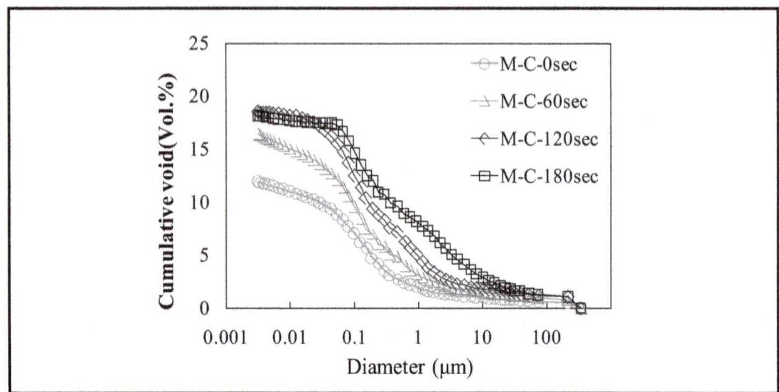

Figure 12. Cumulative void size distribution of SMCA.

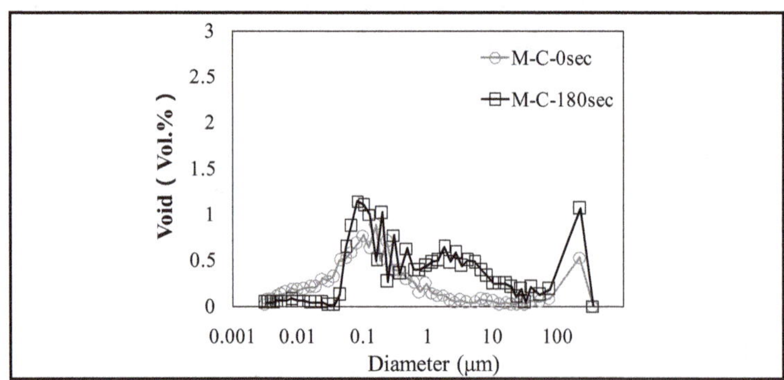

Figure 13. Void size distribution of SMCA.

of the SMCA concrete was slightly higher than that of the OCA concrete. The increase in the concrete porosity [18] [19] because of microwave heating may represent the weakening of the cement paste around the ITZ and aggregate. Thus, the increase in pores in the SMCA concrete because of microwave heating may result from the evaporation of bound water caused by the decomposition of calcium hydroxide and C-S-H hydrates and the occurrence of micro-cracks in the concrete [20].

5. Performance Review of SMCA Concrete

5.1. Experiment Overview and Method

Concrete is generally composed of cement, water, and fine and coarse aggregates. However, with respect to the SMCA concrete, the surface of the OCA was coated with cement paste. This required that the modified paste be reflected in the mixing design, as shown in **Figure 15**. In the mixing design, the amount of coarse aggregate coated with an admixture can be increased. Thus, the amount of SMCP coated on the coarse aggregate can be excluded from the amounts of cement and water in the mixing ratio, and the amounts of cement, water, and

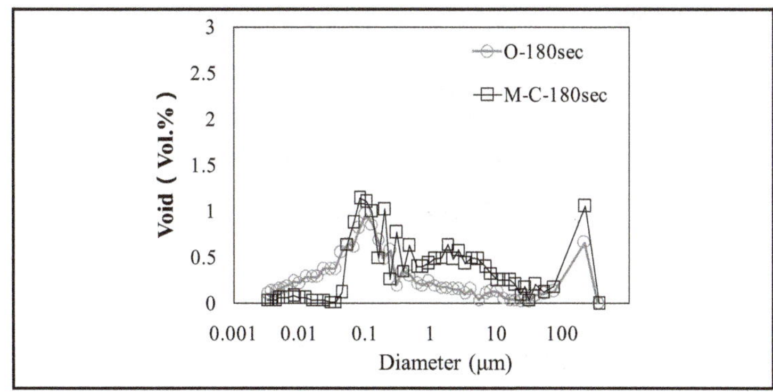

Figure 14. Void size distribution of each aggregate.

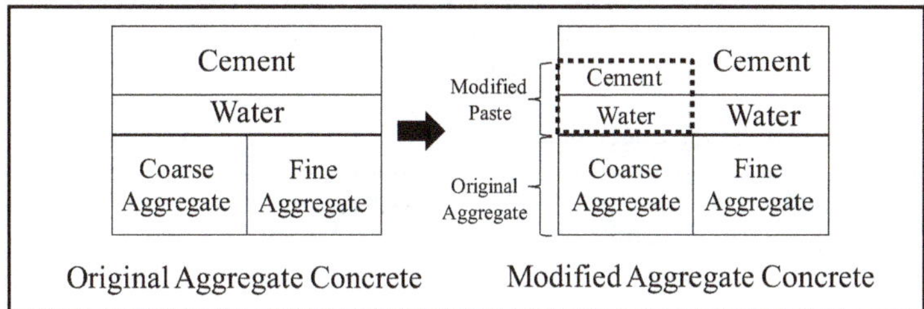

Figure 15. Ratios of constituents in concrete (Note: Materials/m^3) [5].

OCA for the OCA and SMCA concretes can be the same [5]. In this experiment, OCA and SMCA concrete specimens with a water-cement (W/C) ratio of 55% were compared. In order to achieve equal characteristics for the fresh specimens for the comparison, the W/C ratio of the SMCA concrete and amount of chemical admixture were adjusted as shown in **Table 5**. Using the above method, the same amount of cement was applied to the mixing design for the OCA and SMCA. As the binder for the surface admixture, only cement and cement with admixtures (e.g., pozzolanic material) were used with a W/C ratio of 30%. The mixing percentage of the surface admixture is shown in **Table 6**. Based on the mechanical performance and weakening of the SMCA concrete in Sections 3 and 4, the compressive and splitting tensile strengths of the SMCA concrete were examined in this experiment to assess the improvement in the mechanical properties. The recovery rates of RCA by microwave heating (frequency of 2.45 GHz and high-frequency output of 1800 W) for 0, 60, 120, and 180 s were measured to assess the aggregate recovery characteristics.

For the OCA used in this experiment, the parent materials were collected from a rocky mountain, crushed with a jaw crusher, and sorted by diameter from 5 mm to 20 mm. Then, the particle size distribution was adjusted to satisfy the standard diameter prescribed in JIS A 1102 (standard for aggregate sieving test), and the density and water absorption ratio of the OCA and SMCA specimens (only cement and cement with pozzolanic materials, respectively) were measured in accordance with JIS A 1110 (standard for coarse aggregate density and water absorption test). The results of the preliminary experiment showed that the density of OCA with a modified surface was higher than that with no surface modification by 1% - 1.5%, as shown in **Table 7**. This may be because the SMCP with a low W/C ratio contained high-density iron oxide, which resulted in the ITZ microstructure becoming more compact. On the other hand, the water absorption ratio also increased. This was due to the cement paste forming a thin film on the OCA surface, which resulted in the OCA and SMCP absorbing water. However, the OCA and SMCA in this experiment satisfied the criteria prescribed in JIS A 5021 (standard for concrete using recycled aggregate H) for H class of RCA, which can be used for structural aggregate. Thus, the water absorption ratio was not expected to cause any potential problems in this experiment.

5.2. Mechanical Properties of SMCA Concrete

The compressive and splitting tensile strength tests were performed based on JIS A 1108 and JIS A 1113. For the mechanical properties, the compressive and splitting tensile strengths of each SMCA concrete specimen were higher than those of the OCA concrete specimens by 5% - 12%, as shown in **Figure 16** and **Figure 17**. The

Table 5. Mix proportions of concrete.

Type	W/C (%)	Slump (mm)	Air (%)	G_{max} (mm)	Unit weight(kg/m^3)				Admixture (%)
					W	C	S	G	
OCA's concrete (O)	55	180±25	4.5±1.5	20	175	318	833	961	C × 0.5 (a*)
SMCA's concrete (M)					175	311	833	982	C × 0.9 (a*)

Note: O: OCA concrete; M: SMCA concrete; a*: Plasticizer.

Table 6. Mix proportions of modification paste.

W/B (%)	Water (g)	Binder (g)	Fe_2O_3 (g)	Superplasticizer (g)	Table flow (mm)
30	21	70	B × 100%	B × (1.9% - 2.5%)	300

Note: Based on 1 kg of OCA.

Table 7. Type and properties of each coarse aggregate.

Type		Density (g/cm^3)	Water absorption ratio (%)
O		2.66	0.70
M	C	2.69	1.65
	C + FA	2.70	1.63
	C + BFS	2.70	1.65
	C + SF	2.70	1.62

Note: O: OCA (OCA concrete); M: Modified coarse aggregate (SMCA concrete), C: Cement; C + FA: Cement + Fly ash; C + BFS: Cement + Blast furnace slag; C + SF: Cement + Silica fume.

improved strength of the modified aggregate concrete was due to the reinforced physical and chemical bonding between the modified paste and cement matrices, which was caused by the increased mechanical friction resulting from the particle size and shape of the iron oxide and the SMCP coating effect on the ITZ [5]. In addition, the SMCA concrete using pozzolanic materials (silica fume and fly ash) showed an increase in strength of approximately 8% after 7, 14, and 28 days of curing compared to the SMCA concrete using only cement (**Figure 16**). The splitting tensile strength test results showed an improvement in strength of approximately 5% after 7, 14, and 28 days of curing (**Figure 17**). As noted in Section 2, when pozzolanic materials such as silica fume and fly ash were added to the SMCP, the micro-filler effect and pozzolanic reaction caused the ITZ structure to become denser.

5.3. Recovery Properties of SMCA Concrete

As shown in **Figure 18(a)** and **Figure 18(b)**, when the microwave heating times were 0 and 60 s, the RCA recovery rate for the SMCA concrete was equivalent to or slightly lower than that of the OCA concrete. This may be due to the enhanced bonding caused by the SMCP coating between the OCA and cement matrices. On the other hand, when the microwave heating times were 120 and 180 s, the recovery rate for the SMCA concrete became substantially lower than that of the OCA concrete regardless of the admixture (cement only and cement with pozzolanic materials). When the heating time was120 s, the temperature reached 300°C, which is the weakening temperature of cement paste [5] [6]. This was the point at which the recovery rate began to differ. In particular, when the heating time was180 s, the recovery rate of RCA for the SMCA concrete reached nearly 100%; thus, the microwaves were concluded to effectively heat the dielectric material (iron oxide) present in the SMCP. This result could be predicted by the RCA recovery rates for both the OCA and SMCA concrete specimens after 180 s of microwave heating, as shown in **Figure 19**.

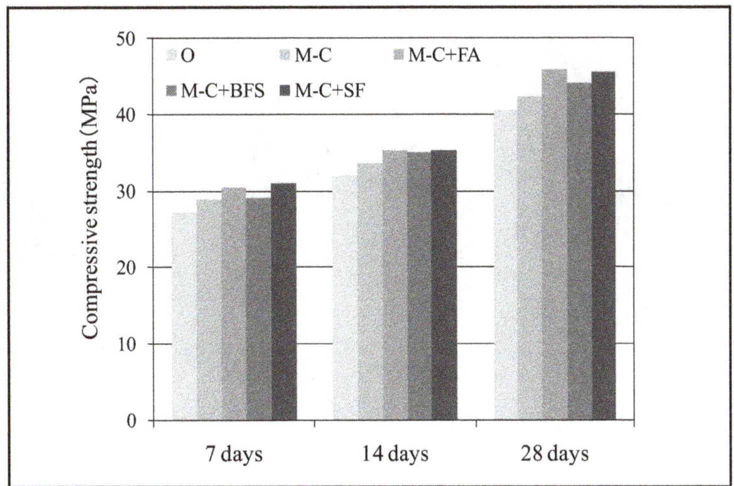

Figure 16. Comparison of strength.

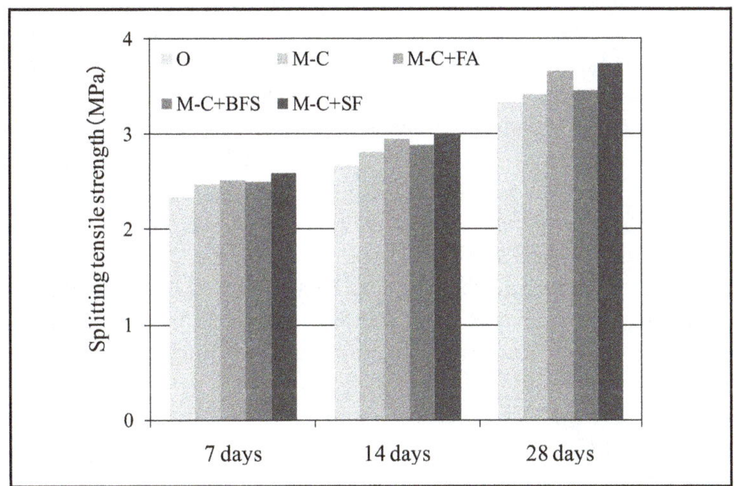

Figure 17. Splitting tensile strength.

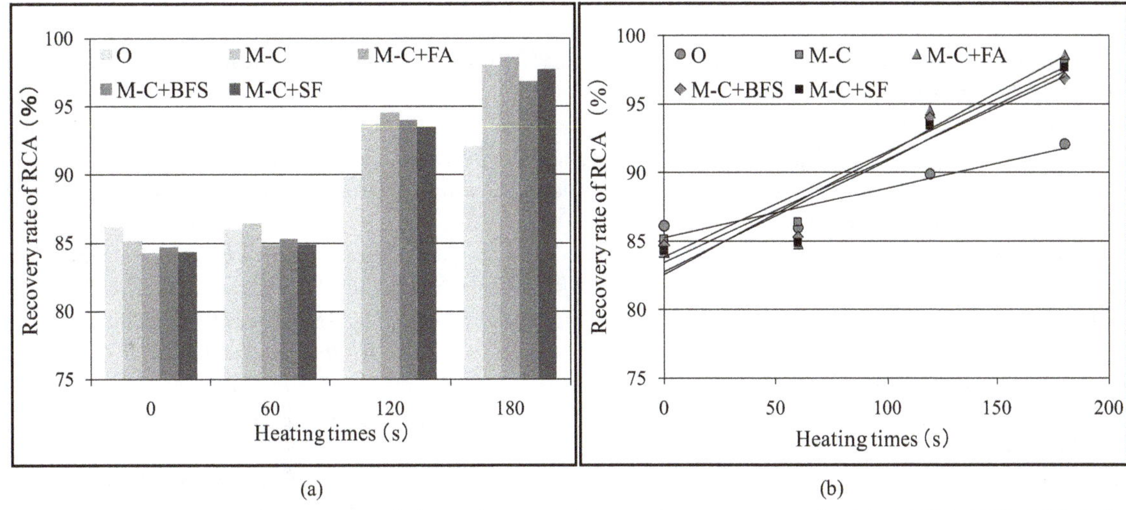

Figure 18. Recovery rate of RCA of each concrete specimen by microwave heating. (a) RCA recovery rates; (b) Comparison of RCA recovery rates.

Figure 19. Comparison of RCA from each type of concrete.

6. Conclusions

The following conclusions were made based on the results of the experiments conducted to examine the mechanical properties and recovery performance of the SMCA concrete with admixtures and microwave heating.

1) When the W/C ratio was 55%, the improvements in the compressive strength and split tensile strength, including the shear bond strength, were confirmed to be due to the SMCP coating; the ITZ structure was made denser by the admixtures (only cement and pozzolanic materials).

2) When the SMCA concrete containing iron oxide (Fe_2O_3), was heated with microwaves, the temperature increased more significantly compared to the OCA concrete. In particular, when the microwave heating time was 180 s, the maximum temperature was over 400°C, and micro-cracks occurred in the concrete along with an increased void volume caused by dehydration of the hydrates. Based on these results, the microwaves effectively heated the iron oxide contained in the SMCP.

3) The RCA recovered after microwave heating for 180 s contained less than 5% of paste and fine aggregate, regardless of the admixture. The recovered RCA was very similar to the OCA, which proves the feasibility of recovering high-quality RCA. Thus, microwave heating was determined to weaken binders containing a dielectric material for efficient recovery of RCA.

The mechanical performance of SMCA can be improved by the use of inorganic materials and microwave heating to effectively recover RCA.

References

[1] Noguchi, T. and Tamura, M. (2001) Concrete Design towards Complete Recycling. *Structural Concrete Journal of the Fib*, **2**, 155-167.

[2] Noguchi, T. (2008) Resource Recycling in Concrete: Present and Future. *Stock Management for Sustainable Urban Regeneration*, **4**, 255-274.

[3] Hendriks, Ch.F. and Janssen, G.M.T. (2001) Construction and Demolition Waste: General Process. *HERON*, **46**, 79-88.

[4] Shima, H., Tateyashiki, H., Matsuhashi, R. and Yoshida, Y. (2005) An Advanced Concrete Recycling Technology and its Applicability Assessment through Input-Output Analysis. *Journal of Advanced Concrete Technology*, **3**, 53-67. http://dx.doi.org/10.3151/jact.3.53

[5] Choi, H.S., Kitagaki, R. and Noguchi, T. (2014) Effective Recycling of Surface Modification Aggregate using Microwave Heating. *Journal of Advanced Concrete Technology*, **12**, 34-45. http://dx.doi.org/10.3151/jact.12.34

[6] Choi, H.S., Kitagaki, R. and Noguchi, T. (2012) A Study on the Completely Recovery of Surface Modification aggregate using Microwave and Effective Utilization. *Proceedings of the 5th ACF International Conference*, Pattaya, October 2012, Session 1-2, ACF2012-0093, 41-46.

[7] Kunio, Y. (2003) A Study on the Manufacturing Technology of High-Quality Recycled Fine Aggregate. *Japan Concrete Institute*, **25**, 1217-1222.

[8] Shima, H. and Tateyashiki, H. (1999) New Technology for Recovering High-Quality Aggregate from Demolished Concrete. *Proceedings of the 5th International Symposium on East Asian Recycling Technology*, The M.M.P.I. in Japan 1999, 106-109.

[9] Tamura, M., Tomosawa, F. and Noguchi, T. (1997) Recycle-Oriented Concrete with Easy-to-Collect Aggregate. *Ce-*

ment Science and Concrete Technology, **51**, 494-499.

[10] Tsujino, M., Noguchi, T., Tamura, M., Kanematsu, M. and Maruyama, I. (2007) Application of Conventionally Recycled Coarse Aggregate to Concrete Structure by Surface Modification Treatment. *Journal of Advanced Concrete Technology*, **5**, 13-25. http://dx.doi.org/10.3151/jact.5.13

[11] Value Engineering Benefits (2010) Concrete Recycling.org. Retrieved 2010-04-05.

[12] Mehta, P.K. and Moneiro, P.J.M. (2006) Concrete: Microstructure, Properties and Materials. McGraw-Hill Companies, New York.

[13] Diamond, S. and Huang, J. (2001) The ITZ in Concrete. *Cement and Concrete Composite*, **23**, 59-64.

[14] Elsharief, A., Cohen, D. and Olek, J. (2003) Influence of Aggregate Size, Water Cement Ratio and Age on the Microstructure of the Interfacial Transition Zone. *Cement and Concrete Research*, **33**, 1837-1849. http://dx.doi.org/10.1016/S0008-8846(03)00205-9

[15] Robin, P.J. and Austin, S.A. (1995) A Unified Failure Envelope from the Evaluation of Concrete Repair Bond Tests. *Magazine of Concrete Research*, **47**, 57-68. http://dx.doi.org/10.1680/macr.1995.47.170.57

[16] Austin, S., Robins, P. and Pan, Y.G. (1999) Shear Bond Testing of Concrete Repair. *Cement and Concrete Research*, **29**, 1067-1076. http://dx.doi.org/10.1016/S0008-8846(99)00088-5

[17] McGill, S.L., *et al.* (1988) The Effects of Power Level on the Microwave Heating of Selected Chemicals and Minerals. *Proceedings of the MRS Symposium*, Nevada, April 1988, 124.

[18] Schneider, U. (1982) Behavior of Concrete at High Temperatures. Deutscher Ausschuss für Stahlbeton, Berlin, 28-33.

[19] Bazant, Z.P. and Kapaln, M.F. (1996) Concrete at High Temperatures: Material Properties and Mathematical Models. Prentice Hall, Upper Saddle River.

[20] Takeo, A., Fukujiro, F., Kuniyuki, T., Kenji, K. and Isao, K. (1999) Mechanical Properties of High-Strength Concrete at High Temperatures. *Architectural Institute of Japan*, **515**, 163-168.

[21] Tsujino, M., Noguchi, T., Kitagaki, R. and Nagai, H. (2010) Completely Recyclable Concrete of Aggregate-Recovery Type by a New Technique Using Aggregate Coating. *Architectural Institute of Japan*, **75**, 17-24.

[22] Tsujino, M., Noguchi, T., Kitagaki, R. and Nagai, H. (2011) Completely Recyclable Concrete of Aggregate-Recovery Type by Using Microwave Heating. *Architectural Institute of Japan*, **76**, 223-229.

7

Determination of N and O-Atoms, of $N_2(A)$ and $N_2(X, v > 13)$ Metastable Molecules and N_2^+ Ion Densities in the Afterglows of Ar-N_2 Microwave Discharges

Andre Ricard, Hayat Zerrouki, Jean-Philippe Sarrette

Laplace, Toulouse, France
Email: ricard@laplace.univ-tlse.fr

Abstract

Early afterglows of Ar-N_2 flowing microwave discharges are characterized by optical emission spectroscopy. The N and O atoms, the $N_2(A)$ and $N_2(X, v > 13)$ metastable molecules and N_2^+ ion densities are determined by optical emission spectroscopy after calibration by NO titration for N and O-atoms and measurements of NO and N_2 band intensities. For an Ar-xN_2 gas mixture with × increasing from 2 to 100% at 4 Torr, 100 Watt and an afterglow time of 3×10^{-3} s at the 5 liter reactor inlet, it is found densities in the ranges of (2 - 6) × 10^{14} cm^{-3} for N-atoms, one order of magnitude lower for $N_2(X, v > 13)$ and for O-atoms (coming from air impurity), of 10^{10} - 10^{11} cm^{-3} for $N_2(A)$ and of 10^8 - 10^9 cm^{-3} for N_2^+.

Keywords

Ar-N_2 Microwave Discharge, Flowing Afterglow, N-Atoms, N_2 Metastables, N_2^+ Ions

1. Introduction

Afterglows of N_2 flowing microwave discharges have been studied at medium gas pressures (1 - 20 Torr) for sterilization of medical instruments by N-atoms [1] [2]. The mentioned project of sterilization in N_2 afterglow is based on N-atom etching of bacteria without oxidation by O-atoms. A part of the present study is to detect the O-atoms from air impurity to appreciate their influence on the sterilization process.

The main part concerns a study of Ar-N_2 gas mixtures to enhance the sterilization process in the early afterglow. The interest of N_2 dilution into Ar is to increase the electron energy in the plasma at constant values of transmitted power and of gas pressure. Superelastic collisions of electrons on the Ar metastable atoms produced in the plasma could enhance the electron energy. It is mentioned here that in the present measurements of flowing afterglow, the Ar metastable atoms have disappeared after collisions on the tube wall (destruction probability of about 1). As a consequence, the excitation transfers of Ar metastable atoms on N_2 can be discarded at a distance of about 1 cm after the discharge end. Another interest of Argon dilution is to maintain the plasma at high gas pressure, up to the atmospheric gas pressure while keeping a plasma power as low than 100 Watt [3].

The early flowing afterglows produced from Ar-N_2 microwave plasmas are presently studied by emission spectroscopy with the same experimental methods as in N_2-H_2 RF afterglow [4] [5], in N_2, N_2-O_2 [6] and in N_2-H_2, Ar-N_2-H_2, Ar-N_2-O_2 microwave early afterglows [6].

The present paper is focused on Ar-N_2 early afterglow by directly introducing the discharge tube of 5 mm dia. inside the 5 litre reactor. By this way, it is expected to add the metastable $N_2(A)$ and $N_2(X, v > 13)$ molecules and N_2^+ ions to the N-atoms in the surface treatments as previously experimented [1] [2]. The studied active species are as in [6] the N and O-atoms, the $N_2(A)$ and $N_2(X, v > 13)$ metastable molecules and N_2^+ ions. The intensities emitted by the N_2 first positive (1st pos.) and N_2 second positive (2nd pos.) systems and by the NO_β bands are measured to obtain the mentioned active specie densities after NO titration to calibrate the N and O-atom densities [6]. The O-atoms are coming from air impurity in the discharge.

2. Experimental Setup and NO Titration

The experimental setup is changed in comparison to the one used in [6]. The dia. 5 mm discharge tube is now directly connected to the 5 litre reactor as shown in **Figure 1**. The Ar-N_2 microwave plasmas is always produced by a surfatron cavity at 2450 MHz, 100 Watt, 1 slm, but lowering the gas pressure from 8 Torr in [6] to 4 Torr to allow a satisfactory diffusion of the afterglow inside the 5 litre reactor.

The plasma is located inside the dia.5 mm tube with a length after the surfatron gap varying from about 5 cm in pure N_2 to 20 cm in the Ar-2%N_2 gas mixture. With a discharge tube length of 30 cm after the surfatron gap, the residence time before the afterglow in the 5 litre reactor is 3×10^{-3} s.

The optical emission spectroscopy across the reactor is performed by means of an optical fiber connected to an Acton Spectra Pro 2500i spectrometer (grating 600 gr/mm) equipped with a Pixis 256E CCD detector (front illuminated 1024 × 256 pixels).

The N-atom density is obtained from the I_{580} measured intensity after calibration by NO titration as described in [6].

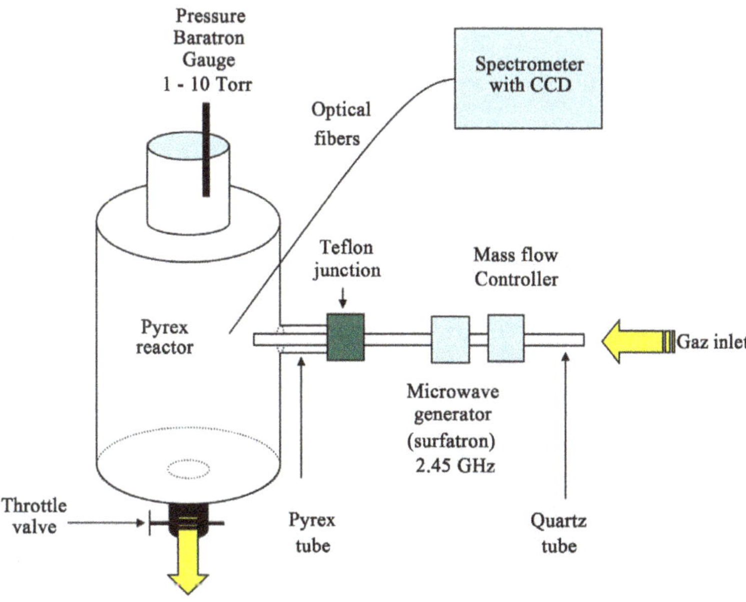

Figure 1. Microwave discharge and post-discharge reactor of 5 liters.

3. The Ar-N2 Early Afterglow

3.1. N-Atom Density

As reported in [6], the pure late afterglow emission is produced by reaction R1 in **Table 1**.

The N_2 (580 nm) band head intensity (I^m_{580}) in arbitrary unit (a.u) was measured for constant parameters of the Acton spectrometer (grating 600 gr/mm, slit of 150 μm, integrating time 1 s).

I^m_{580} is then deduced from reaction R1 with v' = 11 and hυ = hc/λ (580 nm), as follows:

$$I^m_{580} = k_1 [N]^2 \tag{1}$$

with k_1 explicited in [4]-[6].

The reaction R1 produced with an excess of Ar atoms results in a change of the N_2(B, v') distribution as compared to pure N_2 at a given a_{N+N} value. The N + N recombination coefficient a_{N+N} has been calculated in [6] in conditions of pink and late afterglows for Ar-xN_2 gas mixture with x from 2% to 100%.

Equation (1) becomes:

$$a_{N+N} I^m_{580} = k_1 [N]^2 \tag{2}$$

By NO titration,it has been verified the same k_1 value inside the error bars as for pure N_2 [6]:

$k_1 = 0.6\ (+/-\ 0.3)10^{-26}$ cm^6 counts/s with I^m_{580} in counts/s and [N] in cm^{-3}.

It is obtained $a_{N+N} = 0.9$ for pure N_2 and $a_{N+N} = 0.5$ for the Ar-2% N_2 mixture in the 5 litre reactor.

This result indicates that the early afterglow in N_2 is dominated by the N+N recombination as expressed by R1.

The N-atom density is then obtained in the 5 litre reactor by taking into account the change of diameter from 2.1 cm in the tube to 15 cm in the reactor.

It is reported in **Figure 2** the N-atom density variation with the %N_2 into Ar

A slow increase of N-atom density is found in the range 2% - 10% N_2 to reach a constant value of (5 - 6) × 10^{14} cm^{-3} between 10 and 100% N_2. The uncertainty on N-atom density is estimated to be 30% [6].

Table 1. Kinetic reactions in Ar-N_2 afterglow.

Reactions	
$N + N + (Ar\text{-}N_2) \rightarrow N_2(B, v') + (Ar\text{-}N_2)$ $N_2(B, v') \rightarrow N_2(A, v'') + h\upsilon_{580}$	R1
$N + O + (Ar\text{-}N_2) \rightarrow NO(B,0) + (Ar\text{-}N_2)$ $NO(B,0) \rightarrow NO(X,8) + h\upsilon(320\text{ nm})$	R2
$N_2(A) + N_2(A) \rightarrow N_2(C,1) + N_2(X)$ $N_2(C,1) \rightarrow N_2(B,0) + h\upsilon(316\text{ nm})$	R3
$N_2(A) + N_2(A) \rightarrow N_2(B,11) + N_2$	R4
$N_2(A) + N_2(X, v>13) \rightarrow N_2(B,11) + N_2$	R5
$N_2(a') + N_2(a') \rightarrow e + N_2^+(X) + N_2(X)$	R6
$N_2^+(X) + N_2(X>12) \rightarrow N_2^+(B) + N_2(X)$ $N_2^+(B,0) \rightarrow N_2^+(X,0) + hv(391\text{ nm})$	R7
$Ar^+ + N_2 \rightarrow Ar + N_2^+$	R8

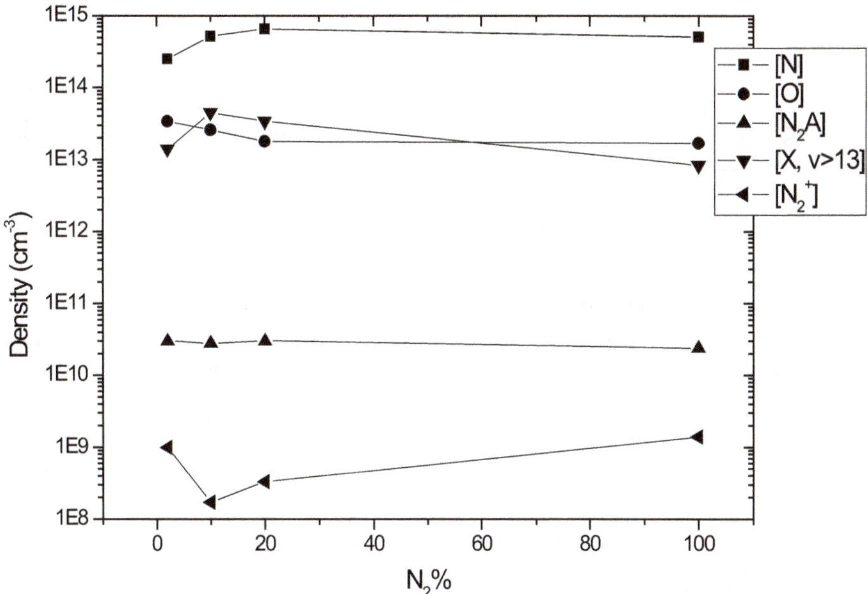

Figure 2. Active species density versus the %N_2 into the Ar-N_2 early afterglow in the 5 litre reactor at 4 Torr, 1 Slm, afterglow time of 3×10^{-3} s, plasma 100 Watt.

3.2. Density of O-Atoms in Impurity in the Ar-N2 Early Afterglow

The NO_β bands are presently observed as a result of the recombination of N and O atoms by reaction R2. In a similar way than for Equation (1), the NO (320 nm) measured band intensity (I_{320}^m) is deduced from reaction R2 as follows:

$$I_{320}^m = k_3[N][O] \quad (3)$$

The coefficients in k_3 are explicited in ref. 6 as for k_1.

The O atom density can be deduced from the N-atom density by considering the $a_{N+N}\cdot\ I_{580}^m/I_{320}^m$ ratio of reactions 2 and 3, as follows:

$$a_{N+N}\cdot I_{580}^m/I_{320}^m = k_4[N]/[O] \quad (4)$$

with $k_4 = k_1/k_3$.

After several NO titration experiments, it was found in [6]: k_4 = 1(+/−0.4). From k_4 obtained by NO titration, the O-atom density in the Ar-N_2 early afterglow inside the reactor was determined by Equation (4) after measurements of $a_{N+N}\,I_{580}^m/I_{320}^m$ and [N] versus the N_2 percent into Ar. The results are reproduced in **Figure 2**. If the uncertainty on N-atom density is estimated to be 30% (see part 3.1), the experimental errors on O-atom density calculated from Equation (4), with the uncertainty on k_4 of 40% is 90% that is near the order of magnitude.

As shown in **Figure 2**, there is a slow decrease of the O-atom density from 3 to 2×10^{13} cm^{-3} between 2% to 100%N_2.

3.3. Density of N2(A) Metastable Molecules

It has been detected the N_2(C, 1→ B, 0) emission at 316 nm near the NO_β emission at 320 nm which is used as in [4]-[6] to determine the density of the N_2(A) metastable molecule.

It is considered that the N_2 2nd positive system in the early afterglow is produced by reaction R3.

The N_2 (316 nm) measured intensity (I_{316}^m) is then given by:

$$I_{316}^m = k_5\left[N_2(A)\right]^2 \quad (5)$$

with k_5 explicited in [6].

From Equations (3) and (5), it comes the following I_{320}^m/I_{316}^m intensity ratio:

$$I^m_{320}/I^m_{316} = k_6[N][O]/[N_2(A)]^2 \quad (6)$$

with $k_6 = k_3/k_5$. The $N_2(A)$ density is then obtained from equation (6) with the N and O atom densities previously determined.

As shown in **Figure 2**, the $N_2(A)$ density kept a constant value in the Ar-N_2 gas mixture. It is estimated that it is obtained the order of magnitude of $N_2(A)$ density in the range 10^{10} - 10^{11} cm^{-3}.

3.4. Density of $N_2(X, v > 13)$ Molecules

The production of $N_2(B, 11)$ by R1 in the early afterglow is less than 1 ($a_{N+N} < 1$).

Other collisional processes in the pink afterglow [7] also excite the $N_2(B)$ states, in addition to reaction R1.

For this other part $(1 - a_{N+N})$, it is considered the reactions R4 and R5 whose rate coefficients are reported in [4]-[6]. The contribution of reactions R4 and R5 on I^m_{580} is then written as follows:

$$(1-a_{N+N})I^m_{580} = k_{R4}[N_2(A)]^2 + k_{R5}[N_2(A)][N_2(X, v > 13)] \quad (7)$$

where k_{R4}, k_{R5} are the rate coefficients of reactions R4, R5. As $a_{N+N}I^m_{580} = k_1[N]^2$, it is deduced:

$$(a_{N+N}/1-a_{N+N})\left(k_{R4}[N_2(A)]^2 + k_{R5}[N_2(A)][N_2(X, v > 13)]\right) = [N]^2 k_1 \quad (8)$$

With the experimental values of a_{N+N} and of N and $N_2(A)$ densities, it is found that $(a_{N+N}/1-a_{N+N})k_{R4}[A]^2$ is about 2 orders of magnitude lower than $[N]^2k_1$.It results that Equation 8 can be simplified as:

$$(a_{N+N}/1-a_{N+N})k_{R5}[N_2(A)][X, v > 13] = [N]^2 k_1 \quad (9)$$

From the obtained values of N-atom and $N_2(A)$ density, it was deduced the values of $[N_2(X, v > 13)]$ as reproduced in **Figure 2**.

It is observed about one order of magnitude lower $N_2(X, v > 13)$ density as compared to N values.

Such values of $[N_2(X, v > 13)]$ can be considered as an estimated value depending on the R5 rate coefficient.

3.5. Density of N_2^+ Ions

The emission of the N_2^+ band at 391 nm is observed in the present early afterglows. It is generally proposed [8] that the N_2^+, 391 nm band is produced in the pink afterglow by reactions R6 and R7.

The I^m_{391} intensity is then expressed as follows:

$$I^m_{391} = k_{10}[N_2^+][N_2, X, v > 12] \quad (10)$$

with $k_{10} = c_{391} \cdot V \cdot A_{391} \cdot k_{R7} \Big/ \left(\upsilon^R_{N_2^+} + [N_2]k^Q_{N_2^+B,0}\right)$, where c_{391} is the spectral response of spectrometer, V is the detected afterglow volume, A_{391} the Einstein coefficient of the N_2^+ (391 nm) transition, k_{R7} the rate coefficient of reaction R7 with $k_{R7} = 4\times10^{-11}\ cm^3 \cdot s^{-1}$ [9], $\upsilon^R_{N_2^+} = 1.7\times10^7\ s^{-1}$ and $k^Q_{N_2^+B,0} = 8.8\times10^{-10}\ cm^3 \cdot s^{-1}$ [10].

By comparing the intensities of I^m_{316} from Equation (5) and I^m_{391} from equation (10), it is calculated:

$$I_{391}/I_{316} = k_{11}\left([N_2^+][N_2, X, v > 12]/[N_2(A)]^2\right) \quad (11)$$

with k_{11} increasing from 6.6 10^{-2} in pure N_2 to 0.17 in Ar-2%N_2.

By assuming the equality $[N_2, X, v>12] = [N_2, X, v > 13]$, it is found a N_2^+ density which decreases from about 10^9 cm^{-3} in pure N_2 to 2×10^8 cm^{-3} in Ar-10%N_2 and increases again to 10^9 cm^{-3} in Ar-2%N_2. To verify that the N_2^+ ions are not coming from the end of a plasma jet at Ar-2%N_2, the measurements have also be performed 5 cm above in the 5 litre reactor, keeping about the same results.

Compared to published data [11] [12], the value of N_2^+ density in pure N_2 appears to be in the same order of magnitude.

4. Interest of Ar-N_2 Gas Mixture for Surface Treatments

It is reported in **Figure 3** the N/N_2, $N_2(X, v > 13)/N_2$, $N_2(A)/N_2$ and N_2^+/N_2 density ratio versus the %N_2 into

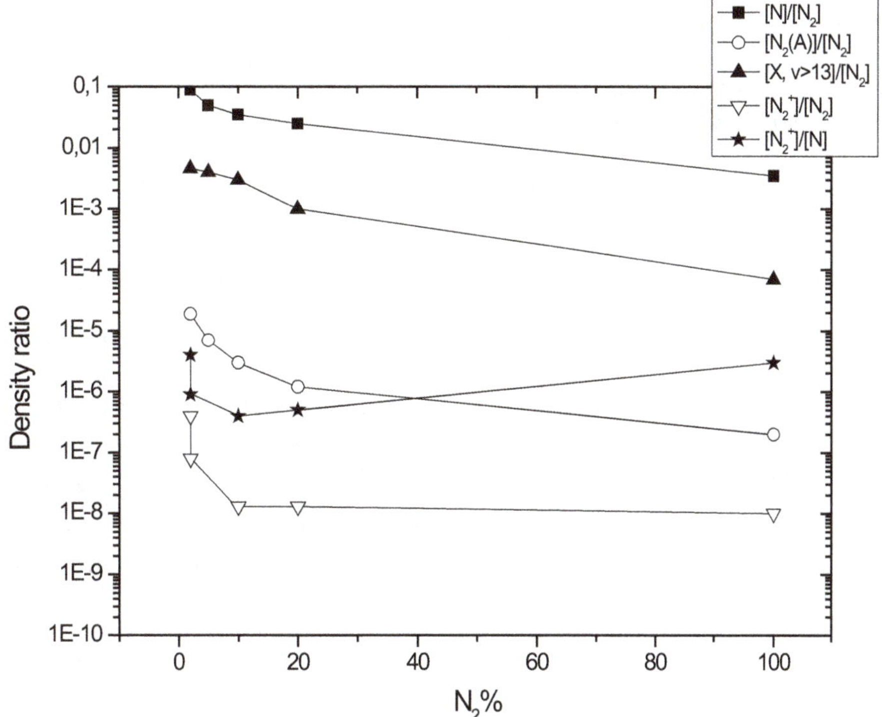

Figure 3. Density ratios of active species on N_2 versus the $\%N_2$ in the Ar-N_2 gas mixtures. In addition N_2^+/N ratio.

Ar. Clearly, there is an interest of low $\%N_2$ to increase the active species density relative to N_2 if it can be considered that the Ar atoms have no influence on the surface processes.

The N_2^+/N_2 density ratio is nearly constant from pure N_2 to Ar-10%N_2 with a new increase with Ar-2%N_2.

There is thus an interest of Ar-xN_2 gas mixtures with x = 2% - 20% for surface treatments with high N, N_2(A, Xv > 13) and N_2^+ density values (see **Figure 2**).

The N_2^+/N density ratio decreased from pure N_2 to Ar-10%N_2 with an increase at Ar-2%N_2 to find again the value in pure N_2.

This increase of N_2^+ density for Ar-2%N_2 could be the result of the charge transfer R8 at the benefit of the N_2^+ ions [13].

5. Conclusions

Densities of N and O atoms (the O-are coming from air impurity), N_2(A) and N_2(X, v > 13) metastable molecules and N_2^+ ions have been determined in Ar-N_2 early afterglows of flowing microwave discharges at 1 slm, 4 Torr, afterglow time of 3×10^{-3} s and 100 W, after NO calibration.

The density of these active species are obtained by comparing the N_2 (580 nm), NO_β (320 nm), N_2 (316 nm) and N_2^+ (391 nm) band intensities and by writing the dominant kinetic equations.

It is found densities in the ranges of $(2 - 6) \times 10^{14}$ cm^{-3} for N-atoms, one order of magnitude lower for both N_2(X, v > 13) and O-atoms (coming from air impurity), of 10^{10} - 10^{11} cm^{-3} for N_2(A) and of 10^8 - 10^9 cm^{-3} for N_2^+.

The densities obtained by these line-ratio measurements are with an uncertainty of 30% for N-atoms and the order of magnitude for O-atoms and N_2(A) metastable molecules. Estimated densities values are obtained for the N_2(X, v > 13) metastable and N_2^+ ions which are depending on the kinetics reaction rates.

It is found that the main interest of N_2 dilution into Ar is to increase the N/N_2 dissociation from 0.5% in N_2 to about 10% in the Ar-2%N_2 which could be of interest for surface reactions of N-atoms with less N_2 molecules. The other N_2(A)/ N_2, N_2(X, v > 13)/N_2 density ratios are also increasing at low $\%N_2$ into Ar. It is not the case for the N_2^+/N_2 and N_2^+/N ratios which are constant or decreasing from pure N_2 up to 10%N_2.

References

[1] Villeger, S., Sarrette, J.P. and Ricard, A. (2005) Synergy between N and O Atom Action and Substrate Temperature in a Sterilization Process Using a Flowing N_2-O_2 Microwave Post-Discharge. *Plasma Process and Polymers*, **2**, 709-711. http://dx.doi.org/10.1002/ppap.200500040

[2] Villeger, S., Sarrette, J.P., Rouffet, B., Cousty, S. and Ricard, A. (2008) Treatment of Flat and Hollow Substrates by a Pure Nitrogen Flowing Post Discharge. Application to Bacterial Decontamination in Low Diameter Tubes. *European Physical Journal Applied Physics*, **42**, 25-32. http://dx.doi.org/10.1051/epjap:2007177

[3] Ricard, A., Gaboriau, F. and Canal, C. (2008) Optical Spectroscopy to Control a Plasma Reactor for Surface Treatments. *Surface and Coatings Technology*, **202**, 5220-5224. http://dx.doi.org/10.1016/j.surfcoat.2008.06.070

[4] Ricard, A., Oh, S.G. and Guerra,V. (2013) Line-Ratio Determination of Atomic Oxygen and N_2 Metastable Absolute Densities in an RF Nitrogen Late Afterglow. *Plasma Sources Science and Technology*, **22**, Article ID: 035009. http://dx.doi.org/10.1088/0963-0252/22/3/035009

[5] Ricard, A. and Oh, S.G. (2014) Densities of Active Species in N_2 and N_2-H_2 RF Pink Afterglow. *Plasma Sources Science and Technology*, **23**, Article ID: 045009. http://dx.doi.org/10.1088/0963-0252/23/4/045009

[6] Zerrouki, H., Ricard, A. and Sarrette, J.P. (2014) Determination of N and O-Atoms, of $N_2(A)$ and $N_2(X, v>13)$ Metastable Molecules and N_2^+ Ion Densities in the Afterglows of N_2-H_2, Ar-N_2-H_2 and Ar-N_2-O_2 Microwave Discharges. *Journal of Physics: Conference Series*, **550**, Article ID: 012045. http://dx.doi.org/10.1088/1742-6596/550/1/012045

[7] Levaton, J. and Amorim, J. (2012) Metastable Atomic Species in the N_2 Flowing Afterglow. *Chemical Physics*, **397**, 9-17. http://dx.doi.org/10.1016/j.chemphys.2011.11.010

[8] Sa, P.A., Guerra, V., Loureiro, J. and Sadeghi, N. (2004) Self-Consistent Kinetic Model of the Short Lived Afterglow in Flowing Nitrogen. *Journal of Physics D: Applied Physics*, **37**, 221-231. http://dx.doi.org/10.1088/0022-3727/37/2/010

[9] Piper, L.G. (1994) Further Observations on the Nitrogen Orange Afterglow. *Journal of Chemical Physics*, **101**, 10229-10236. http://dx.doi.org/10.1063/1.467903

[10] Kang, N., Lee, M., Ricard, A. and Oh, S.G. (2012) Effect of Controlled O_2 Impurities on N_2 Afterglows of RF Discharges. *Current Applied Physics*, **12**, 1448-1453. http://dx.doi.org/10.1016/j.cap.2012.04.009

[11] Sadeghi, N., Foissac, C. and Supiot, P. (2001) Kinetics of $N_2(A)$ Molecules and Ionization Mechanisms in the Afterglow of a Flowing N_2 Microwave Discharge. *Journal of Physics D: Applied Physics*, **34**, 1779-1788. http://dx.doi.org/10.1088/0022-3727/34/12/304

[12] Ferreira, J.A., Stafford, L., Leonelli, R. and Ricard, A. (2014) Electrical Characterization of the Flowing Afterglow of N_2 and N_2/O_2 Microwave Plasmas at Reduced Pressure. *Journal of Applied Physics*, **115**, Article ID: 163303.

[13] Fehsenfeld, F.C., Ferguson, E.E. and Schmeltekopf, A.J. (1966) Thermal Energy Ion-Neutral Reactions Rates VI. Some Ar^+ Charge Transfer Reactions. *Journal of Chemical Physics*, **45**, 404-405. http://dx.doi.org/10.1063/1.1727351

8

The Effect of Light on the CVC of Strained p-n-Junction in a Strong Microwave Field

Muhammadjon Gulomkodirovich Dadamirzaev

Namangan Engineering Pedagogical Institute, Namangan, Uzbekistan
Email: dadamirzaev70@umail.uz

Abstract

For the first time the effect of light on the CVC of strained p-n-junction in a strong microwave field is examined. It is shown that the deformation and the microwave field increase the current through p-n-junction, and the light decreases it. The mechanism of this phenomenon is explained by the fact that under heating of the charge carriers by microwave field the recombination current arises, and under the action of light the generation current arises which are directed oppositely. And under the influence of the deformation the band gap of the semiconductor will be changed.

Keywords

Hot Electrons and Holes, The Microwave Field, p-n-Junction, Light, Photocurrent Lasing and Recombination Currents, Light, Deformation, CVC Strain p-n-Junction

1. Introduction

Research of the effects associated with the heating of the carriers reveal new properties of semiconductor materials and devices, and the physical processes occurring in them that are difficult or even impossible to study in equilibrium conditions of the carriers and the lattice. Interest in the study of hot carriers is caused mainly by the following circumstances. First, the heating of the carriers leads to a change in their energy distribution. This is evident in the well-known kinetic effects, and leads to the emergence of new phenomena unique to the state of the hot carrier's gas. Secondly, the presence of preheated charge carriers leads to a number of features in the behaviour of semiconductors, which are inhomogeneous in their structure, particularly having p-n, n+-n-junctions and others (change of voltage-current characteristics, amplification of current, and the occurrence of additional heat, thermal and photo-thermal emf's in certain conditions and others). Thirdly, the effects caused by the heating of the carriers, are increasingly used in practical application. On the basis of various kinds of instabilities of current accompanied by the heating effect, radio-electronic devices were created, many of whom work in the microwave frequencies.

In recent years, much attention is paid worldwide to research and development of electromechanical transducers and sensors based on homogeneous semiconductors and semiconductor devices. Small and sensitive sensors and sound detectors, working in a wide frequency band, are needed in many areas of technology.

Connection of the mechanical and electrical properties of semiconductors is determined by two major phenomena: the piezoelectric and deformation effects. Piezoelectric effect occurs in the crystals, which are not an observer inversion center. The deformation effect is due to the interaction of electrons with the crystal lattice, which is available in all semiconductors. The physical cause deformation effects are the shift of the energy levels of the semiconductor under the influence of the deformation and the associated change in the spectrum of the current carriers—electrons and holes, depending on the strain.

Changing the current in semiconductor devices under pressure was first observed in 1951, when p-n-transitions to Germany were subjected to uniform compression [1]. Since then, there were many works that investigate physical phenomena in semiconductor devices under pressure, and the possibility of their technical use as electromechanical transducers [2]. [3] analyzed the decay of nickel precipitates in silicon under the influence of a comprehensive hydrostatic pressure. [4] examined the effect of hydrostatic hydrostatic compression (HCV) in the decay rate of a solid solution of Si <Mn> at different temperatures.

[5] studied the change characteristics of the photovoltaic silicon p-n-junctions in a microwave field. It is shown that the barrier height of p-n-junction in a strong microwave field, reducing the barrier illuminated p-n-junction, is proportional to the height of the initial barrier, if the latter is reduced by direct displacement. In [6] the current-voltage characteristic (CVC) of strained p-n-junction in a microwave field is investigated and it has been showed that the deformation increases the current generated in the p-n-junction.

From the foregoing, it follows that the calculation of the current characteristics of p-n junctions in a microwave field does not include the impact of a simultaneous deformation and light on the current-voltage characteristics of p-n-junction in a strong microwave field.

The purpose of this work is to study the effect of light on the CVC of strained p-n-junction in a strong microwave field.

2. Theoretical Calculations of the Effect of Light on the CVC of Strained p-n-Junction in a Strong Microwave Field

In the first look at the CVC of strained p-n-junction with the low power of the microwave when there is no heating of the electrons and holes ($T_e = T_h = T$) in the absence of the effect of light and disturbance potential barrier height ($I_c = 0$ (Is—light current), $U_B = 0$). Under these conditions, for the CVC of p-n-junction we have the following form:

$$\bar{I} = I_{se}(\varepsilon)\left\{\left(\frac{T_e}{T}\right)^{\frac{1}{2}} \exp\left(\frac{e\varphi_0}{kT} - \frac{e(\varphi_0 - U) + eU_B\left|\overline{\cos(\omega t)}\right|}{kT_e}\right) - 1\right\} + I_{sh}(\varepsilon)\left\{\left(\frac{T_h}{T}\right)^{\frac{1}{2}} \exp\left(\frac{e\varphi_0}{kT} - \frac{e(\varphi_0 - U) + eU_B\left|\overline{\cos(\omega t)}\right|}{kT_h}\right) - 1\right\} \quad (1)$$

where: I_{se}; I_{sh}—saturation currents for electrons and holes; φ_0 the height of the potential barrier in the absence of an electromagnetic wave; $\varphi = \varphi_0 - U$; U—a voltage across the diode; $U_B = -\int_0^d E_B dx$—AC voltage of the incident wave created by the barrier diode; T is the temperature of the lattice; k—Boltzmann constant; T_e, and T_h— temperature electrons and holes; E_b—electric field of the wave; e is the charge of the electron.

For silicon p-n-junctions saturation currents is as follows:

$$I_{se}(\varepsilon) = \sqrt{\frac{ek}{\tau_e} \cdot T^{4.4} \cdot 10^9} \cdot \frac{3\times 10^{33} \cdot e^{-\frac{\varepsilon_g(0)+\Delta\varepsilon}{kT}}}{p_p},$$

$$I_{sh}(\varepsilon) = \sqrt{\frac{ek}{\tau_h} \cdot T^{4.7} \cdot 2.5\times 10^8} \cdot \frac{1.5\times 10^{33} \cdot e^{-\frac{\varepsilon_g(0)+\Delta\varepsilon}{kT}}}{n_n} \quad (2)$$

where: $\varepsilon_g(\varepsilon,T)=\varepsilon_g(0)+\Delta\varepsilon$ —permanent deformation $\Delta\varepsilon = 1.27эB$ —the band gap of silicon; n_n and p_p—the concentration of majority carriers; τ_e and τ_z—lifetimes for electrons and holes. Then CVC p-n-junction becomes:

$$I=\left[I_{se}(\varepsilon)+I_{sh}(\varepsilon)\right]\exp\left(\frac{eU}{mkT}-1\right) \quad (3)$$

From this we can derive an expression for the coefficient of imperfection:

$$m=\frac{eU}{kT\ln\left[\frac{I}{\left[I_{se}(\varepsilon)+I_{sh}(\varepsilon)\right]}+1\right]} \quad (4)$$

For the CVC strain p-n-junction with the low power of the microwave when there is a perturbation potential barrier height ($I_c = 0$, $\varepsilon \neq 0$ (ε—deformation); $T_e = T_h = T$; $U_B \neq 0$) of formula (1) can be obtained:

$$\bar{I}=\left(I_{se}(\varepsilon)+I_{sh}(\varepsilon)\right)\left[\exp\left(\frac{e\left(U-U_B\overline{\left|\cos(\omega t)\right|}\right)}{mkT}\right)-1\right] \quad (5)$$

Hence, under these conditions, to m, we have:

$$m=\frac{e\left(U-U_B\overline{\left|\cos(\omega t)\right|}\right)}{kT\ln\left(\frac{I}{I_{se}(\varepsilon)+I_{sh}(\varepsilon)}+1\right)} \quad (6)$$

Again at low powers the microwave if there is a disturbance potential barrier height and the influence of light ($I_c = 0$, $\varepsilon \neq 0$; $T_e = T_h = T$; $U_B \neq 0$). Again at low powers the microwave if there is a disturbance potential barrier height and the influence of light:

$$\bar{I}=\left(I_{se}(\varepsilon)+I_{sh}(\varepsilon)\right)\left[\exp\left(\frac{e\left(U-U_B\overline{\left|\cos(\omega t)\right|}\right)}{mkT}\right)-1\right]-I_c \quad (7)$$

From

$$m=\frac{e\left(U-U_{\mathrm{B}}\overline{\left|\cos(\omega t)\right|}\right)}{kT\ln\left(\frac{I+I_c}{I_{se}(\varepsilon)+I_{sh}(\varepsilon)}+1\right)} \quad (8)$$

At high power microwave energy, when electrons and holes are hot and happening outrage potential barrier height without lighting ($I_ф = 0$, $\varepsilon \neq 0$; $T_e \neq T_h > T$; $U_B \neq 0$). CVC strain for the p-n-junction can be obtained ratio:

$$\begin{aligned}\bar{I}=&I_{se}(\varepsilon)\left\{\left(\frac{T_e}{T}\right)^{\frac{1}{2}}\exp\left(\frac{e\varphi_0}{kT}-\frac{e(\varphi_0-U)+eU_{\mathrm{B}}\overline{\left|\cos(\omega t)\right|}}{kT_e}\right)-1\right\}\\&+I_{sh}(\varepsilon)\left\{\left(\frac{T_h}{T}\right)^{\frac{1}{2}}\exp\left(\frac{e\varphi_0}{kT}-\frac{e(\varphi_0-U)+eU_{\mathrm{B}}\overline{\left|\cos(\omega t)\right|}}{kT_h}\right)-1\right\}\end{aligned} \quad (9)$$

If $\frac{p_p}{n_n} \gg 1$, the major share of investing T_h, then m is determined from the second member (9):

$$m\left(T_h,T,U_B,\varepsilon\right)=\frac{\frac{e}{k}\left(\frac{\varphi_0}{T}-\frac{(\varphi_0-U)+U_B\left|\overline{\cos(\omega t)}\right|}{T_h}\right)}{\ln\left(\left(\frac{I}{I_{sh}(\varepsilon)}+1\right)\left(\frac{T}{T_h}\right)^{\frac{1}{2}}\right)} \tag{10}$$

If $\frac{p_p}{n_n}\ll 1$ m is determined from the second member (9):

$$m\left(T_e,T,U_B,\varepsilon\right)=\frac{\frac{e}{k}\left(\frac{\varphi_0}{T}-\frac{(\varphi_0-U)+U_B\left|\overline{\cos(\omega t)}\right|}{T_e}\right)}{\ln\left(\left(\frac{I}{I_{se}(\varepsilon)}+1\right)\left(\frac{T}{T_e}\right)^{\frac{1}{2}}\right)} \tag{11}$$

For the CVC of strained p-n-junction at high power microwave energy, when electrons and holes are warmed and illuminated p-n-junction, and the place, the height of the potential barrier ($\varepsilon \neq 0$; $T_e \neq T_h > T$; $I_\phi \neq 0$, $U_B \neq 0$) can obtain the following formula:

$$\begin{aligned}\overline{I} &= I_{se}(\varepsilon)\left\{\left(\frac{T_e}{T}\right)^{\frac{1}{2}}\exp\left(\frac{e\varphi_0}{kT}-\frac{e(\varphi_0-U)+eU_B\left|\overline{\cos(\omega t)}\right|}{kT_e}\right)-1\right\}\\ &+I_{sh}(\varepsilon)\left\{\left(\frac{T_h}{T}\right)^{\frac{1}{2}}\exp\left(\frac{e\varphi_0}{kT}-\frac{e(\varphi_0-U)+eU_B\left|\overline{\cos(\omega t)}\right|}{kT_h}\right)-1\right\}-I_c\end{aligned} \tag{12}$$

If $\frac{p_p}{n_n}\gg 1$ m is determined from the second and third members (12):

$$m\left(T_h,T,U_B,I_\phi,\varepsilon\right)=\frac{\frac{e}{k}\left(\frac{\varphi_0}{T}-\frac{(\varphi_0-U)+U_B\left|\overline{\cos(\omega t)}\right|}{T_h}\right)}{\ln\left(\left(\frac{I+I_c}{I_{sh}(\varepsilon)}+1\right)\left(\frac{T}{T_h}\right)^{\frac{1}{2}}\right)} \tag{13}$$

And if $\frac{p_p}{n_n}\ll 1$ it is determined m the first and third members (12):

$$m\left(T_e,T,U_B,I_\phi,\varepsilon\right)=\frac{\frac{e}{k}\left(\frac{\varphi_0}{T}-\frac{(\varphi_0-U)+U_B\left|\overline{\cos(\omega t)}\right|}{T_e}\right)}{\ln\left(\left(\frac{I+I_c}{I_{se}(\varepsilon)}+1\right)\left(\frac{T}{T_e}\right)^{\frac{1}{2}}\right)} \tag{14}$$

From the above analysis it is possible to build a p-n-junction current-voltage characteristics under different conditions (**Figure 1**).

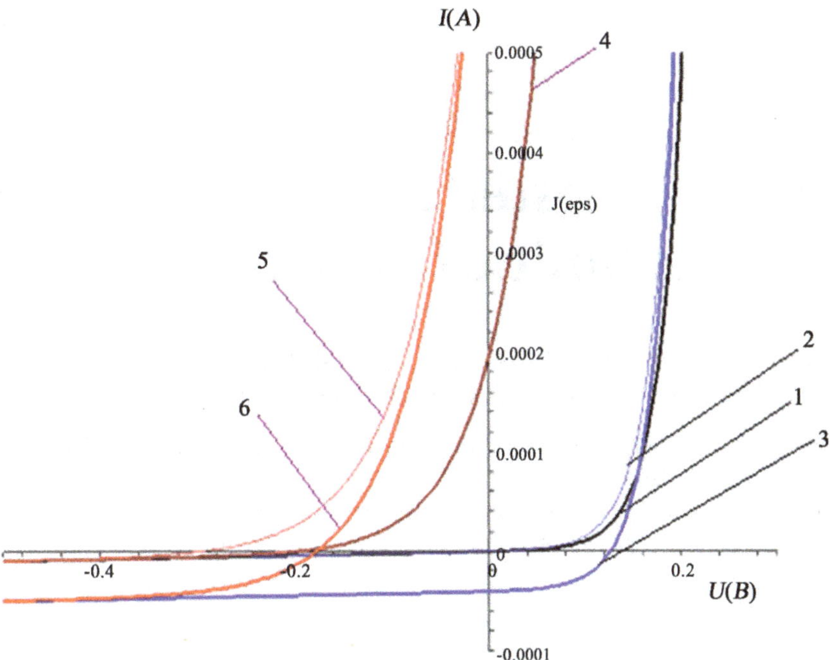

Figure 1. CVC p-n-junction, provided: 1—without deformation, without the microwave field without lighting; 2—in the deformation, the height of the potential barrier (at low powers the microwave) and without light; 3—during deformation, the height of the potential barrier (at low microwave power of the wave) and light; 4—without deformation at high power microwave energy and without lighting; 5—in the deformation at high power microwave energy and without illumination; 6—in the deformation at high power microwave energy and lighting.

3. Conclusion

Based on these studies, the following conclusions can be stated: if light, deformation and microwave field act to the p-n-junction, the deformation and the microwave field increase the current through the p-n-junction, and the light decreases it. The mechanism of this phenomenon is explained by the fact that under heating of the charge carriers by microwave field the recombination current arises, and under the action of light the generation current arises which are directed oppositely. And under the influence of the deformation the band gap of the semiconductor will be changed. As a result, the energy of the electrons and holes will increase, respectively; recombination current also will increase, whose direction corresponds to the direction of the major charge carriers.

References

[1] Hall, H.H., Bardeen, J. and Pearson, G.L. (1951) *Physical Review*, **84**, 129-132. http://dx.doi.org/10.1103/PhysRev.84.129

[2] Polyakova, A.L. (1972) *Acoustic Magazine*, **18**, 1-22.

[3] Zaynabidinov, S.Z., Turaev, A.R., Fistul, V.I. and Khodzhaev, M.D. (1989) *FTP*, **23**, 2118-2121.

[4] Bahadyrhanov, M.K., Abduraimov, A. and Iliev, X.M. (1988) *FTP*, **22**, 123-128.

[5] Ablyazimova, N.A., Veynger, A.I. and Food, V.S. (1992) *Physics and Engineering of St. Petersburg Semiconductors*, **26**, 1041-1047.

[6] Dadamirzaev, M.G. (2015) *Journal of Modern Physics-USA*, **6**, 176-180. http://dx.doi.org/10.4236/jmp.2015.62023

9

Microwave Assisted Liberation of High Phosphorus Oolitic Iron Ore

Mamdouh Omran[1,2*], Timo Fabritius[1], Nagui Abdel-Khalek[2], Mortada El-Aref[3], Abd El-Hamid Elmanawi[3], Mahmoud Nasr[2], Ahmed Elmahdy[2]

[1]Laboratory of Process Metallurgy Research Group, Process and Environmental Engineering Department, University of Oulu, Oulu, Finland
[2]Central Metallurgical Research and Development Institute, Cairo, Egypt
[3]Geology Department, Faculty of Science, Cairo University, Cairo, Egypt
Email: [*]Mamdouh.omran@oulu.fi

Abstract

The influence of microwave treatment on the liberation of iron ore from the high phosphorus oolitic iron ore from Aswan region, Egypt was studied. The effect of microwave power, exposure time and grain size on the liberation of iron ore was investigated. The microfractures and cracks of the samples were characterized before and after microwave treatments. The heating rate of high phosphorus oolitic iron ore was studied. Crystallinity of hematite was characterized before and after microwave pretreatment. The results indicated that intergranular fractures formed between the gangues (fluorapatite and chamosite) and hematite after microwave treatment, leading to improved liberation of iron ore and a significant reduction in comminution energy. Percentages of fraction ≤ −0.125 mm increased from 46.6% to 59.76% with increased exposure time from 0 to 60 seconds. The heating rate of iron ore showed that microwave treatment was less efficient at smaller particle sizes for a fixed applied power density. Crystallinity of hematite increased with the microwave exposure time.

Keywords

Microwave Treatment, High Phosphorus Oolitic Iron Ore, Liberation

1. Introduction

Ironstones may (or may not) contain >50% ooids and pisoids. Ooids are spherical or ellipsoidal coated-grains

[*]Corresponding author.

smaller than <2 mm in diameter, which display regular concentric laminae surrounding a central core. Grains similar to ooids, but larger than >2 mm are known as pisoids [1]. The oolitic iron ores are widely spread worldwide, some of which have huge reserves, for instance, Wadi Fatima mine in Saudi Arabia [2], Lorraine mine in France [3], Bell island mine in Canada [4], Dilband mine in Pakistan [5], Xuanhua region in China [6] and Aswan region in Egypt [7] [8].

The main obstacle for using these deposits is the fine dissemination of silica and aluminum minerals and especially the high level of phosphorus content. This difficulty is mainly due to the poor liberation of iron minerals from oolitic gangues. Song *et al.* [9] observed that fine grinding (commonly 1 - 5 μm) required liberating iron minerals from associated gangue minerals. Such fine particles are very difficult to be beneficiated via conventional processes of mineral processing (e.g. flotation and magnetic separation).

Phosphorus removal from the high phosphorus oolitic iron ores has been investigated by several processes, including a) selective flocculation-reverse flotation [10], b) chemical leaching [11] [12], c) microbiological method (bioleaching) [13] [14], and d) metallurgical method (magnetization roasting and reduction) [15] [16]. Although some of these methods achieve the purpose of phosphorus removal, there still have disadvantages for instance low efficiency of dephosphorization, relatively high cost, and low iron recovery.

The development of a successful and economic process to remove phosphorus from the high phosphorus iron ores would significantly extend the reserves of high grade low phosphorus iron ores [17]. There are two main challenges in mineral comminution: energy consumption and mineral recovery [18]. About 1.5% - 2% of the total national energy consumption in the industrial mining countries is attributed to comminution [19]. Wang *et al.* [20] suggested that there are two main reasons for investigating liberation improvement: a) liberation of particles at large size reduces the energy consumption during grinding, and b) very fine grain size is very difficult in physical separation processes and consumes more grinding energy. Liberation at coarse grain size is suitable for physical separation techniques, such as flotation or magnetic separation. So that it is very crucial to focus on techniques that help in particles liberation with minimum power consumption and particle size reduction. Microwave treatment of ores is considered as a potential way for reducing the grinding energy consumption and increasing the liberation and recovery of valuable minerals [21].

Microwave energy is a non-ionizing electromagnetic radiation with frequencies in the range of 300 MHz to 300 GHz. Microwave frequencies include three bands: the ultra high frequency (UHF: 300 MHz to 3 GHz), the super high frequency (SHF: 3 GHz to 30 GHz) and the extremely high frequency (EHF: 30 GHz to 300 GHz) [22] [23].

Microwave treatment improves the liberation of high phosphorus oolitic iron ores through generating intergranular fractures in oolitic iron ores [9]. The difference in the absorption of microwave energy, thermal expansion and dielectric properties of iron and gangue minerals leads to generating intergranular fractures between iron and gangue minerals [24]-[28]. High phosphorus oolitic iron ores are usually composed of hematite, dolomite, clinochlore, quartz and apatite (fluorapatite or hydroxyl fluorapatite). Microwave radiations have significant influence on the microstructure of the oolitic units [16]. Hematite, phosphorite, silicate minerals and other gangues in the ore differ in absorbing microwave energy. These minerals have different thermal expansion and thus thermal stresses are generated on the boundaries among them. When these thermal stresses reach a certain level, cracks and fissures are formed at the boundaries [29]. Jones *et al.* [24] stated that after microwave radiation, intergranular fractures occur around the grain boundaries between absorbent and transparent phases. Amankwah *et al.* [30] observed that differential heating of different minerals phases in an ore results in thermal stress cracking, which makes the ore more amenable to size reduction and results in a decrease in the work index.

Kingman *et al.* [31] [32] studied the influence of microwave radiation on Norwegian ilmenite ores. It was concluded that short, high-power treatments were most effective and led to a reduction in work index of up to 90% and increased recovery of ilmenite, due to the improvement of liberation and magnetic properties of ilmenite ores after microwave treatment. The influence of mineralogy on the responses of ores to microwave radiation was studied by Kingman *et al.* [33]. They concluded that samples with a mixture of "good heaters" in a lattice of "poor heaters" consisting of coarse grain size gave the best response and greatest reduction in work index after microwave treatment. Poorest response could be expected from ores containing highly disseminated, fine-grained minerals.

The aim of this study was to investigate the effect of microwave pretreatment on liberation of iron bearing minerals from phosphorus and other gangues minerals. The effect of different parameters such as microwave power, exposure time and grain size of particles on the heating rate, crystallinity and intergranular fractures of oolitic iron ore will be studied.

2. Experimental and Analytical Methods

2.1. Iron Ore Sample

The high phosphorous oolitic iron ore used in this study was collected from Aswan region, Egypt. The east of Aswan area represents the main occurrence of the Cretaceous oolitic ironstone bands of South Egypt which are confined to clastic successions belonging to the "Nubian" sandstones or "Nubia facies" [7]. **Figure 1** and **Table 1** show the XRD pattern and chemical analysis, respectively, of iron ore used in the tests.

2.2. Microwave Treatment

The samples were treated using a 2.45 GHz microwave oven (sandstorm, model S25CSS11E and cavity dimension 513 mm (D) × 482 mm (W) × 310 mm (H)) with a maximum output power of 900 W. Iron ore samples were treated in the oven for varying exposure times and power densities. Samples were allowed to cool in the microwave oven to room temperature.

Table 1. Chemical composition of high phosphorus oolitic iron ore.

Oxides	Weight %
Fe_2O_3	74.96
SiO_2	7.48
P_2O_5	3.24
CaO	5.44
Al_2O_3	4.47
MnO	0.54
MgO	1.26
Na_2O	0.37
K_2O	0.05
F	0.19

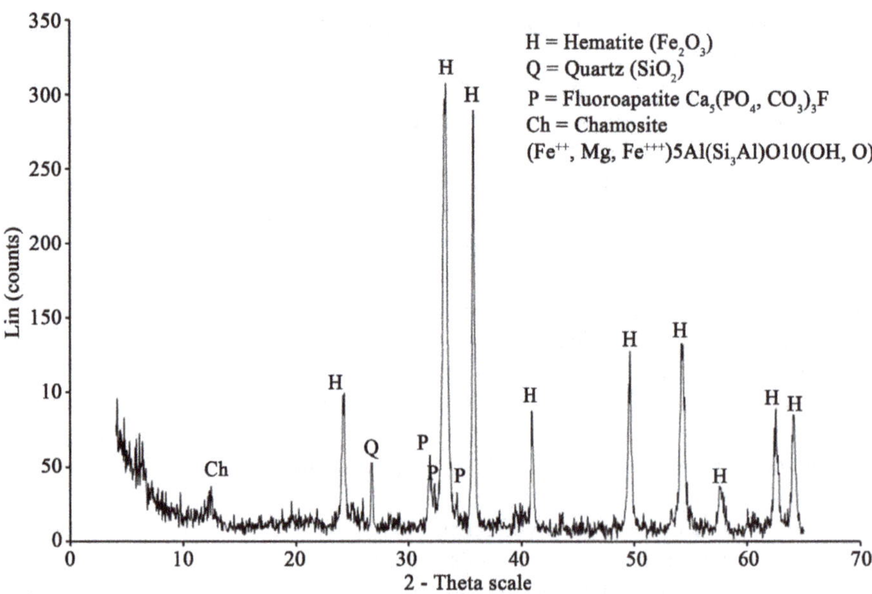

Figure 1. X-ray diffraction chart for high phosphorus oolitic iron ore.

2.3. Analytical Methods

2.3.1. X-Ray Diffraction (XRD)

The bulk mineralogical composition and crystallinity of the iron ore were performed on powdered samples using Siemens D5000 XRD powder diffractometer. The device contains a Cu Kα radiation with a graphite monochromator. The XRD analyses were done using 40 KV and 40 mA.

2.3.2. X-Ray Fluorescence (XRF)

Chemical analyses were performed on whole rock powders by X-ray fluorescence (Bruker AXS S4 Pioneer). The major elements were determined on fused beads (glass disks) in order to minimize matrix effects.

2.3.3. Scanning Electron Microscope (SEM)

The micro-morphological characteristics of the iron ore before and after treatment with microwave were investigated using Zeiss ULTRA plus field emission scanning electron microscope (FESEM) attached to an Energy-dispersive X-ray spectroscopy (EDS) unit for chemical analysis.

2.3.4. Electron Probe Microanalyses (EPMA)

The mineral chemistry of the iron minerals and the element distribution maps within oolites and interstitial spaces between the ferruginous oolites were determined by Electron Probe Microanalyses (EPMA). The EPMA were performed on a Jeol JXA-8200 device with WDS/EDS microanalyzer.

The XRD, XRF, SEM and EPMA analyses were carried out at the Center of microscopy and nanotechnology, University of Oulu, Finland.

3. Results and Discussion

3.1. Mineralogy and Chemistry of the High Phosphorous Oolitic Iron Ore

According to El Sharkawi *et al.* [7] the true oolitic ironstone of Aswan region consists entirely of closely spaced (grain-supported) ferruginous ooids (>95%) with less abundant detrital quartz grains, kaolinitic rock fragments and ferruginous clayey materials "chamosite" (<5%). Ooids are spherical or ellipsoidal coated-grains <2 mm in diameter, which display regular concentric laminae surrounding a central core. These laminae are usually coalesced in group forming zones, which are distinguished by color variation [8].

XRD analysis indicated that hematite is the main iron bearing minerals, whereas quartz, fluorapatite and chamosite are the main gangue minerals **Figure 1**. XRF analysis of the original sample indicated that Fe_2O_3 and P_2O_5 grades are 74.96% and 3.24% respectively **Table 1**. P_2O_5, CaO and F content are related to fluoroapatite, whereas Al_2O_3, MgO and MnO content are related to chamosite. SiO_2 content related to quartz and chamosite. Optical photomicrograph and SEM images of the high phosphorous iron ores show that Fe-bearing minerals occur as ooiltic hematite **Figure 2(A)** and **Figure 2(F)**. Fluoroapatite (phosphorus bearing mineral) occur mainly as fine-grained cement-like materials mixed with iron filling the spaces between ooid grains **Figure 2(C)** and **Figure 2(E)**. Chamosite occurs as rim surrounded the ooid grains **Figure 2(C)** and **Figure 2(D)**.

Distribution of Phosphorus

Figure 3 shows element maps made in the ferruginous ooids and spaces between ooids for Fe, P, Ca, Si and Al. The EDS distribution map of iron shows that iron has higher concentration inside ooids than in the spaces between ooids **Figure 3**. It can be seen that phosphorus and calcium associated closely (particles with high phosphorus content also contained high calcium content) and concentrated in the spaces between ooids, the distribution of P and Ca are related to fluoroapatite **Figure 3**. This indicates that fluoroapatite concentrated mainly in the interstitial spaces between ooids. Silicon clearly detected in the rim zones around the ooids. The distribution of silicon related to both quartz and chamosite. Aluminum related to chamosite and concentrated mainly around ooids and less concentrated inside ooids.

3.2. Heating Rate of the Iron Ore

Different size fractions of iron ore were prepared by crushing and sieving. These fractions are +8 mm, −8 + 4

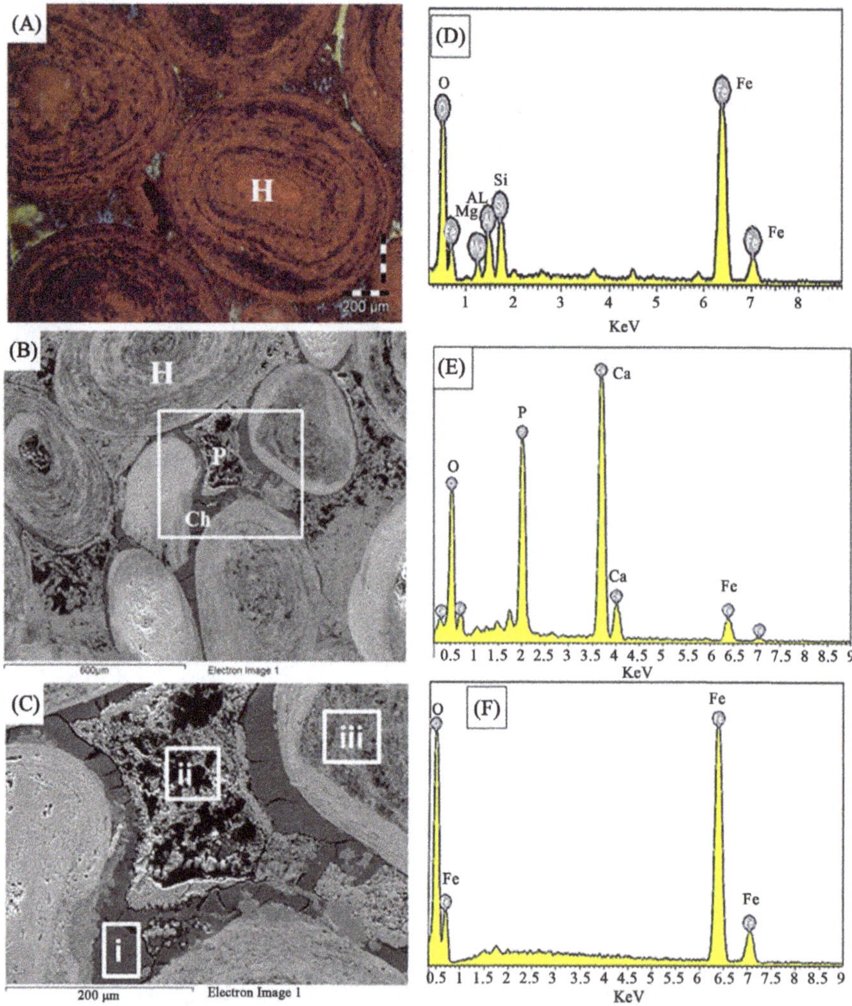

Figure 2. Thin sections and SEM photomicrographs of the high phosphorus oolitic ironstone (A) An optical photomicrograph showing the oolitic iron structure. (B) and (C) SEM photomicrographs showing the matrix between ooids. (D), (E) and (F) EDX analyses of the squared area (i, ii, iii) in **Figure 3(C)** respectively.

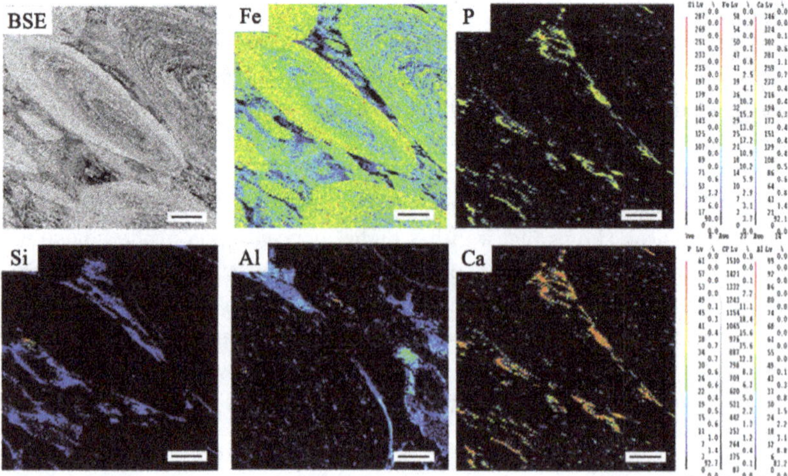

Figure 3. Back-scattered image of the oolitic structure and element maps show the distribution of Fe, P, Si, Al and Ca inside oolitic structure and in the matrix between ooides. Scale bar is 100 µm.

mm, −4 + 2 mm, −2 + 1 mm, −1 + 0.5 mm, −0.5 + 0.25 mm, −0.25 + 0.125 mm, −0.125 + 0.065 mm, −0.065 + 0.032 mm and –0.032 mm. 50 grams of representative samples of different size fractions were used in the tests. Samples were placed in the oven in crucible from pure alumina. The crucible was located in the central position. The sample was treated with microwave radiation at different power levels and exposure times. The temperature of the test sample was measured by quickly inserting thermocouple into the sample after the power was turned off and monitored by a digital display temperature controller **Figure 4** [34]. The measured temperatures are the bulk temperature of the test sample.

The effects of power density, exposure time and particle size on the temperature of the iron ore were studied.

Figure 5 shows that temperature increases with increasing particle size. For example for a +8 mm particle size exposed to a microwave power density of 900 W for 60 s the particle temperature is 546˚C while for the + 0.5 − 1 mm particle size under the same conditions the particle temperature is 485˚C. These small size particles are very important in the process flow sheet: as the size decreases it requires more energy to crush and grind. Also higher power densities and exposure times are required for smaller particle sizes to exhibit the same temperature, and subsequent weakening of particle as large particle.

Figure 5 also shows that with increasing exposure time the temperature of the iron particles increases as expected. The longer exposures time the higher particle temperature. At 90 s exposure time portion of the sample melted, and the measurement of accurate bulk temperature became difficult.

The power density is very important in generation of the temperatures required to thermally damage the rock. **Figure 6** indicates that with increasing power density, the temperature of the iron particles increased. For example + 8 mm particle size exposed to 900 W power density at 50 s radiation time the particle temperature is 420˚C,

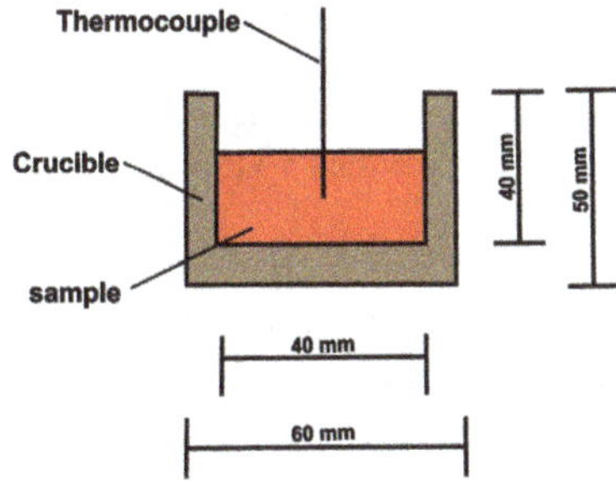

Figure 4. Schematic representation of the crucible and temperature measurement of the sample.

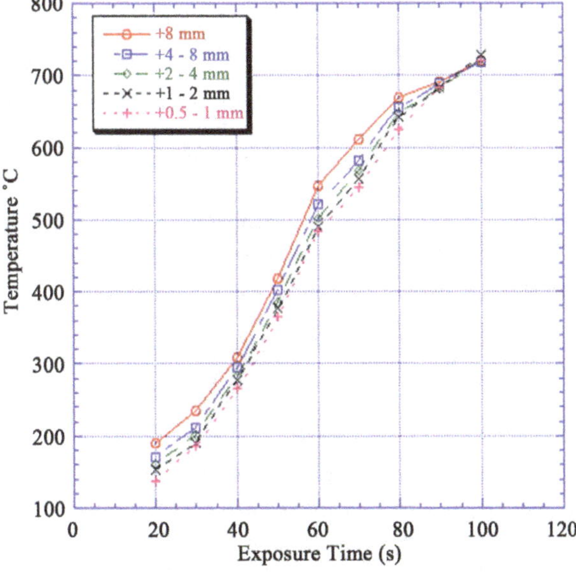

Figure 5. Effect of exposure time on mineral temperature at different particle sizes.

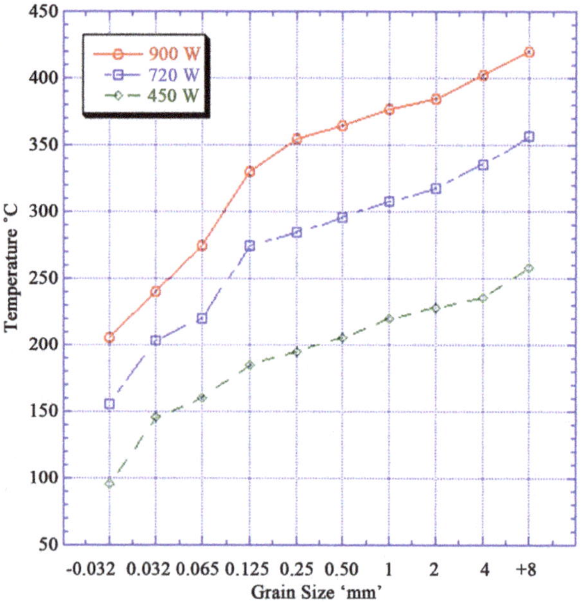

Figure 6. Effect of microwave power density on the mineral temperature (At 50 s radiation time).

while the same particle size exposed to 450 W power density for the same exposure time the particle temperature is 258°C. It can be noted that the power density has a large influence on the temperature and weakening of the iron ores.

Figure 7 shows the XRD analysis of samples before and after microwave treatment. It has been found that the microwave treated sample has peaks sharper than that of untreated. It means the crystallinity increases with increase microwave exposure time, but no phase change. With increasing microwave radiation time, the peak intensity (Crystallinity) of hematite increases. When melting start (at exposure time up to 150 s) the peak intensity of hematite decreases. No phase transformation of hematite after microwave heating was detected. Lack of any phase change after microwave treatment has been also noted by Barani *et al.* [35]. The peaks of chamosite disappear after 50 s exposure time. At this exposure time temperature of sample exceeds 400°C and chamosite decomposed after this temperature and became amorphous.

3.3. Effect of Microwave Radiation on the Liberation of High Phosphorous Oolitic Iron Ore

The oolitic iron ore was observed with the SEM before and after microwave treatment. After that, the images at the same area were compared to find the changes of the ore before and after microwave treatment. The influence of power density, exposure time and grain size on the damage and microfracture of the oolitic iron ore were investigated.

As the previous theoretical studies concerning on microwave treatment of ore, the main cause of the damage after microwave treatment is the thermally-induced tensile stresses, which occurred during the thermal expansion of the absorbent phases, exceeding the tensile strength of the material [21] [24] [26] [35]-[37]. Hematite is an active material to microwave heating, while gangues are inactive materials. **Table 2** gives the heating properties of hematite and gangues minerals with microwave [38] [39]. It was reported that the microwave heating rates for hematite and quartz were 170°C/s and 2°C/s, respectively [23]. When iron ore exposed to microwave radiation, hematite expanded more than quartz this difference on the expanding resulted in the formation of intergranular fractures [9].

Figure 8 SEM images of the oolitic iron ore before and after treatment by microwave radiation. For the short exposure time 40 s and microwave power of 900 W microfractures occur in the matrix around the oolites **Figure 9(A)**. With increased exposure time to 50 s at the same power density these fractures appeared between oolites and matrix (intergranular fractures between oolite and matrix) **Figure 9(B)**. With increased exposure time to 60 s at microwave power of 900 W, the cracks were more localized around the oolites boundaries and almost no damages in the oolites grains **Figure 10**. At this stage, oolites are mostly liberated from the matrix which means

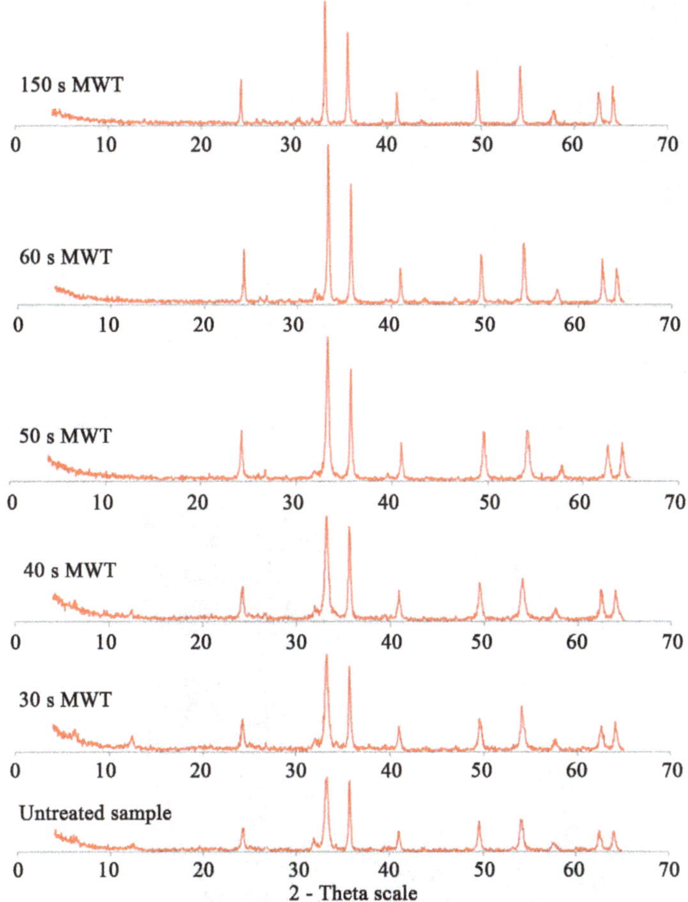

Figure 7. X-ray analysis of microwave treated and untreated samples.

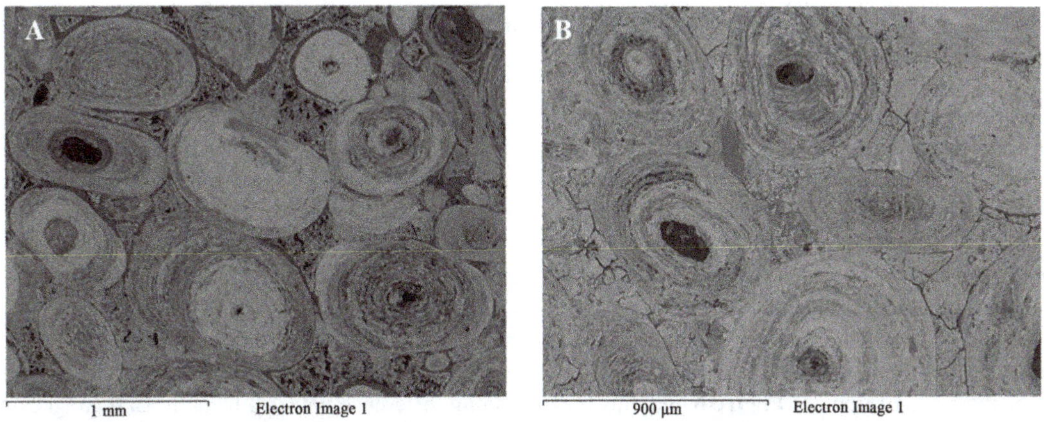

Figure 8. BSE of the Oolitic iron ore (A) Before and (B) After microwave treatment.

Table 2. Heating properties of minerals with microwave radiation.

Mineral	Formula	Microwave heating
Hematite	Fe_2O_3	Heat readily, but no mineral phase change (active)
Quartz	SiO_2	Does not heat (inactive)
Fluroapatite	$Ca_5(PO_4, CO_3)_3F$	Very little or no heat generated
Chamosite	$(Fe^{++}, Mg, Fe^{+++})_5Al(Si_3Al)O_{10}(OH, O)_8$	Very little or no heat generated

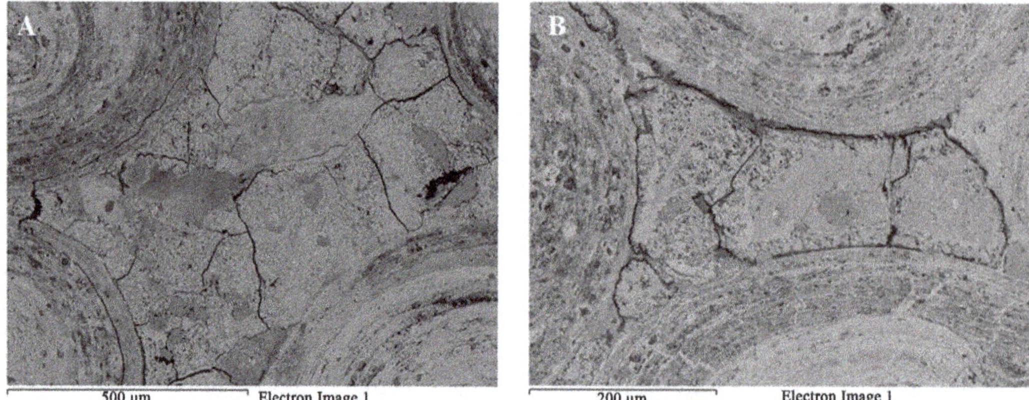

Figure 9. BSE images of the oolitic iron ore after microwave treatment (A) Matrix microfractures; (B) Intergranular fracture between ooids and matrix.

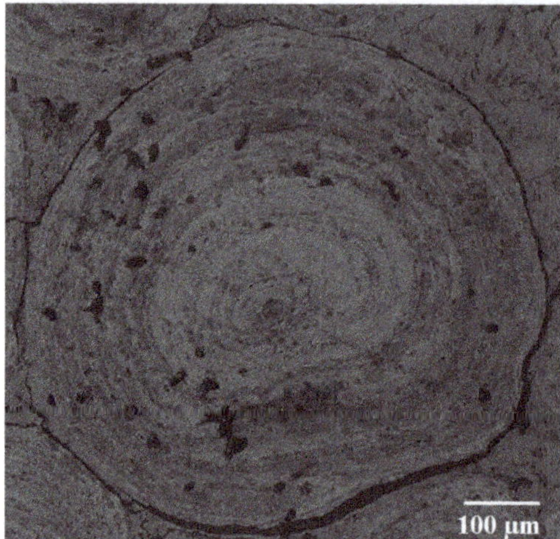

Figure 10. BSE image of completely liberated ooid grain.

that most of phosphorus can be removed **Figure 3**. These localized damages would effectively facilitate liberation of oolites at coarser size and reduce over grinding and slimes losses. At higher exposure time 80 s and microwave power of 900 W intergranular microfractures in the oolitic layers and transgranular fractures in the oolites occurred **Figure 11**. With increased exposure time up to 90 s at the same microwave power, part of the sample melted **Figure 12**. At exposure time up to 150 s the sample completely melted **Figure 13**.

It can be concluded that, as the exposure time increases the fractures increases. The microfracture firstly occurs in the matrix, and then by increasing exposure time intergranular oolite/matrix and transgranular fractures in the oolitic structures occurs.

Figure 14 shows the effect of the microwave power density on the formation of microfractures or intergranular oolite/matix boundaries fractures. At the same exposure time, increasing microwave power would increase the heating rate of iron ore **Figure 6** and thus the expanding difference between hematite and gangues increased. **Figure 14** shows that oolitic iron ore exposed to 900 W and 450 W microwave powers for 50 s radiation times, a significant damage was observed for high microwave power, while only few micro-cracks in the matrix occurred in the low microwave power. Lower microwave power required more exposure times to induce the same damages as higher microwave power. Whittles *et al.* [25] investigated the effect of power density on the microwave treatment of ores and found that the power density is an important factor in microwave treatment of ores. It decreases energy consumption and improves the efficiency.

Figure 15 shows the effect of grain size on the oolites boundaries fractures. The oolitic iron ore of different grain sizes exposed to the same microwave powers and exposure times (at 900 W microwave power and 60 s exposure time). Based on SEM observation, as the grain size decreases, the damage incurred in the samples be-

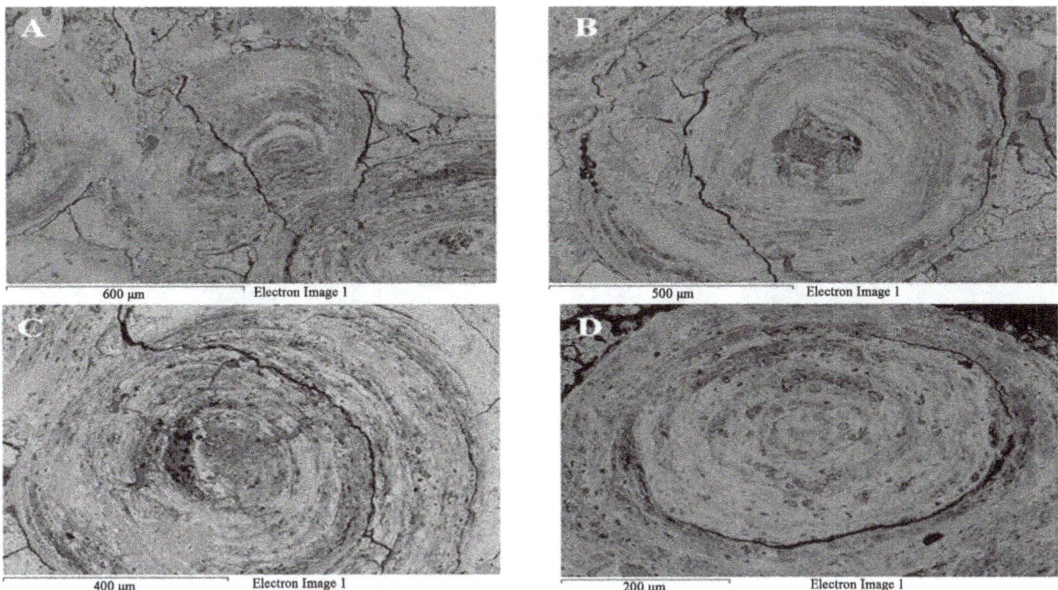

Figure 11. BSE of oolitic iron ore after microwave treatment (A) and (B) Transgranularoolitic fractures; (C) and (D) Intergranular microfractures in oolitic layers.

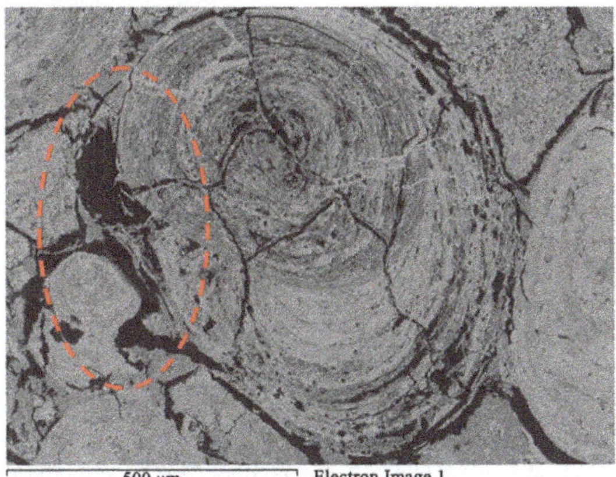

Figure 12. BSE image of microwave treated oolitic iron ore shows that part of the sample was melted.

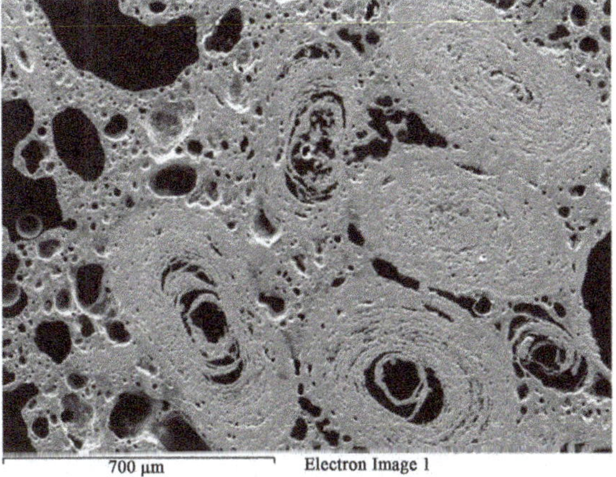

Figure 13. SEM image of the melting oolitic iron ore sample.

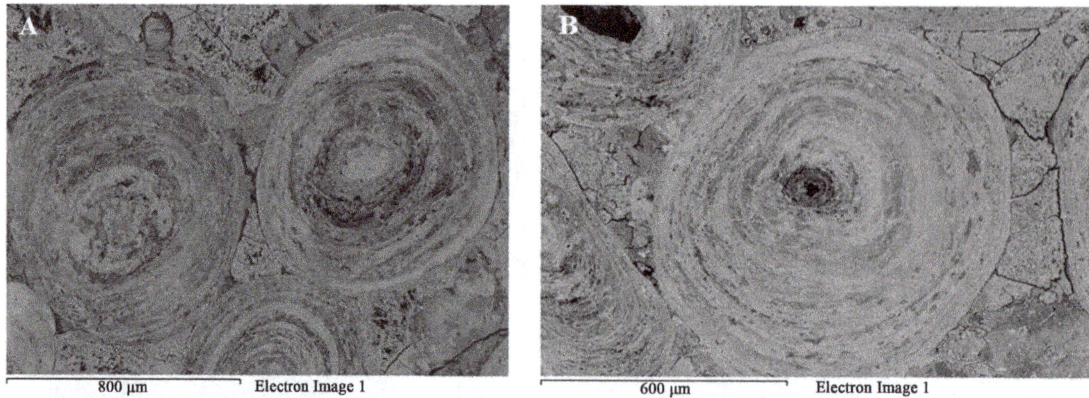

Figure 14. BSE images of oolitic iron ores treated at different microwave powers at 50 s exposure time. (A) 450 W power; (B) 900 W.

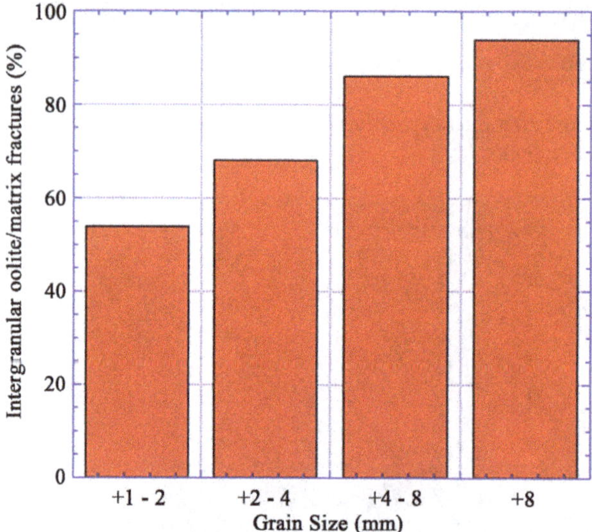

Figure 15. Effect of grain size of the oolitic iron ore on the amount of intergranular oolite/matrix fractures (oolites boundaries damage).

comes lower. Thus a higher energy input is required for the fine grain size to have the same damage as for a coarse grained ore. For example the amount of the ooids boundaries damage for grain sizes +8 mm and + 1 − 2 mm are 94% and 54% respectively.

3.4. Grindability Test

To measure the changes in the grindability of iron ore, the microwave treated and untreated samples were ground for 30 s. 100 grams of the crushed ore sample was first treated in the microwave oven for different exposure times 30, 40, 50 and 60 s at microwave power of 900 W. After grinding, the fraction of less than 0.125 mm of the ground specimen was determined by sieve analysis for both untreated and microwave-treated iron ore samples.

Then calculate grindability % = wt of undersize fraction (0.125)/total wt before grinding ×100

Figure 16 shows that the weight percentage of untreated and microwave treated samples for –0.125 mm size fractions. It is clear from **Figure 16** after 60 s of microwave pretreatment, the increases in the rate of weight percentages of microwave treated iron ore for –0.125 mm particle size to 59.76%, while the weight percent for untreated sample is 46.6%. Microwave radiation displayed more cracks and fractures in iron ores, these fractures occur around the grain boundaries between iron and gangues minerals. A reduction in comminution energy is

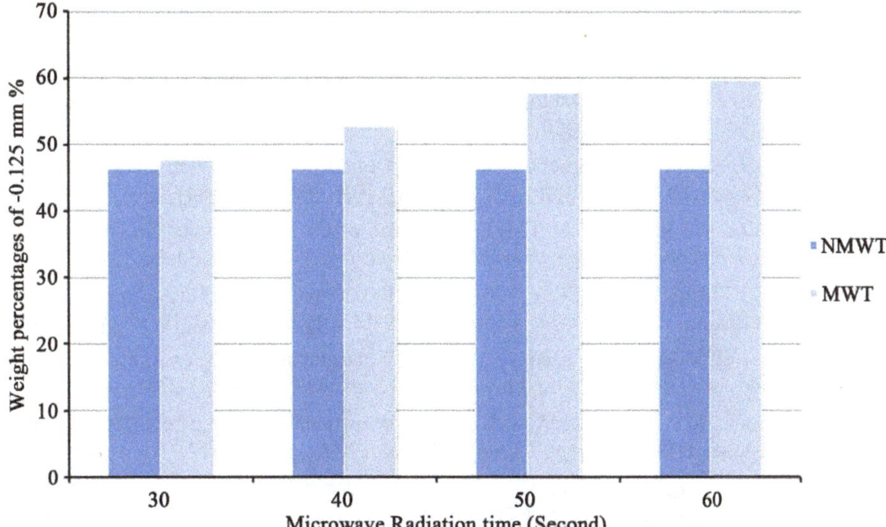

Figure 16. The weight percentage of untreated and microwave treated samples for –0.125 mm size fractions.

possible after microwave treatment.

The particles size distribution of microwave treated and untreated samples show that microwave pretreatment have two advantages:

1) Increased grindability of iron;

2) Reduced very fine size (slimes) produced during grinding. The weight percentages of particles less than 32 micron for untreated and microwave treated sample for 60 s are 1.24% and 0.48% respectively. These slimes are not suitable and interfere with physical separation techniques.

4. Conclusions

The effect of microwave pretreatment on the liberation of high phosphorus oolitic iron ore has been investigated. According to the experiments:

- SEM analysis indicated that intergranular fractures occurred between oolite and matrix (fluoroapatite and chamosite) after microwave treatment, which facilitated minerals liberation from each other at coarse size. The study showed that liberation and intergranular fractures increased by increasing microwave exposure time and grain size.
- According to the results of heating rate of iron ore, particles size is a very important factor. As the size of the particle decreases, more energy is required to raise the temperature of the particles and subsequent weakening and damage of particles. XRD analyses indicate that the peak intensity of hematite increase with increased exposure time.
- The particles size distributions indicate that microwave pretreatment of oolitic iron ore can be applied effectively to enhance the grindabilty and reduced slime production during grinding of iron ore.

Acknowledgments

The authors are very much thankful to (Cultural Affair and Mission Sector, Egypt) and CIMO (Center for International Mobility, Finland) for their financial grant to carry out the present research work. The authors are indebted to Prof. Ali Abdelmotelib from geology Department, Cairo University for his assistance during sample collection from Aswan region, Egypt. The authors would like to thank Mr. Riku Mattila and Mr. Tommi Kokkonen for their technical support throughout their work.

References

[1] Flügel, E. (2010) Microfacies of Carbonate Rocks. Analysis, Interpretation and Application. Springer-Verlag, Berlin.

[2] Manieh, A.A. (1984) Oolite Liberation of Oolitic Iron Ore, Wadi Fatima, Saudi Arabia. *International Journal of Mineral Processing*, **13**, 187-192. http://dx.doi.org/10.1016/0301-7516(84)90002-4

[3] Champetier, Y., Hamdadou, E. and Hamdadou, M. (1987) Examples of Biogenic Support of Mineralization in Two Oolitic Iron Ores—Lorraine (France) and Garadjebilet (Algeria). *Sedimentary Geology*, **51**, 249-255. http://dx.doi.org/10.1016/0037-0738(87)90050-9

[4] Ozdemir, O. and Deutsch, E.R. (1984) Magnetic Properties of Oolitic Iron Ore on Bell Island, New Found Land. *Earth and Planetary Science Letters*, **69**, 427-441. http://dx.doi.org/10.1016/0012-821X(84)90201-2

[5] Abro, M.M., Pathan, A.G. and Mallah, A.H. (2011) Liberation of Oolitic Hematite Grains from Iron Ore, Dilband Mines Pakistan. *Mehran University Research, Journal of Engineering Technology*, **30**, 329-338.

[6] Li, K., Ni, W., Zhu, M., Zheng, M. and Li, Y. (2011) Iron Extraction from Oolitic Iron Ore by a Deep Reduction Process. *Journal of Iron and Steel Research International*, **18**, 9-13. http://dx.doi.org/10.1016/S1006-706X(11)60096-4

[7] El Sharkawi, M.A., El Aref, M.M. and Mesaed, A.A. (1996) Stratigraphic setting and Paleoenvironment of the Conician-Santonian Ironstones of Aswan, South Egypt. *Geological Society of Egypt*, 243-278

[8] El Aref, M.M., El Sharkawi, M.A. and Mesaed, A.A. (1996) Depositional and Diagenetic Microfabric Evolution of the Cretaceous Oolitic Ironstone of Aswan, Egypt. *Geological Society of Egypt*, 279-312.

[9] Song, S., Campos-Toro, E. F. and Valdivieso, A. L. (2013) Formation of Micro-Fractures on an Oolitic Iron Ore under Microwave Treatment and its Effect on Selective Fragmentation. *Journal of Powder Technology*, **243**, 155-160. http://dx.doi.org/10.1016/j.powtec.2013.03.049

[10] Ji, J. (2003) Study on Dephosphorization Technology for High-Phosphorus Iron Ore. Mining & Metallurgy. **12**, 33-37.

[11] Xia, W.T., Ren, Z.D. and Gao, Y.F. (2011) Removal of Phosphorus from High Phosphorus Iron Ores by Selective HCl Leaching Method. *International Journal of Iron and Steel Research*, **18**, 1-4. http://dx.doi.org/10.1016/S1006-706X(11)60055-1

[12] Cheng, C.Y., Misra, V.N., Clough, J. and Muni, R. (1999) Dephosphorisation of Western Australian Iron Ore by Hydrometallurgical Process. *Minerals Engineering*, **12**, 1083-1092. http://dx.doi.org/10.1016/S0892-6875(99)00093-X

[13] Wang, J.C., Shen, S.B., Kang, J.H., Li, H.X. and Guo, Z.C. (2010) Effect of Ore Solid Concentration on the Bioleaching of Phosphorus from High-Phosphorus Iron Ores Using Indigenous Sulfur-Oxidizing Bacteria from Municipal Wastewater. *Process Biochemistry*, **45**, 1624-1631.

[14] Delvasto, P., Valverde, A., Ballester, A., Munoz, J.A., Gonzalez, F. and Blazquez, M.L. (2008) Diversity and Activity of Phosphate Bioleaching Bacteria from a High-Phosphorus Iron Ore. *Hydrometallurgy*, **92**, 124-129. http://dx.doi.org/10.1016/j.hydromet.2008.02.007

[15] Yu, Y.F. and Qi, C.Y. (2011) Magnetizing Roasting Mechanism and Effective Ore Dressing Process for Oolitic Hematite Ore. *Journal of Wuhan University of Technology. Materials Science Ed.*, **26**, 176-181. http://dx.doi.org/10.1007/s11595-011-0192-6

[16] Tang, H.Q., Guo, Z.C. and Zhao, Z.L. (2010) Phosphorus Removal of High Phosphorus Iron Ore by Gas-Based Reduction and Melt Separation. *International Journal of Iron and Steel Research*, **17**, 1-6. http://dx.doi.org/10.1016/S1006-706X(10)60133-1

[17] Fisher-White, M.J., Lovel, R.R. and Sparrow, G.J. (2012) Phosphorus Removal from Goethitic Iron Ore with a Low Temperature Heat Treatment and a Caustic Leach. *ISIJ International*, **52**, 797-803.

[18] Kumar, P., Sahoo, B.K., De, S., Kar, D.D., Chakraborty, S. and Meikap, B.C. (2010) Iron Ore Grindabilityim-Provement by Microwave Pretreatment. *Journal of Industrial and Engineering Chemistry*, **16**, 805-812. http://dx.doi.org/10.1016/j.jiec.2010.05.008

[19] Tromans, D. (2008) Mineral Comminution: Energy Efficiency Considerations. *Minerals Engineering*, **21**, 613-620. http://dx.doi.org/10.1016/j.mineng.2007.12.003

[20] Wang, E., Shi, F. and Manlapig, E. (2012) Mineral Liberation by High Voltage Pulses and Conventional Comminution with Same Specific Energy Levels. *Minerals Engineering*, **27-28**, 28-36. http://dx.doi.org/10.1016/j.mineng.2011.12.005

[21] Ali, A.Y. and Bradshaw, S.M. (2009) Quantifying Damage around Grain Boundaries in Microwave Treated Ores. *Chemical Engineering and Processing: Process Intensification*, **48**, 1566-1573. http://dx.doi.org/10.1016/j.cep.2009.09.001

[22] Roussy, G. and Pearce, J.A. (1995) Foundations and Industrial Applications of Microwave and Radiofrequency Fields-Physical and Chemical Processes, Chapters 10, 11, 12. Wiley, Hoboken.

[23] Haque, K.E. (1999) Microwave Energy for Mineral Treatment Processes—A Brief Review. *International Journal of Mineral Processing*, **57**, 1-24. http://dx.doi.org/10.1016/S0301-7516(99)00009-5

[24] Jones, D.A., Kingman, S.W., Whittles, D.N. and Lowndes, I.S. (2005) Understanding Microwave Assisted Breakage.

Minerals Engineering, **18**, 659-669. http://dx.doi.org/10.1016/j.mineng.2004.10.011

[25] Whittles, D.N., Kingman, S.W. and Reddish, D.J. (2003) Application of Numerical Modelling for Prediction of the Influence of Power Density on Microwave-Assisted Breakage. *International Journal of Mineral Processing*, **68**, 71-91. http://dx.doi.org/10.1016/S0301-7516(02)00049-2

[26] Jones, D.A., Kingman, S.W., Whittles, D.N. and Lowndes, I.S. (2007) The Influence of Microwave Energy Delivery Method on Strength Reduction in Ore Samples. *Chemical Engineering and Processing: Process Intensification*, **46**, 291-299. http://dx.doi.org/10.1016/j.cep.2006.06.009

[27] Fitzgibbon, K. and Veasey, T. (1990) Thermally Assisted Liberation—A Review. *Minerals Engineering*, **3**, 181-185. http://dx.doi.org/10.1016/0892-6875(90)90090-X

[28] Kingman, S.W. and Rowson, S.A. (1998) Microwave Treatment of Minerals—A Review. *Minerals Engineering*, **11**, 1081-1087. http://dx.doi.org/10.1016/S0892-6875(98)00094-6

[29] Tang, H.Q., Wang, J.W., Guo, Z. and Ou, T. (2013) Intensifying Gaseous Reduction of High Phosphorus Iron Ore Fines by Microwave Pretreatment. *International Journal of Iron and Steel Research*, **20**, 17-23. http://dx.doi.org/10.1016/S1006-706X(13)60091-6

[30] Amankwah, R.K., Khan, A.U., Pickles, C.A. and Yen, W.T. (2005) Improved Grindability and Gold Liberation by Microwave Pretreatment of a Free Milling Gold Ore. *Mineral Processing and Extractive Metallurgy*, **114**, 30-36. http://dx.doi.org/10.1179/037195505X28447

[31] Kingman, S.W., Corfield, G. and Rowson, N.A. (1999) Effect of Microwave Radiation upon the Mineralogy and Magnetic Processing of a Massive Norwegian Ilmenite. *Magnetic and Electrical Separation*, **9**, 131-148.

[32] Kingman, S.W. and Rowson, N.A. (2000) The Effect of Microwave Radiation on the Magnetic Properties of Minerals. *Journal of Microwave Power and Electromagnetic Energy*, **35**, 141-150.

[33] Kingman, S.W., Vorster, W. and Rowson, N.A. (2000) The Influence of Mineralogy on Microwave Assisted Grinding. *Minerals Engineering*, **13**, 313-327. http://dx.doi.org/10.1016/S0892-6875(00)00010-8

[34] Aguilar-Garib, J.A. (2011) Thermal Microwave Processing of Materials. In: Grundas, S., Ed., *Advances in Induction and Microwave Heating of Mineral and Organic Materials*, InTech. http://www.intechopen.com/books/advances-in-induction-and-microwaveheating-of-mineral-and-organic-materials/thermal-microwave-processing-of-materials

[35] Barani, K., Koleini, S.M.J. and Rezaei, B. (2011) Magnetic Properties of an Iron Ore Sample after Microwave Heating. *Separation and Purification Technology*, **76**, 331-336. http://dx.doi.org/10.1016/j.seppur.2010.11.001

[36] Ali, A.Y. and Bradshaw, S.M. (2010) Bonded Particle Modelling of Microwave Induced Damage in Ore Particles. *Minerals Engineering*, **23**, 780-790. http://dx.doi.org/10.1016/j.mineng.2010.05.019

[37] Salsman, J.B., Williamson, R.L., Tolley, W.K. and Rice, D.A. (1996) Short Pulse Microwave Treatment of Disseminated Sulphide Ores. *Minerals Engineering*, **9**, 43-54. http://dx.doi.org/10.1016/0892-6875(95)00130-1

[38] Chen, T.T., Dutrizac, J.E., Haque, K.E., Wyslouzil, W. and Kashyap, S. (1984) The Relative Transparency of Minerals to Microwave Radiation. *Canadian Metallurgical Quarterly*, **23**, 349-351. http://dx.doi.org/10.1179/cmq.1984.23.3.349

[39] Kobusheshe, J. (2010) Microwave Enhanced Processing of Ores. Ph.D. Thesis, the University of Nottingham, Nottingham.

10

The Nonideality Coefficient of Current-Voltage Characteristics for Asymmetric p-n-Junctions in a Microwave Field

Gafur Gulyamov[1,2], Muhammadjon Gulomkodirovich Dadamirzaev[1,2], Hasan Yusupovich Mavlyanov[1,2]

[1]Namangan Engineering Pedagogical Institute, Namangan, Uzbekistan
[2]Namangan State University, Namangan, Uzbekistan
Email: dadamirzaev70@umail.uz

Abstract

It is shown that the nonideality coefficient *m* actually depends on the electron temperature T_e, and the hole temperature T_h. We get more general expression for the nonideality coefficient, taking into account the concentration of electrons and holes, as well as their temperature, coefficient and diffusion length, the temperature of the phonons, the applied voltage, and the height of the potential barrier.

Keywords

Hot Electrons, The Microwave Field, The Open Circuit Voltage, Short Circuit Current, Current-Voltage Characteristics of p-n-Junction, The Nonideality Coefficient

1. Introduction

The development occupies an important place in microelectronics, semiconductor devices; therefore it requires a deep study of the physical fundamentals of semiconductor devices based on p-n-junction and the impact of external and internal factors on the characteristics of these devices. The main characteristic of a semiconductor diode based on p-n junction is a voltage-current characteristic. The theory of an ideal p-n junction was developed by Shockley [1]. CVC real p-n-junction is different from the ideal. For CVC real p-n-junction from Shockley was introduced the nonideality coefficient [2]. External influences—pressure, light, temperature and other factors affect the nonideality coefficient of Current-Voltage Characteristics for p-n-junction.

In the works of S. P. Ashmontas [3] [4], the effect of the heating of electrons was examined in the p-n-junction under the influence of the microwave field by the nonideality coefficient.

The nonideality coefficient of CVC silicon p-n-junction in a microwave field was studied in the works of A. I. Veynger and others [5] [6], where the abnormally high currents and electromotive force (EMF) had been observed.

However, the above-mentioned work is not considered dependence of the nonideality coefficient of the simultaneous heating of electrons and holes.

The aim of this work is to study the effect of heating of electrons and holes by a nonideality coefficient of CVC p-n-junction.

2. Theoretical Studies of Nonideality Coefficient of CVC Asymmetric p-n-Junction in a Microwave Field

The resulting expression of your total current passing through the diode consists of electron and hole currents and is defined as follows [7] [8]:

$$\begin{aligned}\bar{I} &= \frac{eD_e n_p}{L_e}\left\{\left(\frac{T_e}{T}\right)^{\frac{1}{2}}\exp\left(\frac{e\varphi_0}{kT}-\frac{e(\varphi_0-U)}{kT_e}\right)\int_0^{2\pi}\exp\left(-\frac{eU_B\cos(\omega t)}{kT_e}\right)\frac{\mathrm{d}(\omega t)}{2\pi}-1\right\}\\ &+\frac{eD_h p_n}{L_h}\left\{\left(\frac{T_h}{T}\right)^{\frac{1}{2}}\exp\left(\frac{e\varphi_0}{kT}-\frac{e(\varphi_0-U)}{kT_h}\right)\int_0^{2\pi}\exp\left(-\frac{eU_B\cos(\omega t)}{kT_h}\right)\frac{\mathrm{d}(\omega t)}{2\pi}-1\right\}\end{aligned} \tag{1}$$

where, $I_{se}=\frac{eD_e n_p}{L_e}$; $I_{sh}=\frac{eD_h p_n}{L_h}$ are the saturation currents for electrons and holes; φ_0: height of the potential barrier in the absence of an electromagnetic wave; $\varphi=\varphi_0-U$; U: a voltage across the diode; $U_B=-\int_0^d E_B\mathrm{d}x$: AC voltage of the incident wave created by the barrier diode; T is the temperature of the lattice; k: Boltzmann constant; T_e and T_h: the temperatures of electrons and holes; E_b—electric field of the wave; e is the charge of an electron; D_e and D_h—diffusion coefficients of electrons and holes, L_e and L_h—their diffusion length; n_p and p_n—the concentration of minority carriers.

Using the mean value theorem [9]:

$$\int_a^b f(x)\mathrm{d}x = f(\xi)\int_a^b \mathrm{d}x = f(\xi)(a-b),$$

we received for the considered case

$$f(U_B)=\left|\frac{1}{2\pi}\int_0^{2\pi}\exp\left(-\frac{eU_B\cos(\omega t)}{kT_e}\right)\mathrm{d}(\omega t)\right|\approx\left|\exp\left(-\frac{eU_B\left|\cos(\omega t)\right|}{kT_e}\right)\right|\gg 1$$

of formula (1) we have to CVC p-n-junction following expression:

$$\begin{aligned}\bar{I} &= \frac{eD_e n_p}{L_e}\left\{\left(\frac{T_e}{T}\right)^{\frac{1}{2}}\exp\left(\frac{e\varphi_0}{kT}-\frac{e(\varphi_0-U)+eU_B\left|\cos(\omega t)\right|}{kT_e}\right)-1\right\}\\ &+\frac{eD_h p_n}{L_h}\left\{\left(\frac{T_h}{T}\right)^{\frac{1}{2}}\exp\left(\frac{e\varphi_0}{kT}-\frac{e(\varphi_0-U)+eU_B\left|\cos(\omega t)\right|}{kT_h}\right)-1\right\}.\end{aligned} \tag{2}$$

At low wave power ($T_e=T_h=T$; $U_B\neq 0$), when there is only outrage the potential barrier height, using the formula (2) obtain the CVC p-n-junction:

$$\bar{I}=\left(I_{se}+I_{sh}\right)\left[\exp\left(\frac{e\left(U-U_{B}\left|\overline{\cos(\omega t)}\right|\right)}{mkT}\right)-1\right], \quad (3)$$

to take a logarithm of the expression (3), while taking into account the perturbation of the potential barrier height for the nonideality coefficient of current-voltage characteristics for p-n-junctions find the following formula:

$$m=\frac{e\left[U-U_{B}\left|\overline{\cos(\omega t)}\right|\right]}{kT\ln\left(\frac{I}{I_{se}+I_{sh}}+1\right)}. \quad (4)$$

High power microwave energy heated electrons and holes have different temperatures. In the works of A. I. Vengera and other experiments were conducted in a highly asymmetrical p-n-junction, where $p_p \gg n_n$ and $T_e > T_h > T$. However, under these conditions, the main share of the total current flowing through the p-n-junction, make hot holes [7]. Then the CVC p-n-junction consists of a second portion of the formula (2) and the first member can be neglected:

$$\bar{I}=\frac{eD_{h}p_{n}}{L_{h}}\left\{\left(\frac{T_{h}}{T}\right)^{\frac{1}{2}}\exp\left(\frac{e\varphi_{0}}{mkT}-\frac{e(\varphi_{0}-U)+eU_{B}\left|\overline{\cos(\omega t)}\right|}{mkT_{h}}\right)-1\right\}. \quad (5)$$

Hence, for the nonideality coefficient we obtain the following expression:

$$m(T_{h},T,U_{B})=\frac{e\left\{\varphi_{0}T_{h}-T\left[\varphi_{0}-U+U_{B}\left|\overline{\cos(\omega t)}\right|\right]\right\}}{\ln\left(\left(\frac{I_{h}}{I_{sh}}+1\right)\left(\frac{T}{T_{h}}\right)^{\frac{1}{2}}\right)kTT_{h}}. \quad (6)$$

It is seen that the nonideality coefficient is mainly dependent on the temperature of the holes.

If $p_p \ll n_n$, then a major share to the total current flowing through the p-n junction, to introduce hot electrons. Then, similarly to the formula (6) we get:

$$m(T_{e},T,U_{B})=\frac{e\left\{\varphi_{0}T_{e}-T\left[\varphi_{0}-U+U_{B}\left|\overline{\cos(\omega t)}\right|\right]\right\}}{\ln\left(\left(\frac{I}{I_{se}}+1\right)\left(\frac{T}{T_{e}}\right)^{\frac{1}{2}}\right)kTT_{e}}. \quad (7)$$

This expression depends mainly on electron temperature.

According to the analysis of the results shows that in the special case when $T_e = T_h = T$; $U_B = 0$ from (1) to obtain an expression for the nonideality coefficient of current-voltage characteristics of p-n-junction where the charge carriers are not warmed up.

We get more general expression for the nonideality coefficient, which takes into account the asymmetry of concentration, temperature, diffusion coefficients, diffusion length for electrons and holes, the temperature of the phonons, and the applied voltage, the height of the potential barrier. Analyses show that the nonideality coefficient strongly influenced by the concentration of charge carriers and the temperature perturbation potential barrier height, which are consistent with experimental results. In addition to the analytical expressions ((4), (6), (7)) on the basis of formula (1) can be transcendental equation coefficient imperfection independent simultaneous heating of electrons and holes. In addition to the analytical expressions ((4), (6), (7)) on the basis of formula (1) we received a transcendental equation for the nonideality coefficient, which depends on the simultaneous heating of electrons and holes.

$$\bar{j} = \frac{eD_e n_p}{L_e}\left\{\left(\frac{T_e}{T}\right)^{\frac{1}{2}} \exp\left(\frac{e\varphi_0}{mkT} - \frac{e(\varphi_0 - U)}{mkT_e}\right)\int_0^{2\pi} \exp\left(-\frac{eU_B\cos(\omega t)}{mkT_e}\right)\frac{d(\omega t)}{2\pi} - 1\right\}$$
$$+\frac{eD_h p_n}{L_h}\left\{\left(\frac{T_h}{T}\right)^{\frac{1}{2}} \exp\left(\frac{e\varphi_0}{mkT} - \frac{e(\varphi_0 - U)}{mkT_h}\right)\int_0^{2\pi} \exp\left(-\frac{eU_B\cos(\omega t)}{mkT_h}\right)\frac{d(\omega t)}{2\pi} - 1\right\}. \quad (8)$$

Using this formula, for different values of m we can construct CVC p-n-junction (**Figures 1-3**).

The graphs show that the nonideality coefficient is more sensitive to the microwave field and the current-voltage characteristics of p-n-junction is shifted to higher direct voltage with an increase in the nonideality coefficient.

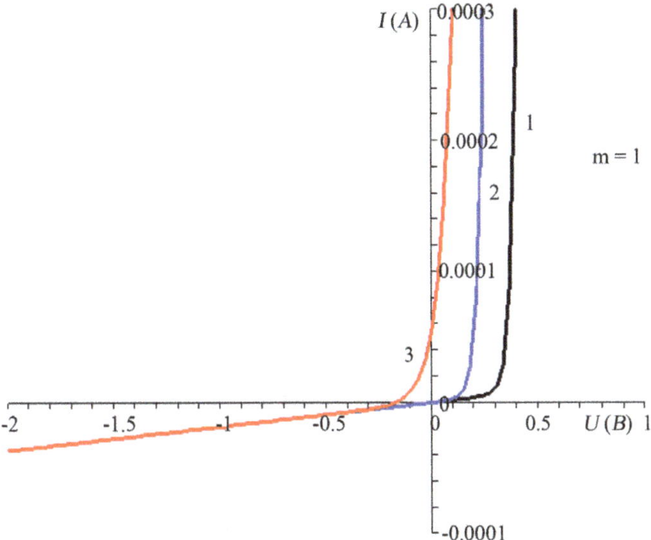

Figure 1. CVC p-n-junction, provided: 1—without the microwave field, 2—at low microwave powers wave and 3—at high power microwave energy and $m = 1$.

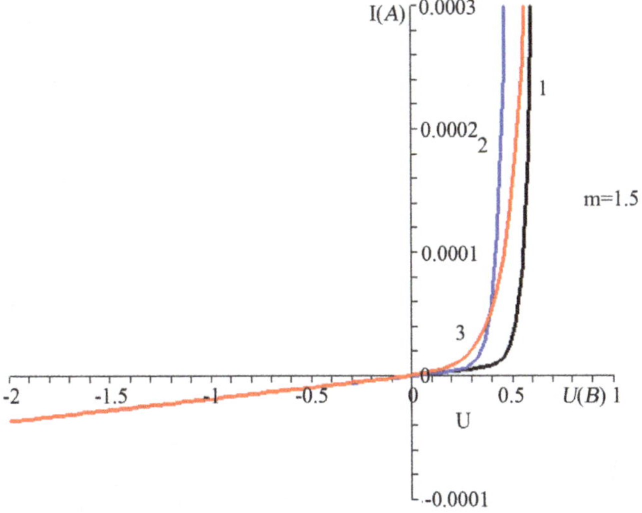

Figure 2. CVC p-n-junction, provided: 1—without the microwave field, 2—at low microwave power wave, and 3—at high power micro-wave energy and $m = 1.5$.

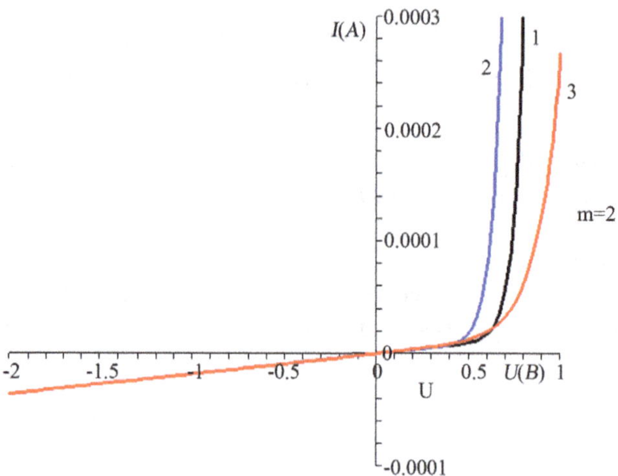

Figure 3. CVC p-n-junction, provided: 1—without the microwave field, 2—at low microwave powers wave, and 3—at high power micro-wave energy and $m = 2$.

3. Conclusions

The results show that the nonideality coefficient m really depends on the electron temperature and the temperature of the holes.

With $p_p \gg n_n$ the nonideality coefficient is determined by the temperature of the hole, and when $p_p \ll n_n$ electron temperature. We obtain a formula the nonideality coefficient for different conditions. It is shown that the nonideality coefficient m is more sensitive to the power of the microwave field. We get more general expression for the nonideality coefficient, taking into account the concentration of electrons and holes, as well as their temperature coefficient and diffusion length, the temperature of the phonons, the applied voltage, and the height of the potential barrier.

We get more general expression for the nonideality coefficient, taking into account the concentration of electrons and holes, as well as their temperature, diffusion coefficient and diffusion length, temperature phonons, applied voltage and indignation potential barrier height.

References

[1] Shockley, W. (1953) Theory of Electrical Semiconductors. Foreign Literature Publishing House, Moscow, 186 p.

[2] Sah, C.T., Noyce, R.H. and Shockly, W. (1957) Carrier Generation and Recombination in p-n-Junctions and p-n-Junctions Characteristics. *Proceedings of the IRE*, **45**, 1228-1243. http://dx.doi.org/10.1109/JRPROC.1957.278528

[3] Ashmontas, S.P., Olekas, A.P. and Shimulis, A.I. (1985) Effect of Warming up of the Charge Carriers in the Form of Current-Voltage Characteristics of p-n-Junction Germanium. *Semiconductors*, **29**, 807-809.

[4] Ashmontas, S.P., Olekas, A.P. and Shimulis, A.I. (1980) Temperature Anomalies Schottky Barrier Diodes Ni-n-Si. *Lithuanian Physical Collection*, **20**, 39-46.

[5] Veynger, A.I. and Sargsiyan, M.P. (1980) The Kinetics of the Thermopower Arising in the p-n-Junction with Heating Carrier. *Semiconductors*, **14**, 2020-2027.

[6] Ablyazimova, N.A., Veynger, A.I. and Pitanov, V.S. (1992) The Impact of a Strong Microwave Field in the Photovoltaic Properties of Silicon p-n-Junctions. *Semiconductors*, **26**, 1041-1047.

[7] Dadamirzaev, M.G. (2011) Heating of Charge Carriers and Rectification of Current in Asymmetrical p-n Junction in a Microwave Field. *Semiconductors*, **45**, 288-291. http://dx.doi.org/10.1134/S1063782611030092

[8] Dadamirzaev, M.G. (2015) Influence of Deformation on CVC p-n-Junction in a Strong Microwave Field. *Journal of Modern Physics*, **6**, 176-180. http://dx.doi.org/10.4236/jmp.2015.62023

[9] Smirnov, V.I. (1974) Course of Higher Mathematician. Vol. 1, Nauka, Moscow, 480 p.

11

Comparative Study on Conventional Sintering with Microwave Sintering and Vacuum Sintering of Y_2O_3-Al_2O_3-ZrO_2 Ceramics

Mayur Shukla[1,2], Sumana Ghosh[2*], Nandadulal Dandapat[2], Ashis K. Mandal[2], Vamsi K. Balla[2]

[1]Academy of Scientific and Innovative Research (AcSIR), CSIR-Central Glass and Ceramic Research Institute, Kolkata, India
[2]CSIR-Central Glass and Ceramic Research Institute (CSIR-CGCRI), Kolkata, India
Email: *sumana@cgcri.res.in

Abstract

The present investigation demonstrated the comparative studies carried out on conventional, microwave and vacuum sintering of alumina added yttria stabilized zirconia (YSZ). The conventional, microwave and vacuum sintered specimens were characterized by density measurement, XRD, SEM with EDX analysis and hardness evaluation. Microwave sintering was proved to be the best efficient sintering technique with respect to energy and time savings. Enhanced densification was observed for the microwave and vacuum sintered specimens at lower temperatures compared to the conventionally sintered ones. Further, it was observed that the particle size had significant influence on the enhancement of densification. The microwave sintered specimen showed the highest hardness compared to conventional and vacuum sintered specimens.

Keywords

Ceramics, Zirconia, Sintering, Density, Microstructure, Hardness

1. Introduction

Zirconia (ZrO_2) ceramics possess excellent properties such as high fracture toughness, high hardness and wear

*Corresponding author.

resistance, chemical inertness, low thermal conductivity and ionic conductivity, which allows its use in a range of applications including precision ball valve balls, high-density ball-mill grinding media, rollers and guides for metal tube forming, thread guides, pump seals, oxygen sensors, and solid oxide fuel cell membranes. However, tailored microstructure can be produced through controlled processing of partially stabilized zirconia (PSZ) with alkaline earth or rare earth oxide additions e.g. yttria (Y_2O_3) so that transformation toughening can be achieved by tetragonal ZrO_2 (t-ZrO_2) to monoclinic ZrO_2 (m-ZrO_2) phase transformation [1].

Recently, interest has been growing in the use of microwave energy to sinter ceramic compacts. In conventional thermal processing, energy is transferred to the material through conduction and radiation of heat from the surface. In contrast, microwave energy is delivered directly to the material through molecular interaction with the electromagnetic field. This interaction leads to various beneficial effects, which includes rapid volumetric heating, shorter sintering times, lower sintering temperatures and selective heating. The energy savings are in the range of 25% - 95%. In addition, it is well established that densification of a variety of ceramic materials is enhanced by microwave sintering. Higher densities are being achieved by microwave heating at lower temperatures than that obtained by conventional radiant heating [2]-[7]. Conventional sintering of zirconia at high temperatures above 1600°C results in tetragonal to monoclinic phase transformation on cooling, leading to destruction of the specimen on account of grain enlargement. Sintering at low oxygen partial pressure causes stabilization of high temperature phase of zirconia similar to oxide additives such as Y_2O_3, CaO, MgO, etc. Therefore, stabilization of tetragonal zirconia as well as high densification can be achieved by vacuum sintering method [8] [9].

Few studies were made to investigate the vacuum sintering effect on the phase composition and properties of ceramics compared to the conventionally sintered ones. It was reported that vacuum sintering of plasma-chemical 3 wt.% yttria added zirconia at high temperatures stabilized the tetragonal phase of zirconia in association with obtaining high densification [10] [11]. Sablina *et al.* [12] showed that density increased with increasing temperature and tetragonal to monoclinic phase transformation did not take place on cooling in case of vacuum sintering of ZrO_2-Y_2O_3 and ZrO_2-Y_2O_3-Al_2O_3 based ceramic specimens. Wilson and Kunz [13] evaluated microwave sintering of partially stabilized ZrO_2. They placed ZrO_2 material in SiC susceptors that absorbed microwave energy and transferred heat to the specimen. After initial heating, the ZrO_2 material could then absorb microwave energy and got heated. They showed that cracking of the ZrO_2 material occurred by ultra-rapid heating and observed similar physical properties for conventional and microwave sintered ZrO_2. Janney *et al.* [14] sintered 8 mol% ytttria stabilized zirconia (YSZ) using SiC rod as susceptor. Further, Nightingale and Dunne [15] studied the density and grain growth of 3Y-TZP sintered in the conventional and microwave ovens. Sintering of zirconia ceramics using microwaves of various frequencies (2.45 - 60 GHz) has been already studied [5]. Goldstein *et al.* [16] sintered YSZ by microwave directly. They indicated that microwave sintered specimen had a smaller grain size. Upadhyaya *et al.* [17] studied sintering and grain growth of 3Y-TZP and 3Y-TZP with the addition of TiO_2 and MnO_2 in order to improve the microwave coupling at a given temperature. Wang *et al.* [18] used microwave/conventional hybrid heating technique for various ceramics including zirconia. However, grain growth was enhanced during microwave/conventional hybrid heating compared with conventional heating and thereby, suggesting acceleration of the diffusion processes by microwave/conventional hybrid heating. Matsui *et al.* [19] reported that small amount of Al_2O_3 enhanced the densification rate because of the decrease in the activation energy with the change in the diffusion mechanism from grain boundary diffusion (GBD) to volume diffusion (VD).

In the present investigation, comparative studies have been performed between conventional, microwave and vacuum sintering of alumina added 8 wt.% YSZ ceramics. The objective of this paper is to get full dense YSZ at lower temperature with superior hardness.

2. Experimental

Commercial 8 mol% Y_2O_3-ZrO_2 powder (YSZ, Metallizing Equipment Co. Private Limited, India, particle size 45 ± 10 μm) was mixed with 0.5%, 1%, 1.5% and 2% of alumina (Al_2O_3, Alcoa, USA; 99.99% purity). Alumina was added to enhance the sintering of YSZ. YSZ and Al_2O_3 Powders were milled in a planetary mill to obtain homogenous powder mixture. Powders were then isostatically pressed (EPSI NV, SO, 10,036 Belgium) at 150 MPa to produce disk-shaped green compacts. The green powder compacts were dried at 100°C for 5 h and calcined in an electrical furnace (ELECTROHEAT, Model No.EN170QT, Naskar & Co., Howrah, India) at 1200°C for 1 h, then cut and finally sintered at 1600˚C for various period of time ranging from 2 h to 10 h. Heating and cooling cycle has been shown in **Figure 1**.

Another set of specimens were sintered in a microwave furnace (Enerzi Microwave Systems Pvt. Limited, Bangalore, India) at 1500˚C for 20 min. Heating and cooling time was 25˚C/min. A multimode microwave furnace with a magnetron having frequency of 2.45 GHz and maximum output power of 3 kW was used for the sintering. SiC powder was used as a susceptor to initiate coupling of microwave with the specimen. The specimens were placed in an alumina disc insulated by microwave transparent casket insulating box. The top cover of the insulating box had a hole of 20 mm diameter to monitor temperature through a non-contact IR pyrometer. Temperature measurement accuracy was ±0.3% of the measured value +1˚C with adjustable emissivity (ε: 0.1 - 1.0). Third set of specimens were sintered in a vacuum furnace (Hindhivac Private Limited, Bangalore, India) with a vacuum of 5×10^{-6} mbar at 1500˚C for 20 min. Heating and cooling rate was 10˚C/min and 7˚C/min, respectively. Different temperature programs for conventional, microwave and vacuum sintering were selected to establish the superiority of the microwave and vacuum sintering techniques compared to conventional sintering technique.

The surfaces of sintered specimens were ground using a grinding machine (BAINLINE Belt Linishing machine, Chennai Metco. Limited, Chennai, India) and then polished in a polishing machine (Leco Corporation, USA) with 6 µm, 3 µm and 0.25 µm diamond pastes (Buehler USA). The bulk density of the sintered specimen was measured by Archimedes' principle. Polished specimens were thermally etched at 1300˚C for 30 min in an electrical furnace to reveal the microstructure. In order to examine the phase assemblages of the densified bodies X-ray diffraction was performed (PW 1710, Philips Research Laboratory, Eindhoven, Netherlands) with Cu Kα radiation (45 kV, 35 mA). Microstructural observations were performed by scanning electron microscopy (SEM) (Phenom Pro-X, Netherlands) and elemental composition was determined by energy dispersive X-ray (EDX) analysis (Phenom Pro-X, Netherlands). Microhardness was evaluated by a Vickers hardness tester (ESEWAY, 410 series, Bowers group, U.K.) at a load of 100 g with 30 s loading/unloading time. For a particular type of specimen, five specimens were examined. Considerable numbers of data were taken to avoid any error in hardness measurement.

Similarly, commercial nano-sized 8 wt.% Y_2O_3-ZrO_2 powder (YSZ, TOSOH CORPORATION, Japan, particle size 40 ± 20 nm) was conventionally sintered at 1450˚C for 2 h in a conventional furnace. Heating rate was 3˚C/min and cooling rate was 2˚C/min up to 1000˚C and 4˚C/min up to 600˚C. Furnace cooling to room temperature was conducted after 600˚C. The sintered nano-YSZ specimen was subsequently characterized in the same manner. Conventional sintering of nano-YSZ specimen was conducted in order to establish the effect of particle size.

3. Results and Discussion

Figure 1 shows that conventional sintering method required ~13 h for the total sintering operation of 0.5 wt.% alumina added YSZ specimen whereas total processing time was ~3 h for the microwave sintered specimen. Total processing time for vacuum sintered similar specimen was ~7 h, which was intermediate among the three processing techniques. **Figure 2(a)** shows the density of YSZ sintered at 1600˚C for different period of time with the increase of alumina addition. It was observed that highest density could be achieved by the addition of 0.5 wt.% Al_2O_3 to YSZ. Therefore, 0.5Al_2O_3-YSZ composition was selected as optimum composition for continuing further investigation. **Figure 2(b)** shows the density values for 0.5 wt.% Al_2O_3 added YSZ specimen at 1600˚C as a function of soaking time. The 0.5 Al_2O_3-YSZ specimen had 80% density after sintering at 1600˚C for 10 h duration whereas 75% density was obtained for the similar specimen sintered at 1600˚C for 8 h. The microwave sintered 0.5 Al_2O_3-YSZ specimen at 1500˚C for 20 min showed 75% density while it was 70% density in the case of vacuum sintered (1500˚C, 20 min) 0.5Al_2O_3-YSZ specimen. It can be said that both microwave and vacuum sintering resulted enhanced densification at lower sintering temperature within shorter processing time leading to energy and time savings. However, highest energy and time savings was observed with the microwave sintering operation. Further, the density of vacuum sintered 0.5 Al_2O_3-YSZ specimen was comparable with that of the microwave sintered 0.5 Al_2O_3-YSZ specimen.

Figure 3 represents the XRD patterns of conventional, microwave and vacuum sintered 0.5 Al_2O_3-YSZ specimens. The crystalline phases were identified as tetragonal zirconia (t-ZrO_2), monoclinic zirconia (m-ZrO_2), Y_2O_3 and Al_2O_3 in all the cases. The major crystalline phase was t-ZrO_2 whereas m-ZrO_2, Y_2O_3 and Al_2O_3 were

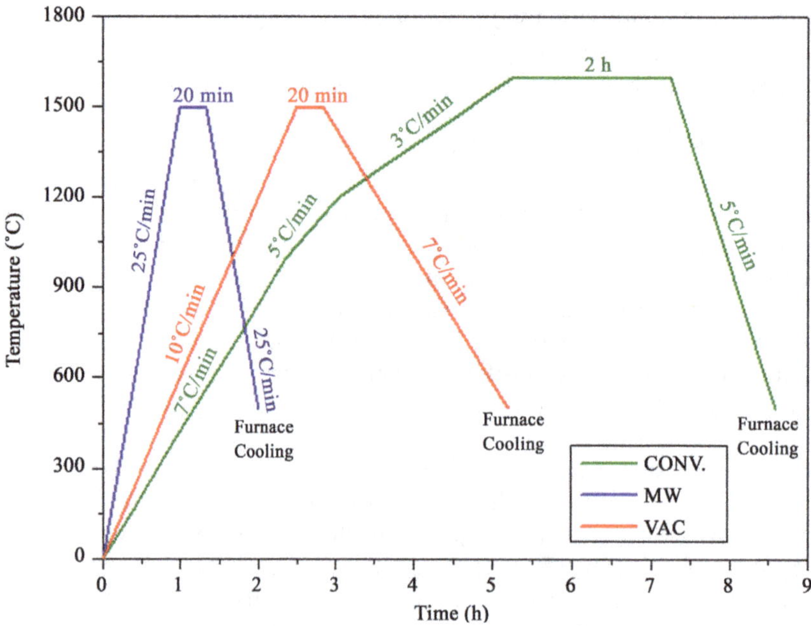

Figure 1. Temperature versus time plots for conventional, microwave and vacuum sintering operations.

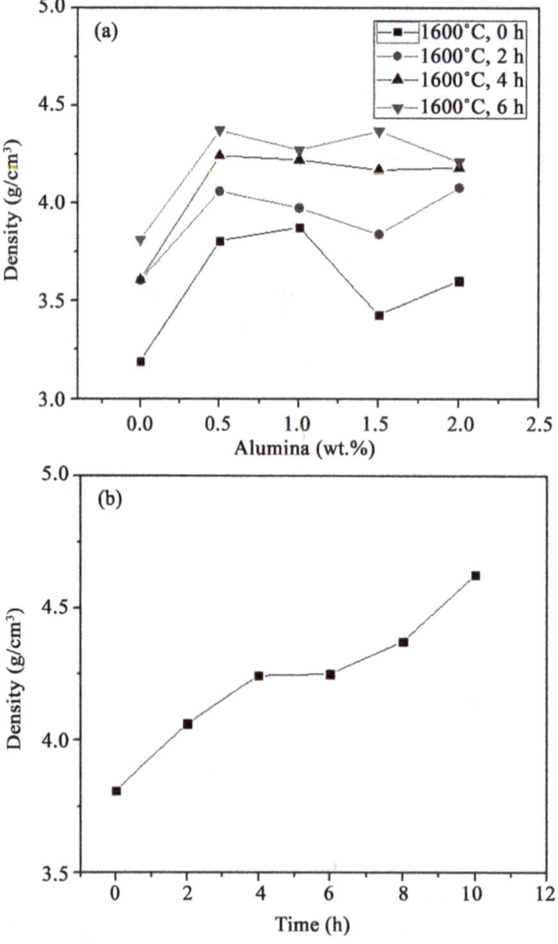

Figure 2. (a) Density of YSZ sintered at 1600°C for different period of time as a function of alumina addition and (b) density versus dwell time plot for YSZ with 0.5 wt. % alumina addition.

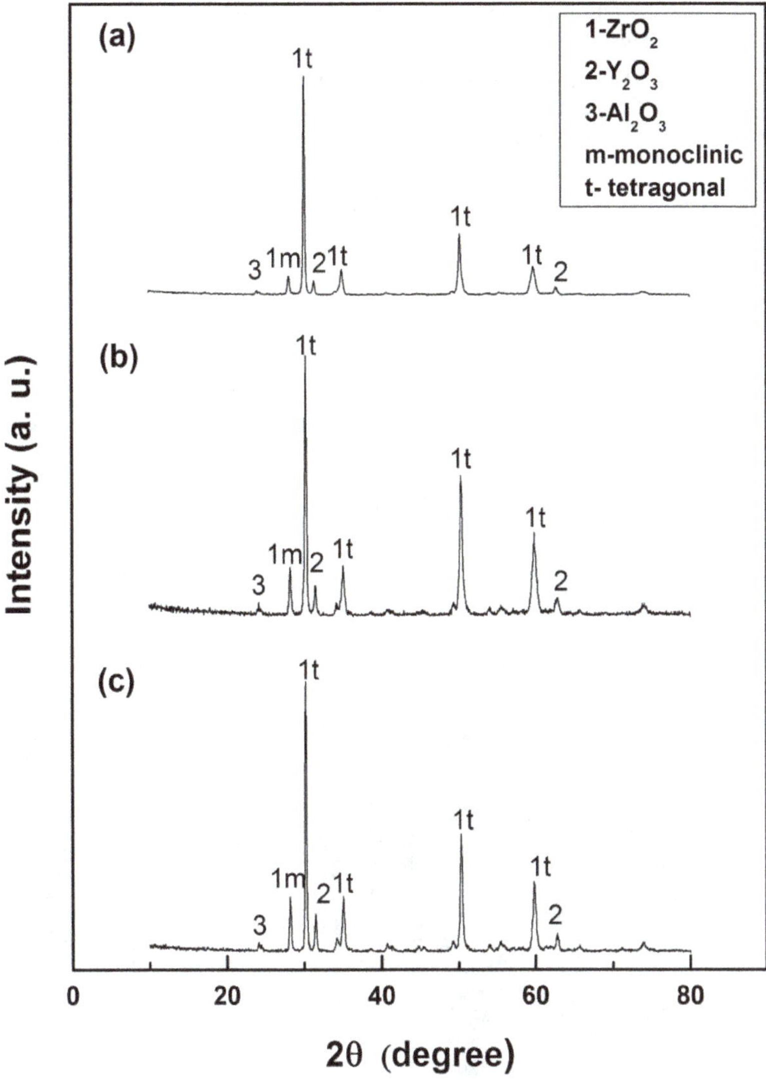

Figure 3. XRD analysis of (a) conventional, (b) microwave and (c) vacuum sintered $0.5Al_2O_3$-YSZ specimens.

present as minor phases. Therefore, transformation toughening can be obtained for the sintered specimen as a consequence of t-ZrO_2 to m-ZrO_2 phase transformation. **Figure 4** shows the microstructures of conventional, microwave and vacuum sintered 0.5 Al_2O_3-YSZ specimens.

Conventionally sintered specimen demonstrated highest grain growth having grain size of 2.5 ± 0.25 μm (**Figure 4(a)**, **Figure 4(b)**). The microwave sintered specimen had finest grain size of 0.8 ± 0.1 μm (**Figure 4(c)**, **Figure 4(d)**). Whereas the vacuum sintered specimen showed a grain size of 1.2 ± 0.2 μm (**Figure 4(e)**, **Figure 4(f)**). In addition, significant difference in the morphologies of the conventional, microwave and vacuum sintered specimens was observed. The microwave sintered specimen displayed uniform grain size. In contrast, the conventional and vacuum sintered specimens showed non-uniformity in the grain size.

Present study showed that 100% dense YSZ specimen was achieved by 2 h soaking at 1450°C in the conventional route through particle size tuning of the starting powder from micron-size to nanometer-size range. **Figure 5(a)** shows the typical XRD analysis of nano-YSZ specimen conventionally sintered at 1450°C for 2 h, which was identical to those obtained for conventional, microwave and vacuum sintered micron-sized 0.5 Al_2O_3-YSZ specimens. The total processing time was analogous to that maintained for conventional sintering of micron-sized 0.5 Al_2O_3-YSZ specimen. The SEM image and corresponding EDX analysis have been shown in **Figure 5(b)**. The SEM microstructure showed that the grain morphology of nano-YSZ specimen was spherical. The EDX pattern confirmed the presence of Zr and O elements only.

Table 1 shows the density and hardness values for conventional, microwave and vacuum sintered 0.5 Al_2O_3-YSZ specimens (micron-sized) and conventionally sintered nano-sized YSZ specimen. Conventionally sintered nano-sized YSZ specimen exhibited 100% density at lowest temperature with highest hardness *i.e.* 1345 ± 25 VHN (13.19 ± 0.25 GPa). The hardness of conventionally sintered 0.5 Al_2O_3-YSZ specimen for 10 h duration was 660 ± 20 VHN (6.47 ± 0.20 GPa) while it was 433 ± 15 VHN (4.25 ± 0.15 GPa) in case of similar specimen conventionally sintered for 8 h. The hardness of microwave sintered (1500°C, 20 min) 0.5 Al_2O_3-YSZ specimen was 440 ± 17 VHN (4.32 ± 0.17 GPa) whereas it was 417 ± 18 VHN (4.09 ± 0.18 GPa) in case of vacuum sintered (1500°C, 20 min) specimen.

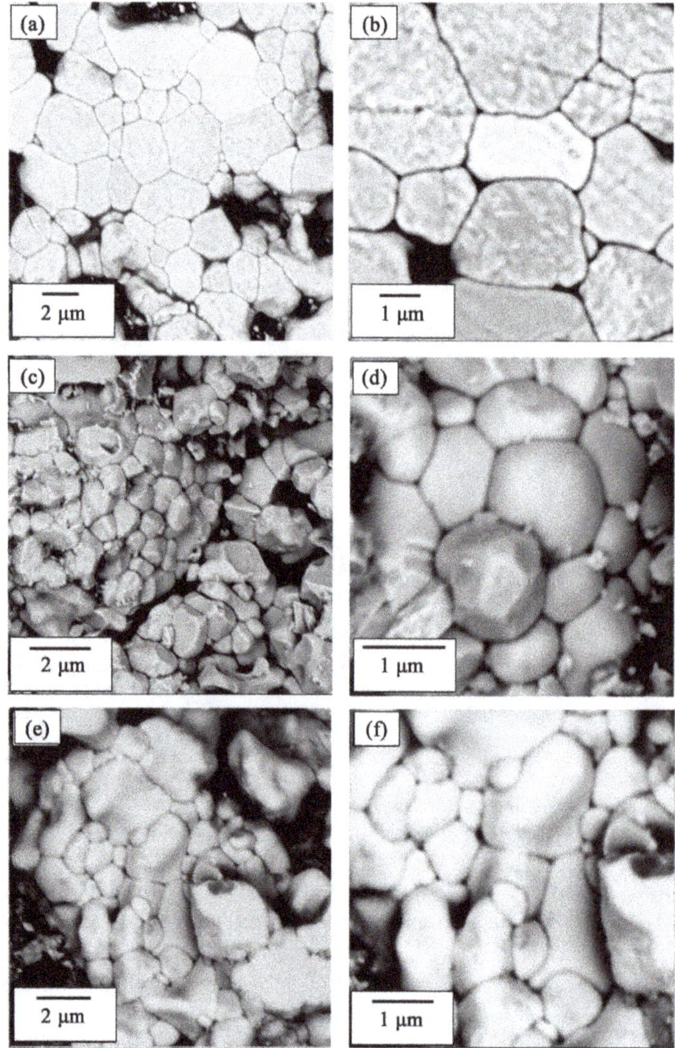

Figure 4. SEM images of (a)-(b) conventional, (c)-(d) microwave and (e)-(f) vacuum sintered 0.5 Al_2O_3-YSZ specimens.

Table 1. Properties of conventional, microwave and vacuum sintered zirconia.

Processing Method	Properties			
	Particle Size	Density (g/cm^3)	Hardness (VHN)	Hardness (GPa)
Conventional (1600°C, 10 h)	45 ± 10 μm	4.8 (80%)	660 ± 20	6.47 ± 0.20
Conventional (1600°C, 8 h)	45 ± 10 μm	4.5 (75%)	433 ± 15	4.25 ± 0.15
Microwave (1500°C, 20 min)	45 ± 10 μm	4.5 (75%)	440 ± 17	4.32 ± 0.17
Vacuum (1500°C, 20 min)	45 ± 10 μm	4.2 (70%)	417 ± 18	4.09 ± 0.18
Conventional (1450°C, 2 h)	40 ± 20 nm	6.0 (100%)	1345 ± 25	13.19 ± 0.25

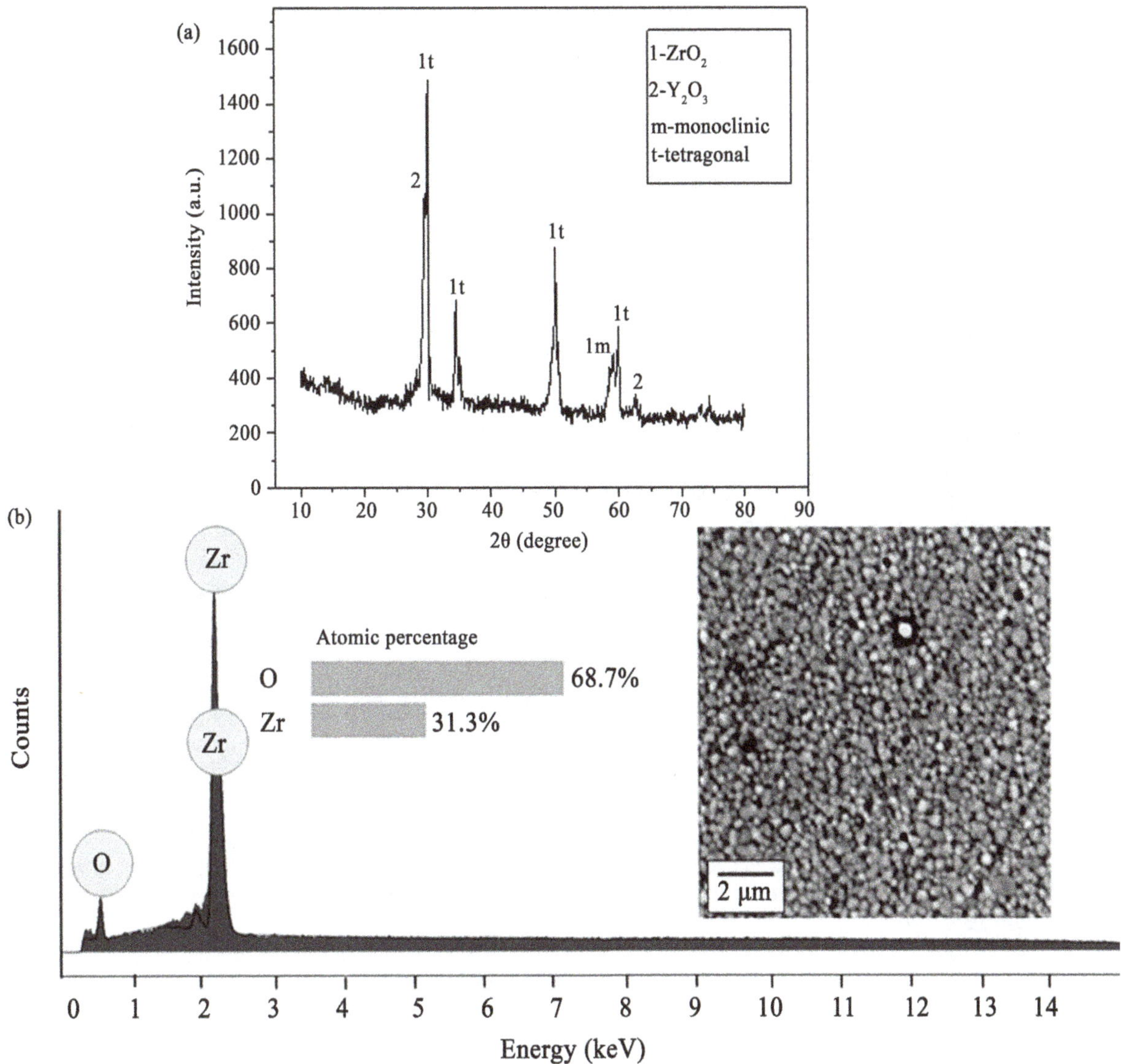

Figure 5. Typical (a) XRD plot and (b) EDX pattern in association with SEM image (shown in inset) of sintered nano-YSZ specimen.

4. Conclusion

The present investigation showed that microwave and vacuum sintering techniques offered enhanced sintering at lower temperatures compared to conventional sintering technique in case of the micron-sized 0.5 Al_2O_3-YSZ specimen. However, microwave sintering technique appeared as the best sintering technique in comparison to the conventional and vacuum sintering techniques. Further, conventional sintering of nano-YSZ specimen imparted 100% density at 1450°C for 2 h soaking. In contrast, conventionally sintered micron-sized 0.5 Al_2O_3-YSZ specimen revealed 80% density after 10 h soaking at 1600°C. The current study also established the particle size effect on densification of the conventionally sintered YSZ specimen. Thus, the present study showed some possible means through which hard zirconia based ceramics can be manufactured for suitable applications in dentistry.

Acknowledgements

The present work was financially supported by CSIR, India under 12 FYP network project titled "Very High Power Microwave Tubes: Design and Development Capabilities (MTDDC)", Grant No. PSC0101.

References

[1] Hannink, R.H.J., Kelly, P.M. and Muddle, B.C. (2000) Transformation Toughening in Zirconia-Containing Ceramics. *Journal of the American Ceramic Society*, **83**, 461-487. http://dx.doi.org/10.1111/j.1151-2916.2000.tb01221.x

[2] Sheppard, L.M. (1988) Manufacturing Ceramics with Microwaves: The Potential for Economical Production. *American Ceramic Society Bulletin*, **67**, 1656-1661.

[3] Katz, J.D. (1992) Microwave Sintering of Ceramics. *Annual Review of Materials Science*, **22**, 153-170. http://dx.doi.org/10.1146/annurev.ms.22.080192.001101

[4] Clark, D.E., Folz, D.C., Schulz, R.L., Fathi, Z. and Cozzi, A.D. (1993) Recent Developments in Microwave Processing of Ceramics. *Materials Research Bulletin*, **18**, 41-46.

[5] Thostenson, E.T. and Chou, T.W. (1999) Microwave Processing: Fundamentals and Applications. *Composites Part A: Applied Science and Manufacturing*, **30**, 1055-1071. http://dx.doi.org/10.1016/S1359-835X(99)00020-2

[6] Weller, M. and Schubert, H. (1986) Internal Friction, Dielectric Loss, and Ionic Conductivity of Tetragonal ZrO_2-3% Y_2O_3 (Y-TZP). *Journal of the American Ceramic Society*, **69**, 573-577.

[7] Kenkre, V.M. (1991) Theory of Microwave Interactions with Ceramics. *Ceramic Transactions*, **21**, 69-80.

[8] Ruh, R. and Garrett, H.J. (1967) Non-Stoichiometry of ZrO_2 and Its Relation to Tetragonal-Cubic Inversion in ZrO_2 *Journal of the American Ceramic Society*, **50**, 257-261. http://dx.doi.org/10.1111/j.1151-2916.1967.tb15099.x

[9] Ramaswamy, P. and Agrawal, D.C. (1987) Effect of Sintering Zirconia with Calcia in Very Low Partial Pressure of Oxygen. *Journal of Materials Science*, **22**, 1243-1248. http://dx.doi.org/10.1007/BF01233116

[10] Savchenko, N.L., Sablina, T.Y., Poletika, T.M., Artish, A.S. and Kul'kov, S.N. (1993) Phase Composition and Mechanical Properties of Zirconium Dioxide Based Ceramic Obtained by High Temperature Sintering in a Vacuum. *Powder Metallurgy and Metal Ceramics*, **32**, 9-10.

[11] Savchenko, N.L., Sablina, T.Y., Poletika, T.M., Artish, A.S. and Kul'kov, S.N. (1994) High Temperature Vacuum Sintering of Plasmochemical Powders Based on ZrO_2. *Powder Metallurgy and Metal Ceramics*, **33**, 1-2.

[12] Sablina, T.Y., Savchenko, N.L., Mel'nikov, A.G. and Kul'kov, S.N. (1994) Vacuum Sintering of a Ceramic Based on Zirconium Dioxide. *Glass and Ceramics*, **51**, 198-201. http://dx.doi.org/10.1007/BF00682584

[13] Wilson, J. and Kunz, S.M. (1988) Microwave Sintering of Partially Stabilized Zirconia. *Journal of the American Ceramic Society*, **71**, C40-C41.

[14] Janney, M.A., Calhoun, C.L. and Kimrey, H.D. (1992) Microwave Sintering of Solid Oxide Fuel Cell Materials: I, Zirconia-8 mol% Yttria. *Journal of the American Ceramic Society*, **75**, 341-346. http://dx.doi.org/10.1111/j.1151-2916.1992.tb08184.x

[15] Nightingale, S.A., Dunne, D.P. and Worner, H.K. (1996) Sintering and Grain Growth of 3 mol% Yttria Zirconia in a Microwave Field. *Journal of Materials Science*, **31**, 5039-5043. http://dx.doi.org/10.1007/BF00355903

[16] Goldstein, A., Travitzky, N., Singurindy, A. and Kravchik, M.J. (1999) Direct Microwave Sintering of Yttria-Stabilized Zirconia at 2.45 GHz. *Journal of the European Ceramic Society*, **19**, 2067-2072. http://dx.doi.org/10.1016/S0955-2219(99)00020-5

[17] Upadhyaya, D.D., Ghosh, A., Gurumurthy, K.R. and Prasad R. (2001) Microwave Sintering of Cubic Zirconia. *Ceramics International*, **27**, 415-418. http://dx.doi.org/10.1016/S0272-8842(00)00096-1

[18] Wang, J., *et al.* (2006) Evidence for the Microwave Effect during Hybrid Sintering. *Journal of the American Ceramic Society*, **89**, 1977-1984.

[19] Matsui, K., Yamakawa, T., Uehara, M., Enomoto, N. and Hojo, J. (2008) Sintering Mechanism of Fine Zirconia Powders with Alumina Added by Powder Mixing and Chemical Processes. *Journal of Materials Science*, **43**, 2745-2753. http://dx.doi.org/10.1007/s10853-008-2493-5

12

Microwave Design and Performance of PTB 10 V Circuits for the Programmable Josephson Voltage Standard

Franz Müller, Thomas Scheller, Jinni Lee, Ralf Behr, Luis Palafox, Marco Schubert, Johannes Kohlmann

Physikalisch-Technische Bundesanstalt (PTB), Braunschweig, Germany
Email: franz.mueller@ptb.de

Abstract

At Physikalisch-Technische Bundesanstalt (PTB), superconducting 10 V circuits for the programmable Josephson voltage standard (PJVS) are routinely manufactured on the basis of Nb_xSi_{1-x} barrier junctions. This paper describes in detail the basic design principles for an operating frequency of 70 GHz. It starts with single junctions, discusses their insertion into microstriplines and closes with the whole microwave circuit containing 69,632 Nb_xSi_{1-x} barrier junctions arranged over 128 microstriplines connected in parallel. The microwave attenuation of this junction type is a key parameter for the 10 V design and we report its experimental determination. Special attention has been devoted to subarrays with just a few Josephson junctions in one of the outermost striplines. The arrangement of these subarrays determines the optimum performance of the complete 10 V series array. The high performance of programmable 10 V circuits fabricated at PTB is characterized by measured operating margins in all subarrays of more than 1 mA centered at the same dc bias current. The observed modulation of the current margins, when changing the frequency around 70 GHz, is explained by microwave reflections caused by the rf waveguide inside the cryoprobe. We determine the current margins in dependence of the output power of a microwave synthesizer and show that 60 mW is sufficient to achieve current margins larger than 1 mA.

Keywords

Josephson Voltage Standard, Programmable Josephson Voltage Standard (PJVS), SNS Josephson Junction, Nb_xSi_{1-x} Josephson Junction

1. Introduction

Nowadays programmable Josephson voltage standards (PJVS) [1] with output voltages up to 10 V replace more and more conventional 10 V Josephson voltage standard (JVS) systems operated between 70 and 75 GHz [2]. PJVSs are based on series arrays of overdamped Josephson junctions (JJs) divided into smaller independently biased subarrays. The overdamped Josephson junctions are externally or intrinsically shunted and exhibit a current-voltage characteristic (IVC) with a negligible hysteresis. In contrast to underdamped and therefore hysteretic SIS (S = superconductor, I = insulator) junctions used for conventional JVS systems, overdamped Josephson junctions display single-valued Shapiro steps under microwave irradiation. Each of these constant voltage steps, including $U = 0$, can be attributed to a definite dc bias current. This special property opens up new areas of application beyond simple dc metrology covered by the conventional JVS. Today, the PJVS constitutes, together with the Josephson arbitrary waveform synthesizer (JAWS) based on pulse-driven Josephson junctions [3], the basis of modern ac voltage metrology. PJVS circuits representing the current state of technology are using Nb_xSi_{1-x} barrier junctions. This junction type allows operating frequencies from 15 to 70 GHz by changing the Nb content [4]. For this reason, it complies perfectly with the strategy of PTB to retain an operating frequency of 70 GHz, or, in other words, to continue to use the microwave equipment in operation for the conventional JVS at most National Metrology Institutes (NMI).

Section 2 of this paper describes the search for a suitable Josephson junction type and for a normal-conducting barrier material (N) in SNS (S = superconductor) junctions that are required for a 70 GHz design. Thereafter, the basic design principles starting from single junctions, their integration into microstriplines and completion of the whole microwave circuit for 70 GHz will be discussed. Section 3 concentrates on experimental results obtained for different design variants regarding the arrangement of the smallest subarrays. The performance of programmable 10 V circuits fabricated at PTB is characterized by the measured current margins for all relevant array segments. Furthermore, we present measured current margins in dependence of the frequency in a small band around 70 GHz and discuss the requirements for the microwave power and their impact on the use of cryocoolers. A summary and some conclusions are given in Section 4. With this paper, the authors intend to provide an overview of the properties of 10 V PJVS circuits made by PTB and a useful guideline for their use. The main metrological applications of the 10 V circuits fabricated at PTB have been described in detail in a review paper by Behr *et al.* [5].

2. 70 GHz Design of 10 V Circuits for the PJVS

2.1. Development of Robust Josephson Junctions Suitable for 70 GHz Operation

To ensure optimum operation of a Josephson voltage standard, the product of critical current I_c and normal resistance: $R_n : I_c R_n = V_c$ (characteristic voltage) of the used Josephson junctions has to be matched to the microwave frequency, *i.e.* it should be $V_c \approx 145\ \mu V$ for 70 GHz (cf. 2.2.) [6].

While the proof-of-concept experiment [1] was still executed with externally shunted SIS junctions operated at 70 GHz, shortly thereafter the National Institute for Standards and Technology (NIST) favoured intrinsically shunted SNS junctions enabling larger Shapiro steps, *i.e.* a better adjustability of the operating point and larger noise immunity [7]. Due to the relatively low resistivity of the normal-conducting barrier materials commonly used and available at that time, e.g. AuPd, HfTi, or later on, TiN_x, microwave design and operating frequency were fixed to the range from 15 to 20 GHz and were maintained in later developments with other barrier materials. In contrast, PTB searched from the beginning for a junction technology which would allow retaining the operating frequency of 70 GHz used for conventional JVSs. A 70 GHz drive reduces the number of junctions necessary to reach a certain voltage level by a factor of 4 to 5, *i.e.* 70,000 JJs instead of 320,000 JJs for 10 V. The entire microwave distribution network for 70 GHz can be completely fabricated on chip. In this way, the requirements for fabrication technology and microwave design are more relaxed.

Motivated by the reliable Nb-Al technology successfully established throughout the 1990s for the fabrication of conventional 10 V SIS JVSs at PTB [8] [9], the intrinsically shunted SINIS Josephson junctions (S = Nb, I = AlO_x, N = Al) composed of two very transparent AlO_x barriers quickly moved into the focus of the PTB researchers. The double-barrier structure of this junction type can be easily tuned to 70 GHz. As early as in 2000, PTB reported first programmable 10 V PJVS circuits with SINIS junctions for 70 GHz drive [10]. However, their limited resolution due to six subarrays and a step width of 200 μA restricted their applications to fast and simplified dc voltage calibrations. Continuous progress in fabrication technology and several design optimization steps led six years later to the first-ever 10 V PJVS circuit really suitable for ac applications [11]. These 10 V circuits with 69,632 series-connected SINIS JJs offered a high resolution and sufficient noise immunity because they were divided into 16 subarrays and the step width of all array segments was larger than 0.6 mA. However, wide adoption in ac metrology suffered from the low fabrication yield, because a few of the very thin insulating AlO_x barriers $(d \leq 1\ \mathrm{nm})$ of the SINIS junctions were obviously damaged by plasma-induced charging effects occurring during several fabrication steps (dry etching) of the microwave circuits [12].

It is well known that SNS junctions with normal-conducting (N) barriers, which are usually relatively thick ($d > 10$ nm), are very robust and free of these yield problems connected to very thin insulating layers with pinholes. For this reason, PTB was continuously and simultaneously to its SINIS activities, seeking a SNS junction type that can be simply implemented as a drop-in replacement in existing 70 GHz SINIS designs of PJVS circuits. Finally, amorphous Nb_xSi_{1-x}, "rediscovered" by the NIST [4], was the appropriate SNS-like normal-conducting material. The resistivity of Nb_xSi_{1-x}, already known as a candidate for JJs since 1987 [13], can be tuned by its Nb content from insulating to metallic and represents an ideal N-barrier for SNS junctions. NIST first used this material for PJVS circuits driven around 15 GHz. In close cooperation between PTB (design, circuit fabrication) and NIST (trilayer deposition) the first 10 V PJVS circuits with 69,632 Nb_xSi_{1-x} barrier junctions operable at 70 GHz were realized in 2008 [12]. Two years later, NIST reported their first own 10 V arrays with about 300,000 triple-stacked junctions and the same barrier material, tuned however for operating frequencies from 16 to 20 GHz [14]. Another type of a 10 V PJVS circuit with double-stacked NbN-TiN_x-NbN junctions operated at 16 GHz and 10 K was developed by the Japanese National Institute of Advanced Industrial Science and Technology (AIST) [15].

After the purchase of a special sputtering cluster tool allowing the automated deposition of the barrier by co-sputtering of Nb and Si in a separate chamber, PTB has been fabricating 10 V circuits with Nb_xSi_{1-x} barrier junctions routinely and completely in-house since 2010. Details of this fabrication process are described elsewhere [12] [16]. Using this technology and benefiting from the advantages of the 70 GHz design mentioned above, we also fabricated the first-ever 20 V array for PJVSs simply by using 69,632 double stacked JJs instead of the single junctions in the 10 V design [17].

The NbSi technology has significantly improved the yield of PJVS circuits. For the first time, 10 V circuits with no defects, *i.e.* no "missing (shorted) junctions", as compared to 10 V SINIS arrays [11], could be manufactured in a complex fabrication process. 10 V arrays with Nb_xSi_{1-x} barrier junctions show a much better performance than those with SINIS junctions. A current width of about 1 mA for the 1st Shapiro steps of all subarrays is typical for the 10 V circuits fabricated at PTB (cf. 3.2.).

2.2. Single Josephson Junctions with Nb_xSi_{1-x} Barriers Optimized for 70 GHz

The underlying physics of Nb-Nb_xSi_{1-x}-Nb Josephson junctions has been thoroughly investigated by Scheller [18]. Major properties of the semiconducting Nb_xSi_{1-x} tunneling barrier for different doping levels of Nb, its influence on the $1/f$ -noise and on the specific capacitance have been summarized by Müller *et al.* [17]. By controlling the Nb content "*x*" and the thickness "*d*" of the Nb_xSi_{1-x} barrier, it is possible to realize two limiting cases representing different junction types: 1) An SIS-like junction with a nearly maximum hysteresis in the IVC for $x = 0$ (pure semiconductor) and $d \approx 5$ nm, or on the other hand, 2) an SNS-like junction with no hysteresis in the IVC for $x \geq 20\%$ (fully degenerate semiconductor) and $d \geq 20$ nm. Junctions designed for an operating frequency of 70 GHz occupy a position between these limiting cases: They present a small (almost negligible) hysteresis and are strictly speaking neither SIS nor SNS junctions. In the following they are called Nb_xSi_{1-x} barrier junctions. Their barrier is typically characterized by $x \approx 10\%$, $d \approx 10.5$ nm and a specific capacitance of about 380 fF/μm^2. This barrier composition enables a characteristic voltage of $V_c = 145\ \mu\mathrm{V}$ (with R_n determined at $2I_c$). Such a value for V_c is optimal for an operating frequency $f = 70$ GHz as the normalized frequency $\Omega = nf/(2eV_c) \approx 1$ [6]. Under this condition, V_c may also be derived from the second Josephson equation describing the ac Josephson effect (cf. e.g. [19]):

$$2eU_n = nhf \tag{1}$$

setting the integer $n = 1$ and $U = V_c$ (e = elementary charge, h = Planck's constant). On the other hand, it can be deduced from Equation (1) that a single Josephson junction under microwave irradiation of 70 GHz generates constant voltage steps corresponding to multiples of ±145 μV. Typically, PJVSs only use Shapiro steps of 1st (±145 μV) and 0th (0 V) order [1]. The current width, in the following also called current margin, of the 1st Shapiro step reaches a theoretical maximum when the $I_c R_n$ product is adjusted to 145 μV. As current density j_c and resistivity of a junction are up to a certain degree independently adjustable by the co-sputtering of Nb and Si, it is possible to obtain the desired $I_c R_n$ for a given (designed) junction area A. To select reasonable values of j_c and A, it should be taken into consideration that the value of the critical current $I_c \approx j_c A$ determines the necessary rf power (together with the stripline impedance). Based on our experience with SINIS designs [12] where a maximum critical current of 3 mA was reasonable, we set $I_c^{\max} \leq 5\ \text{mA}$ for the present 10 V design with Nb_xSi_{1-x} barrier junctions. The width "w" of the junction, perpendicular to the extension of the array, is an important parameter for the microwave impedance of the microstripline. For the present 10 V design a value of $w = 20\ \mu\text{m}$ was chosen. The length "l" of the junction, which should be "short" to limit the length of the array, was fixed to $l = 6\ \mu\text{m}$. For these geometrical design parameters resulting in $A = wl = 120\ \mu\text{m}^2$ of a single junction, it is possible to adjust current density and critical voltage to about $j_c \approx 4\ \text{kA/cm}^2$ and $I_c R_n \approx 145\ \mu\text{V}$, thus enabling optimum formation of Shapiro steps.

2.3. 70 GHz Design of Microstriplines with Embedded Nb_xSi_{1-x} Barrier Junctions

A simple approach to increase the voltage from 145 µV up to 10 V is the series connection of many junctions. In order to ensure that the "quantized" voltages U_1 of each junction add to a much larger quantized voltage level, the series-connected JJs should have nearly the same shape of the IVC under microwave irradiation, *i.e.* Shapiro steps with the same current width and positioned at the same place of the IVC. This requires that nearly the same microwave current is injected into each junction, apart from a small scattering of the junction parameters. Therefore, microwave designs for series arrays used in all types of JVSs had to ensure a nearly homogeneous distribution of the rf power along the array. Superconducting microstriplines with embedded Josephson junction arrays can provide the required homogeneity [20] and were therefore used as a basic design element in most conventional 10 V JVS circuits with hysteretic (underdamped) SIS junctions driven at 70 GHz and above. The insertion of nonhysteretic (overdamped) junctions in superconducting striplines was therefore the first logical step in creating a 70 GHz design for a PJVS. However, a new and surprising effect was detected in this case [21]: In contrast to highly-hysteretic SIS junctions, arrays of embedded nonhysteretic junctions (SINIS, SNS) show self-oscillation coupling effects if their dc bias exceeds the critical current. These coupling effects [12] [22] occur in "active" striplines with $I_{dc} > I_c$, and partly compensate the relatively large microwave attenuation expected for the series connection of many JJs. The attenuation can be estimated according to Kautz for "passive" striplines $(I_{dc} < I_c)$, a model that was originally successfully developed for embedded SIS junctions with so-called "zero-current" Shapiro steps $(I_{dc} \approx 0)$ [23]. The maximum number of Josephson junctions that can be inserted in a single microstripline is the key parameter of the microwave design. For this reason, we determined experimentally the microwave attenuation for a stripline with embedded Nb_xSi_{1-x} barrier junctions. The scheme of the setup used in the measurements is shown in **Figure 1**. The setup allows varying the microwave power supplied by means of an adjustable attenuator at the top of the cryoprobe. First we measure the step width of IVC_{out} with no external attenuation $(D = 0)$. The attenuation of 1500 JJs is given by D when IVC_{in} matches the previous IVC_{out}. In the "passive" case $(I_{dc} = 0)$ the measured attenuation per junction is about 0.022 dB (theoretically: 0.04 dB). Coupling effects occur in "active" striplines $(I_{dc} > I_c)$ and reduce this value to 0.009 dB/JJ. The deviation between experimental and theoretical value for the passive case is possibly due to the influence of the Si substrate on the microwave propagation. As a result, the microwave design has to take into consideration that short but inactive $(I_{dc} = 0)$ array segments can present a higher attenuation than longer and active subarrays. On the other hand, experience previously collected with SIS arrays for conventional JVSs [8] suggest that the maximum microwave attenuation should not exceed 5 dB to ensure a sufficiently homogeneous distribution of the rf power. As a result, the maximum number of Nb_xSi_{1-x} barrier junctions in a single and actively driven stripline should be about 550. This number corresponds to an earlier result found for SINIS arrays that up to 500 junctions the maximum current width of the 1st Shapiro step is only slightly reduced (to 80% of I_c) [24]. The present microwave design is realized on basis of the so-called inverted microstripline structure (**Figure 2**), which was already successfully used for SIS and afterwards for SINIS circuits [25]. In this design, the superconducting ground plane is placed on top instead of underneath the array. Later on (cf. 2.4.), it will become apparent, that such an inverted microstripline substantially simplifies the microwave design for the whole circuit.

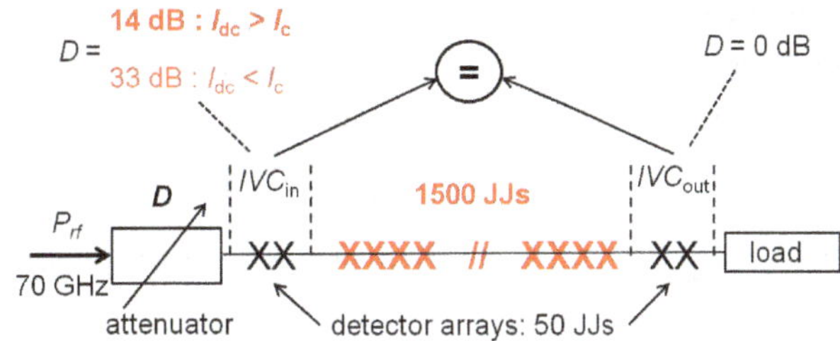

Figure 1. Schematic setup for microwave attenuation measurements.

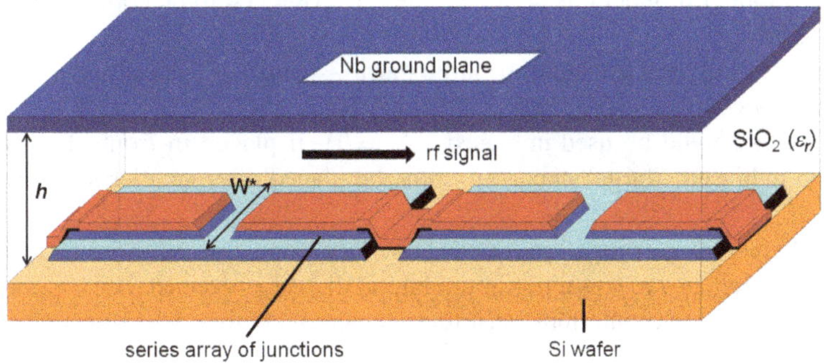

Figure 2. Scheme of an inverted stripline structure with embedded JJs.

Moreover, it allows the direct deposition of the junction trilayer on a clean and polished wafer surface. It is known that a rough surface, due to a previous etching process or the deposition of a thick dielectric film, degrades the quality of the trilayer, especially the sharpness of its interfaces. Last but not least, a Nb ground plane on top of the circuit protects the junction array very well. The microwave impedance "Z" of a single stripline (**Figure 2**) is an important design parameter that determines the necessary rf power to generate Shapiro steps of optimum size. For the design values: $h = 1.2\ \mu\text{m}$, $w^* = 24\ \mu\text{m}$ $\left(w^* \geq w\right)$ and $\varepsilon_r \approx 4.5$ for SiO_2 [26], the characteristic impedance of an "undisturbed" microstripline (neglecting the influence of the substrate and without junctions) becomes approximately $Z \approx 9\ \Omega$. Knowledge of this impedance allows a rough estimation of the rf power P that should be at least coupled into each "stripline-array", according to $P = Zi^2/2$ with $i \approx 2I_c$ (i: rf current) [27]. For $I_c \approx 5\ \text{mA}$, a single microstripline requires approximately $P \approx 0.5\ \text{mW}$.

2.4. 10 V Circuit Design with 69,632 Nb_xSi_{1-x} Barrier Junctions

The second generation of programmable 10 V circuits developed and fabricated at PTB [11] is subdivided into 16 array segments whose junction numbers follow an almost binary sequence. The sequence is exactly binary at the beginning, 2^m (m: 0, 1, 2, 3) giving the numbers 1, 2, 4, and 8. The next segment with 17 junctions breaks this series and starts a new "binary" sequence according to 17×2^p (p: 0, 1, ···), resulting in: 17, 34, 68, 136, 272, and so on. To approach the 10 V level as close as possible, we need a largest subarray with 34,816 junctions $(p = 11)$, leading to a total of 69,630 junctions in the array. The fundamental Josephson equation specially formulated for the programmable version of the Josephson voltage standard becomes:

$$U(t) = nM(t)f/K_{J-90} \tag{2}$$

with $n = 0$, or ± 1 and $K_{J-90} = 483597.9\ \text{GHz/V}$ (Josephson constant). For $f = 70\ \text{GHz}$ and $M = 69,630$, Equation (2) delivers a maximum voltage amplitude of ±10.0788 V which is very close to the 10 V level. To realize for special measurement purposes two equal halves of the circuit (each with 34,816 junctions), two extra single junctions have been added. Therefore, the total number of junctions amounts to 69,632.

For the robust operation of rapidly programmable circuits with a "time-variant" number of junctions $M(t)$, the current margins of each voltage step $U(t)$ generated by the activated array segments should exceed a certain minimum. From our experience and also supported by other researchers [28], a value of at least 0.8 mA is desirable. The current width of a Shapiro step of order n: ΔI_n generated by each junction (embedded in a stripline) under microwave injection is proportional to the critical current I_c and varies with the rf current $i \sim P^{1/2}$ [27] according to the total amount of a Bessel function J_n :

$$\Delta I_n = 2I_c \left| J_n \left(P^{1/2} \right) \right| \tag{3}$$

All 69,632 JJs of the 10 V design have to be distributed over microstriplines rf-driven in parallel, taking into account the following restrictions:

- A binary number 2^N $(N = 1, 2, \cdots)$ of parallel-connected striplines should be used for geometrical reasons and for equal distribution of the rf power coupled by the antenna (**Figure 3**). In order to keep the necessary rf power for the whole circuit low, N should be as small as possible.
- As the maximum number of junctions in one stripline is evaluated to be about 550 (cf. 2.3.), we concluded to use 128 (2^7) striplines.
- To minimize their microwave attenuation, large "array-striplines" should always be biased collectively to benefit from the previously described self-oscillation effect.
- Only small segments should be used unbiased $(I_{dc} = 0)$ if placed in front of active subarrays. Due to the attenuation when unbiased, "large" subarrays must be placed after smaller segments, relative to the microwave propagation.
- The active circuit area comprising all junctions should be concentrated as much as possible to minimize the parameter spread, which also means a minimum lateral distance between the striplines.
- With the exception of the two outermost striplines, superconducting dc connections can only be made to the ends of the striplines.
- To realize at the same time a series connection of all junctions and a parallel rf power supply for each stripline, *dc* blocks have to be inserted at the output of the rf-distributing network (microwave splitters).

The 10 V microwave design schematically depicted in **Figure 3** is the result of these conditions. One 10 V chip covers an area of 24 mm × 10 mm. A finline antenna [29], built by the Nb ground plane and the Nb base metallization of the trilayer, captures microwave power from a rectangular waveguide. The rf power is divided by microwave splitters and fed over capacitors (dc blocks) into the array-striplines connected in parallel. All microstriplines are extended by homogenous Nb strips beyond the array length in order to reach the microwave termination and to match all lengths. The microwave termination of each array stripline is necessary to avoid an inhomogeneous rf power distribution by standing waves. The use of a AuPd load as part of the circuit ground is a special feature of the PTB design. This resistive ground plane, together with the Nb strips which lie underneath and which are connected to the ends of the junction arrays, forms the continuation of the inverted microstripline structure. It seems to be the only reasonable design variant that enables the necessary superconducting connection between the arrays and the 23 bonding pads arranged at the right part of the circuit. The alternative design variant, using AuPd strips and a superconducting ground plane extending over the whole circuit, would make it nearly impossible to establish superconducting connections between dc pads and all array segments.

The distribution of the 69,632 Nb_xSi_{1-x} barrier junctions over the 128 microstriplines is detailed in **Table 1**. From the "low potential dc pad", connected to the end of the 1st row (**Figure 3**), the smallest array segments are originally arranged in decreasing (regular) order of junctions (68, 34, 17, 8, 4, 2, 1, 1, 1), before beginning with the 2nd row, the "natural" quasi-binary sequence is continued: 136, 272, 544, etc. For comparative studies, a small number of chips with a reversed order of the smallest segments in the 1st stripline have also been fabricated. Unbiased junctions in front of active array segments only occur in the 1st row of the 128 striplines, *i.e.* in an extreme case (reversed order) 135 passive junctions lie in front of one active Josephson junction.

3. Experimental Investigations on 10 V PJVS Circuits with Nb_xSi_{1-x} Barrier Junctions

3.1. Measuring Setup and Instructions for Use

The 10 V chips (**Figure 4**) are glued on special carriers and electrically connected by wire bonding. For an optimum microwave coupling the antenna is inserted to 70% of its length in the slot of a WR12 waveguide. The use of a commercially available microwave synthesizer for frequencies around 70 GHz with a maximum output power of about 160 mW simplifies substantially the operation of the programmable circuit. This computer-controlled synthesizer only requires a 10 MHz reference signal without the need of a phase lock loop. To

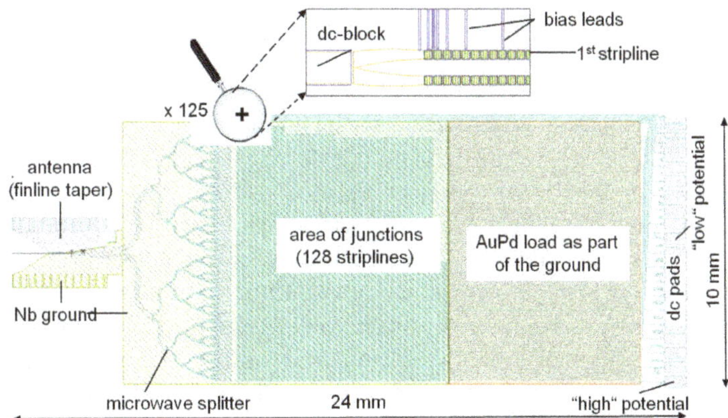

Figure 3. Microwave design of a programmable 10 V circuit operated at 70 GHz. The magnification shows the 1st stripline with the smallest segments arranged in "regular" order.

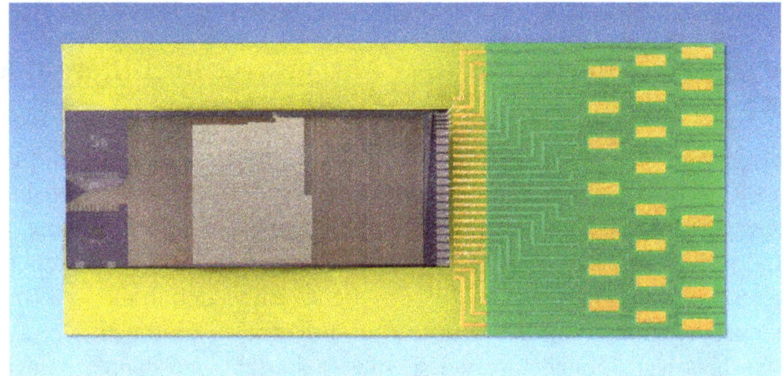

Figure 4. Photograph of a 10 V PJVS circuit for operation at 70 GHz mounted on a special chip carrier.

Table 1. Distribution of 69,632 Josephson junctions over 128 parallel-connected striplines in a programmable 10 V circuit with 16 subarrays.

Subarray N	# Stripline	Number of junctions M	V_N at 70 GHz (mV)
1	1	68	9.84
2	1	34	4.92
3	1	17	2.46
4	1	8	1.16
5	1	4	0.579
6	1	2	0.289
7(3x)	1	1, 1, 1	0.145
8	2	136	19.69
9	3 + 4	272 (2 × 136)	39.37
10	5 + 6	544 (2 × 272)	78.74
11	7 + 8	1088 (2 × 544)	157.49
12	9 to 12	2176 (4 × 544)	314.97
13	13 to 20	4352 (8 × 544)	629.94
14	21 to 36	8704 (16 × 544)	1259.89
15	37 to 68	17,408 (32 × 544)	2 519.78
16	69 to 128	34,816 (34 × 582 + 26 × 578)	5 039.56
Entire Circuit	1 to 128	69,632	10 079.12

realize fast programmability, e.g. for the stepwise synthesis of an ac signal of low frequency (<2 kHz) [5], each of the 16 segments of the binary-divided 10 V array can be separately driven by its own rapidly switchable bias source. However, the IVCs presented in this paper have been recorded in most cases by using only one computer-controlled current source (resolution: 20 μA) and a digital voltmeter.

Immediately after cooling down the array and before starting the operation, it is advised to check whether the circuit is free of magnetic flux. For this purpose, the IVC of the whole circuit without microwave injection should be visualized with high resolution using an oscilloscope. In case of trapped flux, one or more junctions display reduced critical currents. If there is a heater integrated in the cryoprobe, the array can be conveniently warmed up to remove the flux [30]. Otherwise one has to lift up the cryoprobe before slowly lowering it again (cooling down). Unfortunately, it is possible that even after repeated warming up and cooling down the flux remains trapped. This case occurs most likely after a quick warming up so that the circuit is covered by a lot of condensed water. Tiny magnetic particles from the surrounding magnetic shield or from other parts (screws) used for fixing the cryoprobe are possibly transported by the condensing atmosphere and/or flowing He gas. It is therefore advantageous to cover the inner surface of the cylindrical shield with a plastic material and to avoid screws made from alloys containing ferromagnetic components. To remove magnetic contaminants from the surface of the circuit, it is advised to do this with large precaution in a clean room under microscopic observation.

3.2. Current Margins and Effect of Parasitic Microwaves for Different Arrangements of the Smallest Subarrays

The 70 GHz design with a nonuniform distribution of the junctions over the 128 parallel-connected striplines (**Table 1**) represents a trade-off between homogeneous microwave supply and geometrical constraints for the "binary" array segments. Each segment with more than 68 JJs occupies entirely one or more striplines, which are therefore always "active" when the considered segment is biased on the first-order Shapiro step $(I_{bias} > I_c)$. In contrast, the smallest subarrays (1, 1, 1, 2, 4, ···, 68) are all embedded in the first (outermost) stripline. In this case, the microwave propagation along the stripline can be affected by inactive array segments. Our observations indicate that the bias leads connecting the smallest subarrays can couple parasitic microwaves from the outside. The magnetic shield surrounding the cryoprobe sample space contains parasitic microwave radiation because the transition from the waveguide to the circuit antenna is not at all ideal and a certain amount of rf power leaks from the waveguide into the silicon wafer [26] and to the cavity of the cylindrical cryoperm shield. Parasitic microwave power injected in such a way can be superimposed on the "regularly" propagating microwave inside the stripline.

For a rough evaluation of possible effects on the behavior of the circuits, we compare the number of "excellent" 10 V circuits with different designs A, B and C for the 1st stripline and D with a special ground plane. The investigated circuits come from 8 wafers which were processed within a period of two and a half years. Each wafer contained two chips of each of the designs A, B, C and D. Those 10 V circuits corresponding to the so-called "1 mA criterion" will be referred to as "excellent", as they show a 10 V Shapiro step (all array segments activated) of 1 mA or larger under microwave irradiation. **Figure 5** displays, for an example, the IVC of an excellent 10 V circuit with design A. **Table 2** lists the features of the used designs A, B, C and D. There are different arrangements of the smallest subarrays: "regular" (1, 1, 1, 2, ···, 68) or "reversed" (68, 34, ···, 1, 1, 1), each relative to the feeding point of the rf power (stripline input). Additionally, one can distinguish between a direct connection of the subarrays containing 1, 1, 1, 2 and 4 junctions and a connection over inserted sections of a homogeneous microstripline. The latter design modification was made to investigate whether very close bias leads (cf. **Figure 3**) couple parasitic microwaves in a different manner than more distant bias leads. We also tested chips with slots in the ground plane between the striplines (design D) aiming to reduce the flux trapping. A comparison of the numbers of excellent 10 V circuits resulting from each design variant (**Table 2**) suggests that the rather small modifications made are not really relevant for the behavior of the entire circuit. At this point it should be mentioned that characterization and classification of the fabricated 10 V circuits in "excellent" and "good to very good" ($\Delta I_1 = 0.8$ to 1 mA) has been accomplished in most cases by measuring only the whole circuit, *i.e.* with all subarrays activated by one current source. Only in a few cases the large segments including the entire 1st stripline $(\Sigma(N_1 \text{ to } \mathrm{N}_7))$ have been measured too. For that reason, one of the motivations for this paper was also a more thorough investigation of the smallest subarrays $(N_1 \text{ to } N_7)$ and to demonstrate how

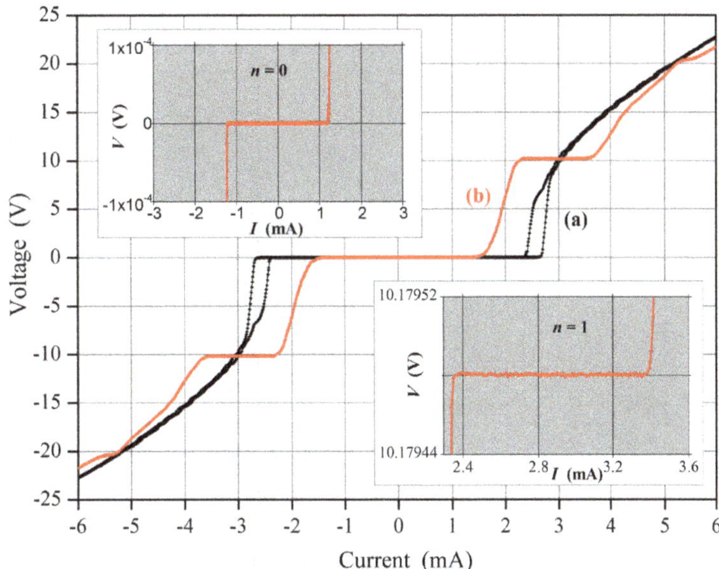

Figure 5. IVC of a 10 V PJVS circuit classified as "excellent" with 69,632 Nb_xSi_{1-x} barrier junctions. (a) without microwaves, (b) with microwaves at 70.7 GHz and approximately 60 mW at the antenna. The insets show the 10 V step $(n=1)$ and the zero-voltage step $(n=0)$ with high resolution. Junction parameters: $I_c = 2.7\ \text{mA}$; $I_cR_n = 155\ \mu\text{V}\ (\text{at}\ 2I_c)$.

Table 2. Design variants of 10 V circuits and the number of excellent 10 V circuits fabricated within two and a half years (for explanation see text).

Design variants	A	B	C	D
order of smallest segments	regular	regular	reversed	regular
connection between subarrays with 1, 1, 1, 2 and 4 junctions	direct	separated	separated	direct
ground plane	homogeneous	homogeneous	homogeneous	slots between striplines
number of excellent 10 V circuits	6	4	4	6

their microwave properties depend on the design of the 1st stripline.

The practical use of programmable voltage standards, especially if fast switchable dc bias sources are used, requires that both the 0th and the 1st constant voltage step of each array segment should be large enough to ensure a good adjustability of the dc operating point and sufficient noise immunity. As a result, the zero-order Shapiro step has to be considered too, especially for the smallest subarrays in the 1st stripline. As explained in 2.4, the subarrays closer to the microwave feeding point have the smaller number of junctions (except of design C) and can be easily overdriven. **Figure 6**, representative for the designs B and C, shows the current widths of the Shapiro steps with $n=0$ and +1 for all array segments in the 1st stripline.

For better comparability and considering thermal voltages, the microwave-induced steps of constant voltage are described by their relative voltage as a function of the dc bias current. **Figure 6** illustrates that the current widths of the steps corresponding to $M=1,\cdots,68$ scatter more for 10 V circuits of design C. This observation demonstrates a better homogeneity of the microwave power along the 1st junction stripline in the case of a regular order (design B) of the smallest array segments. **Figure 6(a)** displays, furthermore, that the array segment N_1 with $M=68$, lying at the end of the stripline and after 68 unbiased junctions, has a minimum step width of $\Delta I_1 = 1.8\ \text{mA}$ or in relative units of $\Delta I_1/2I_c \approx 0.32$. Design C (**Figure 6(b)**) in contrast, enables "only" a step width of 1.3 mA $(\Delta I_1/2I_c \approx 0.16)$ for the single junction (N_7) at the end of the 1st stripline. Despite these differences, both designs fulfill, in principle, the 1 mA criterion for the smallest segments, *i.e.* they are

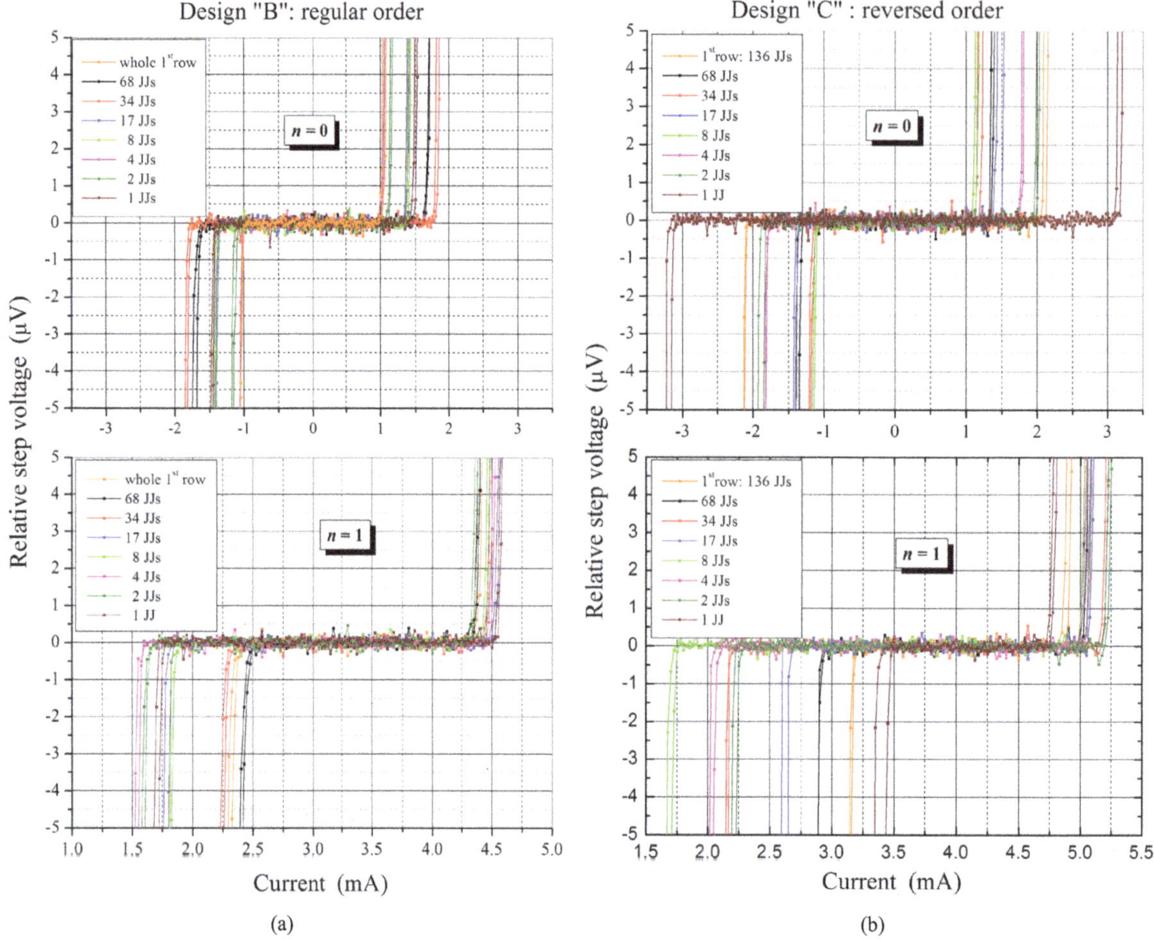

Figure 6. Current widths of the Shapiro steps $n = 0$ and $n = +1$ of the smallest array segments N_1 to N_7. (a) 10 V PJVS circuit with design B. Junction parameters: $I_c = 2.85$ mA; $I_cR_n = 152\ \mu\text{V}$ (at $2I_c$); (b) 10 V PJVS circuit with design C. Junction parameters: $I_c = 4$ mA; $I_cR_n = 180\ \mu\text{V}$ (at $2I_c$).

well suited for ac applications. However, the 1 mA criterion is difficult to meet for the 0th step in some cases, when the critical current is "small" $(I_c \leq 3\text{ mA})$. Such a situation can arise if parasitic microwave is coupled into the 1st stripline or if the load is incorrectly matched. In both cases the rf power distribution becomes inhomogeneous and may lead to reduced 0th steps (below 1 mA) in one or more segments without significant changes to the 1st step. **Figure 7** shows the Bessel functions of 0th and 1st order (J_0, J_1) in dependence on the microwave current $(\sim P^{1/2})$ and explains this effect: In a certain range of the microwave power P (hatched region) the 0th step $\Delta I_0 \approx 2I_c |J_0(x)|$ is quickly reduced, whereas the width of the 1st Shapiro step $\Delta I_1 \approx 2I_c |J_1(x)|$ remains nearly unchanged.

In most cases, a slight change of the array position relative to the surrounding magnetic shield can lead to a significant change in the 0th step, so that $\Delta I_1 > 1$ mA can be achieved for all subarrays in the 1st stripline. This behavior is most pronounced for circuits with design D. The comparative evaluation of 5 wafers with critical currents smaller than 5 mA delivers significantly smaller values for $\Delta I_0 / 2I_c$ while forming a 10 V step for circuits with the designs A and D. This is in our opinion due to a larger amount of parasitic microwave coupled into the 1st stripline from outside over the bias leads (**Figure 3**) and possibly favored by slots in the ground plane (design D). The aim to reduce the flux trapping (cf. 3.1.) by means of a ground plane with slots between the striplines has not been realized.

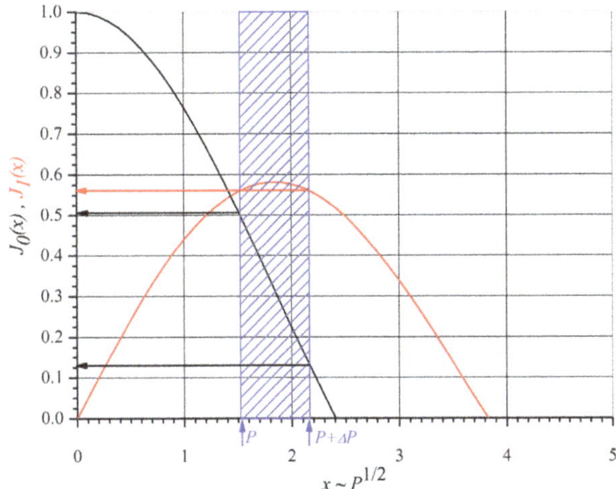

Figure 7. Plot of the Bessel functions of the 1st kind: $J_0(x)$ and $J_1(x)$. The argument of the functions is proportional to the injected microwave current $i \sim P^{1/2}$. The hatched region and arrow lines illustrate the drastic change in the 0th step widths for a change in microwave power ΔP which leaves the width of the 1st step unchanged.

In summary, one can conclude from these investigations that the design variants A and B with a regular order of the smallest subarrays are obviously most suitable for the operation of a programmable voltage standard, in which design B, having enough space between all lateral dc bias leads, is less sensitive to coupling of parasitic microwave.

An outstanding feature of excellent 10 V circuits fabricated at PTB is that both small and large array segments can be operated on the 1st Shapiro step with the same dc bias current. This is a surprising property considering the unequal distribution of nearly 70,000 junctions over the 128 striplines connected in parallel. **Figure 8(a)** demonstrates for a bias current of 6 mA that the current margins at each of the 16 voltage levels are at least 1 mA. In addition, we have also checked, for reasons explained previously, the current width ΔI_0 of the zero-order steps for the same power level. **Figure 8(b)** displays these steps for the smallest segments from N_1 to N_7 (regularly arranged). An investigation of the larger subarrays (N_8 to N_{16}) with regard to ΔI_0 is unnecessary because their microwave-induced reduction of the critical current is comparatively small, which results without exception in large values $\Delta I_0 > I_c$.

Whereas **Figure 8** is typical for 10 V circuits with relatively large critical currents (compared to the design value of 5 mA), analogous microwave behavior could already be demonstrated earlier for circuits with an I_c of about 3 mA [5]. The fact that circuits with low $I_c\,(I_c < 3\text{ mA})$ have similarly good properties $(\Delta I_1 > 1\text{ mA})$ as those with a "large" one $(I_c \geq 5\text{ mA})$ is a second outstanding feature of our 10 V samples.

The thermal load generated by these circuits in operation at 10 V is therefore lower than 100 mW. This value has to be compared with the values reported in the literature for other designs of 10 V PJVSs: 350 mW at 4.2 K [31] or about 470 mW at 10 K [15].

3.3. Performance of Programmable 10 V Circuits with Nb_xSi_{1-x} Barrier Junctions in a Narrow Band Close to 70 GHz

Since the used microwave source (cf. 3.1.) delivers the largest amount of microwave power in a frequency range from 70 to 70.7 GHz, we restricted our investigations to this frequency band. Exact statements regarding the broadband properties of our circuits with Nb_xSi_{1-x} barrier junctions are therefore not possible. However, from earlier investigations of 10 V SINIS circuits with basically the same microwave design except for the geometry and current density of the junctions, we assume that 10 V circuits with Nb_xSi_{1-x} barrier junctions should work as well at 75 GHz, but would require more microwave power [11]. In any case, it should be noted that for optimal operation of the PJVS circuit the critical voltage of the junctions has to be adjusted according to $I_cR_n = fh/(2e)$ close to the selected operating frequency. In principle, it is difficult to decide whether the design of the circuit, including mismatches, or the design of the single junction (electrical parameters: V_c, j_c) is the reason for a special frequency behavior. The microwave response of 10 V circuits shows a lot of resonances

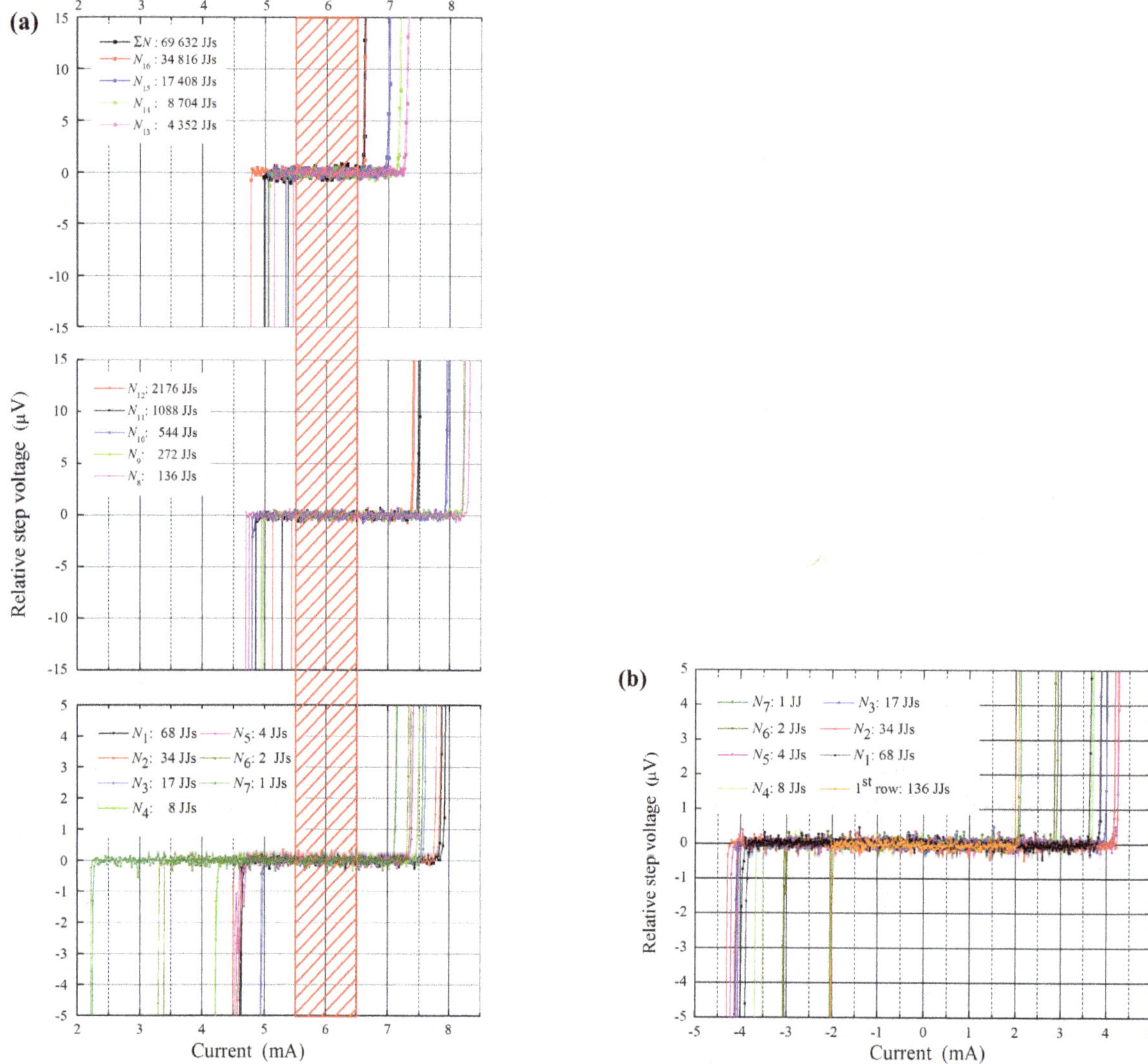

Figure 8. Current widths of all relevant Shapiro steps of a 10 V PJVS circuit (design A). Junction parameters: $I_c = 6\text{ mA}$; $I_c R_n = 175\ \mu\text{V}$ at $(2I_c)$. (a) $n = 1$ for all array segments N_1 to N_{16}. The hatched region illustrates a current margin of 1 mA for all array segments biased at 6 mA; (b) $n = 0$ for the array segments N_1 to N_7 in the 1st stripline.

if the frequency of the synthesizer is changed by small steps from 69.6 and 70.9 GHz. **Figure 9** depicts the current margins corresponding to 0 V $(n = 0)$ and to 10 V $(n = 1)$ for two different rf power levels if all subarrays are activated by a single current source. The dashed line marks the region above which robust operation as a rapidly programmable voltage standard is possible. It can be seen that there are numerous frequencies where excellent operating conditions are fulfilled, *i.e.* where the 1 mA criterion holds. Obviously there is a regular modulation of the current margins best visible in **Figure 9(a)**. We found an explanation for this phenomenon by investigating the measuring probe by means of a network analyzer. If the waveguide at the cold end of the measuring probe used to immerse the sample holder into a liquid helium dewar was shorted, resonances with a period of about 100 MHz could be observed in reflection. This period is clearly visible in **Figure 9(a)** and is correlated to the length of the waveguide used for the measuring probe. The waveguide is approximately 1.25 m long and there are two reflections for the microwave, first at the WR12 waveguide input flange with a thin foil of PTFE to prevent helium gas losses, and second at the end of the WR12 waveguide in the cold part. Multiples of the half wavelength match in this length and cause the frequency modulation of the current margins.

Figure 10 compares the frequency behavior of the two halves of a 10 V circuit. The largest array segment

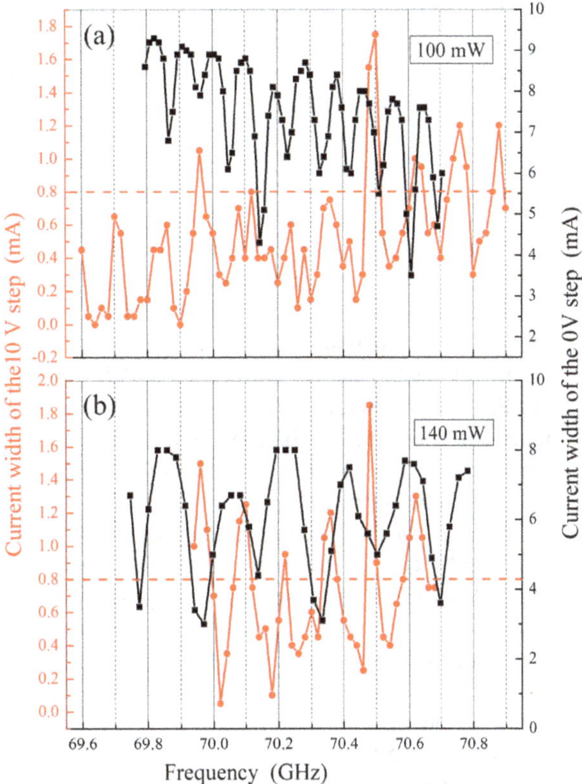

Figure 9. Current width of the 10 V and 0 V steps as a function of the frequency (in a small band at approximately 70 GHz) for two different output power levels of a microwave synthesizer: (a) = 100 mW and (b) = 140 mW. Junction parameters: $I_c = 6$ mA ; $I_c R_n = 175$ μV (at $2I_c$) .

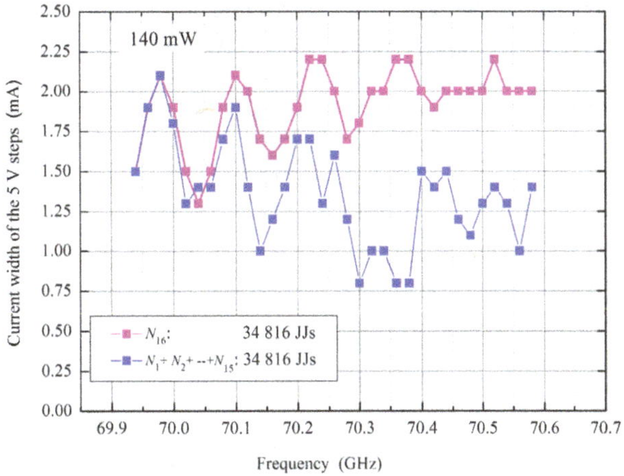

Figure 10. Current width of the 5 V steps generated by the two halves of a programmable 10 V circuit as function of the frequency in a small band at approximately 70 GHz. Junction parameters: $I_c = 6$ mA ; $I_c R_n = 175$ μV (at $2I_c$) .

N_{16} (34,816 JJs) with an almost homogeneous distribution of the junctions over the striplines #69 to #128 (**Table 1**) displays at nearly all frequencies for the 5 V step $(n=1)$ larger current margins than the other half of the circuit (sum of segments N_1 to N_{15}) which is characterized by an inhomogeneous distribution of the junctions over the striplines #1 to #68. Nevertheless for some frequencies just below 70 GHz, the current margins of both halves are in principle identical. For such a frequency, the current margins at 5 V $(n=1)$ and at 0 V $(n=0)$ have been determined as a function of the rf power (synthesizer output) in case the whole circuit is driven by a single current source. **Figure 11** demonstrates that 10 V circuits, even with a relatively large critical current, can be driven by a microwave power of about 60 mW, provided a suitable frequency has been selected. Therefore commercial Gunn diodes with a maximum output power of 80 to100 mW can be used in principle too. However, it seems more difficult to look for a suitable frequency with a Gunn diode than with a computer-controlled microwave synthesizer.

4. Summary and Conclusion

This paper presents the state of the art for 10 V programmable Josephson voltage standards (PJVS) at PTB. The technological progress from the appearance of the fundamental idea in 1995 to the current state of technology, characterized by the use of Nb_xSi_{1-x} barrier junctions, is described. The barrier of this special kind of Josephson junctions can be tuned for a wide range of driving frequencies. The operation of the arrays in the same microwave frequency range that has already been used for conventional JVSs, around 70 GHz, strikes us as the most obvious and promising way for PJVSs. We therefore decided on this frequency range very early even before the reappearance of Josephson junctions with Nb_xSi_{1-x} barriers. The paper underlines the advantages of a 70 GHz drive and discusses step by step the creation of a 10 V design based on striplines with embedded Nb_xSi_{1-x} barrier junctions connected in parallel for the microwaves.

Different designs for the arrangement of the junctions in the outermost stripline and the nature of the ground (cover) plane have been experimentally compared by evaluating the current margins for the steps $(n=0)$ and $(n=1)$. In addition, the susceptibility to flux trapping and to coupling of parasitic microwaves has been investigated. Optimum 10 V circuits for PJVSs have homogeneous (gapless) ground planes and a regular order of the smallest array segments with enough space between them (design B). It is shown that a complete and safe classification of the circuit performance is only possible by evaluating the current margins for the steps of order $(n=0)$ and $(n=1)$ of all array segments in the 1st stripline.

We show that the PTB circuits classified as "excellent" can be operated with the same dc bias current in all array segments while keeping the current margins larger than 1 mA. The resonances in the current margins of the PJVS circuits observed in a small frequency band around 70 GHz have been related to the length of the waveguide used in the measuring probe. In addition, 10 V circuits with small critical currents $(I_c \leq 3\,\text{mA})$ present low thermal loads (<100 mW), due to the combined dc and rf power, and are therefore very advantageous for operation in cryocoolers. Even though a powerful compact microwave synthesizer was used for the experiments described in this paper, the results obtained indicate that also Gunn diodes with an output power of about 100 mW can be used for the routine operation of 10 V PJVSs.

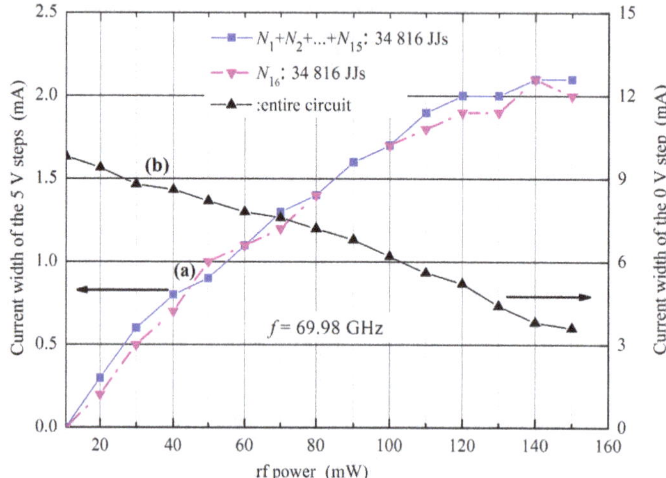

Figure 11. Current width of the 5 V $(n=1)$ and 0 V $(n=0)$ steps generated by the two halves of a programmable 10 V circuit as a function of the output power of a microwave synthesizer operated at 69.98 GHz (cf. **Figure 10**). Junction parameters: $I_c = 6\,\text{mA}$; $I_cR_n = 175\,\mu\text{V}\,(\text{at } 2I_c)$. (a) current widths of the two 5 V steps; (b) current width of the 0 V step of the entire circuit.

The results clearly demonstrate that the fabrication process of arrays with Nb_xSi_{1-x} barrier junctions has achieved the major technological requirements to be robust, reliable, and reproducible. State of the art 10 V circuits are available for routine use in PJVSs that successfully extend the application of Josephson voltage standards more and more from *dc* to *ac*.

Acknowledgements

The authors would like to thank R. Wendisch, T. Weimann, B. Egeling, P. Hinze, K. Störr and P. Duda for technical support, and O. Kieler for helpful discussions.

References

[1] Hamilton, C.A., Burroughs, C.J. and Kautz, R.L. (1995) Josephson D/A Converter with Fundamental Accuracy. *IEEE Transactions on Instrumentation and Measurement*, **44**, 223-225. http://dx.doi.org/10.1109/19.377816

[2] Pöpel, R. (1992) The Josephson Effect and Voltage Standards. *Metrologia*, **29**, 153-174. http://dx.doi.org/10.1088/0026-1394/29/2/005

[3] Benz, S.P. and Hamilton, C.A. (1996) A Pulse-Driven Programmable Josephson Voltage Standard. *Applied Physics Letters*, **68**, 3171-3173. http://dx.doi.org/10.1063/1.115814

[4] Baek, B., Dresselhaus, P.D. and Benz, S.P. (2006) Co-Sputtered Amorphous Nb_xSi_{1-x} Barriers for Josephson-Junction Circuits. *IEEE Transactions on Applied Superconductivity*, **16**, 1966-1970. http://dx.doi.org/10.1109/TASC.2006.881816

[5] Behr, R., Kieler, O., Kohlmann, J., Müller, F. and Palafox, L. (2012) Development and Metrological Applications of Josephson Arrays at PTB. *Measurement Science and Technology*, **23**, 124002-124020. http://dx.doi.org/10.1088/0957-0233/23/12/124002

[6] Kautz, R.L. (1995) Shapiro Steps in Large-Area Metallic-Barrier Josephson Junctions. *Journal of Applied Physics*, **78**, 5811-5819. http://dx.doi.org/10.1063/1.359644

[7] Benz, S.P., Hamilton, C.A., Burroughs, C.J., Harvey, T.E. and Christian, L.A. (1997) Stable 1-Volt Programmable Voltage Standard. *Applied Physics Letters*, **71**, 1866-1868. http://dx.doi.org/10.1063/1.120189

[8] Müller, F., Pöpel, R., Kohlmann, J., Niemeyer, J., Meier, W., Weimann, T., Grimm, L., Dünschede, F.W. and Gutmann, P. (1997) Optimized 1 V and 10 V Josephson Series Arrays. *IEEE Transactions on Instrumentation and Measurement*, **46**, 229-232. http://dx.doi.org/10.1109/19.571819

[9] Müller, F., Schulze, H., Behr, R., Kohlmann, J. and Niemeyer, J. (2001) The Nb-Al Technology at PTB-a Common Base for Different Types of Josephson Voltage Standards. *Physica C*, **354**, 66-70. http://dx.doi.org/10.1016/S0921-4534(01)00037-5

[10] Schulze, H., Behr, R., Kohlmann, J., Müller, F. and Niemeyer, J. (2000) Design and Fabrication of 10 V SINIS Josephson Arrays for Programmable Voltage Standards. *Superconductor Science and Technology*, **13**, 1293-1295. http://dx.doi.org/10.1088/0953-2048/13/9/301

[11] Mueller, F., Behr, R., Palafox, L., Kohlmann, J., Wendisch, R. and Krasnopolin, I. (2007) Improved 10 V SINIS Series Arrays for Applications in AC Voltage Metrology. *IEEE Transactions on Applied Superconductivity*, **17**, 649-652. http://dx.doi.org/10.1109/TASC.2007.898736

[12] Mueller, F., Behr, R., Weimann, T., Palafox, L., Olaya, D., Dresselhaus, P.D. and Benz, S.P. (2009) 1 V and 10 V SNS Programmable Voltage Standards for 70 GHz. *IEEE Transactions on Applied Superconductivity*, **19**, 981-986. http://dx.doi.org/10.1109/TASC.2009.2017911

[13] Barrera, A.S. and Beasley, M.R. (1987) High-Resistance SNS Sandwich-Type Josephson Junctions. *IEEE Transactions on Magnetics*, **23**, 866-868. http://dx.doi.org/10.1109/TMAG.1987.1064867

[14] Dresselhaus, P.D., Elsbury, M., Olaya, D., Burroughs, C.J. and Benz, S.P. (2011) 10 V Programmable Josephson Voltage Standard Circuits Using NbSi-Barrier Junctions. *IEEE Transactions on Applied Superconductivity*, **21**, 693-696. http://dx.doi.org/10.1109/TASC.2010.2079310

[15] Yamamori, H., Ishizaki, M., Shoji, A., Dresselhaus, P.D. and Benz, S.P. (2006) 10 V Programmable Josephson Voltage Standard Circuits Using NbN/TiNx/NbN/TiNx/NbN Double Junction Stacks. *Applied Physics Letters*, **88**, Article ID: 042503. http://dx.doi.org/10.1063/1.2167789

[16] Scheller, T., Mueller, F., Wendisch, R., Kieler, O., Springborn, U., Stoerr, K., Egeling, B., Weimann, T., Palafox, L., Behr, R. and Kohlmann, J. (2012) SNS Junctions for AC Josephson Voltage Standards. *Physics Procedia*, **36**, 48-52. http://dx.doi.org/10.1016/j.phpro.2012.06.128

[17] Müller, F., Scheller, T., Wendisch, R., Behr, R., Kieler, O., Palafox, L. and Kohlmann, J. (2013) NbSi Barrier Junc-

tions Tuned for Metrological Applications up to 70 GHz: 20 V Arrays for Programmable Josephson Voltage Standards. *IEEE Transactions on Applied Superconductivity*, **23**, Article ID: 1101005. http://dx.doi.org/10.1109/TASC.2012.2235895

[18] Scheller, T. (2014) Electrical Properties of Josephson Junctions with Nb_xSi_{1-x} Barriers Customized for AC Voltage Standard Applications. Ph.D. Thesis, Friedrich-Schiller-University, Jena. (in Press)

[19] Likharev, K.K. (1986) Dynamics of Josephson Junctions and Circuits. Gordon and Breach Science Publishers, New York.

[20] Niemeyer, J., Hinken, J.H. and Kautz, R.L. (1984) Microwave-Induced Constant-Voltage Steps at One Volt from a Series Array of Josephson Junctions. *Applied Physics Letters*, **45**, 478-480. http://dx.doi.org/10.1063/1.95222

[21] Behr, R., Schulze, H., Müller, F., Kohlmann, J., Krasnopolin, I. and Niemeyer, J. (1999) Microwave Coupling of SINIS Junctions in a Programmable Josephson Voltage Standard. *Digest of the International Superconducting Electronics Conference, ISEC*'99, Berkeley, 21-24 June 1999, 128-130.

[22] Kim, K.-T., Kim, S.-T., Chong, Y. and Niemeyer, J. (2006) Simulations of Collective Synchronization in Josephson Junction Arrays. *Applied Physics Letters*, **88**, Article ID: 062501. http://dx.doi.org/10.1063/1.2171796

[23] Kautz, R.L. (1992) Design and Operation of Series-Array Josephson Voltage Standards. Metrology at the Frontiers of Physics and Technology. In: Crovini, L. and Quinn, T.J., Eds., North-Holland, Amsterdam, 259-296.

[24] Kieler, O., Behr, R., Müller, F., Schulze, H., Kohlmann, J. and Niemeyer, J. (2002) Improved 1 V Programmable Josephson Voltage Standard Using SINIS Junctions. *Physica C*, **372-376**, 309-311. http://dx.doi.org/10.1016/S0921-4534(02)00657-3

[25] Müller, F., Kohlmann, J., Hebrank, F.X., Weimann, T., Wolf, H. and Niemeyer, J. (1995) Performance of Josephson Array Systems Related to Fabrication Techniques and Design. *IEEE Transactions on Applied Superconductivity*, **5**, 2903-2906. http://dx.doi.org/10.1109/77.403199

[26] Schubert, M., Anders, S., Haertel, E., Wende, G., Hähle, R., Fritzsch, L., Starkloff, M., Springborn, U., Müller, F., Kohlmann, J. and Meyer, H.-G. (2011) Microwave Properties of Microstrip Line Circuits Used for Josephson Voltage Standard Arrays at 70 GHz. *Superconductor Science and Technology*, **24**, Article ID: 085006. http://dx.doi.org/10.1088/0953-2048/24/8/085006

[27] Kautz, R.L. (1994) Quasipotential and the Stability of Phase Lock in Nonhysteretic Josephson Junctions. *Journal of Applied Physics*, **76**, 5538-5544. http://dx.doi.org/10.1063/1.357156

[28] Burroughs, C.J., Dresselhaus, P.D., Rüfenacht, A., Olaya, D., Elsbury, M.M., Tang, Y.H. and Benz, S.P. (2011) NIST 10 V Programmable Josephson Voltage Standard System. *IEEE Transactions on Instrumentation and Measurement*, **6**, 2482-2488. http://dx.doi.org/10.1109/TIM.2010.2101191

[29] Hinken, J.H. (1983) Simplified Analysis and Synthesis of Fin-Line Tapers. *Archiv der Elektronischen Übertragungstechnik*, **37**, 375-380.

[30] Lee, J., Behr, R., Palafox, L., Katkov, A., Schubert, M., Starkloff, M. and Böck, A.C. (2013) An AC Quantum Voltmeter Based on a 10 V Programmable Josephson Array. *Metrologia*, **50**, 612-622. http://dx.doi.org/10.1088/0026-1394/50/6/612

[31] Howe, L., Burroughs, C.J., Dresselhaus, P.D., Benz, S.P. and Schwall, R.E. (2013) Cryogen-Free Operation of 10 V Programmable Josephson Voltage Standards. *IEEE Transactions on Applied Superconductivity*, **23**, Article ID: 1300605. http://dx.doi.org/10.1109/TASC.2012.2230052

13

Study of Correlation Coefficient for Breast Tumor Detection in Microwave Tomography

L. Mohamed, N. Ozawa, Y. Ono, T. Kamiya, Y. Kuwahara*

Graduate School of Engineering, Shizuoka University, Hamamatsu-shi, Japan
Email: *tykuwab@ipc.shizuoka.ac.jp

Abstract

In microwave tomography, it is necessary to increase the amount of diverse observation data for accurate image reconstruction of the dielectric properties of the imaging area. The multi-polarization method has been proposed as a suitable technique for the acquisition of a variety of observation data. While the effectiveness of employing multi-polarization to reconstruct images has been confirmed, the physical considerations related to image reconstruction have not been investigated. In this paper, a compact-sized imaging sensor using multi-polarization for breast cancer detection is presented. An analysis of the correlation coefficient of the received data of adjacent antennas was performed to interpret the imaging results. Numerical simulation results demonstrated that multi-polarization can reconstruct images better compared to single polarizations owing to its low correlation coefficient and condition number.

Keywords

Microwave Tomography, Inverse Scattering, Distorted Born Iterative Method (DBIM), Breast Tumor, Correlation Coefficient

1. Introduction

In Japan, the breast cancer incidence and associated mortality rates of women are lower compared to Western countries. However, it is one of the major causes of death among women, and the incidence rate has increased since 1975, regardless of age group [1]. Therefore, early detection of breast cancer is important because it helps prevent and reduce deaths from breast cancer. In Japan, breast cancer screening by X-ray mammography is recommended for women aged over 40. However, X-ray mammography involves risks, such as X-ray exposure, overlooked tumors owing to the low contrast of breast tissue, and pain during the examination. In addition, a combination of palpation and ultrasound diagnosis is also commonly used in breast cancer screening. Because

*Corresponding author.

the examination results depend on the skill of the examiner, the probability of cancer detection can differ accordingly. For the above reasons, an alternative to current breast cancer screening methods is required.

Recently, studies on the early detection of breast cancer by microwave imaging (MWI) have attracted considerable interest among researchers [2]. This approach to breast cancer detection uses a fundamental principle—when electromagnetic waves are irradiated to the imaging region, scattered waves are generated by the differences in the dielectric properties of the cancer and the normal breast tissue [3]. The MWI method can be classified to ultra-wideband (UWB) radar and tomography. In our laboratory, we have developed a multi-static UWB radar measurement system for the early detection of breast cancer and conducted clinical trials. The results demonstrated that the system can detect cancer that has a clear boundary and is isolated from fibro-glandular tissue. However, if the boundary of the fibro-glandular tissue and the cancer is irregular, the system was unable in some cases to correctly reconstruct the shape of the cancer [4]-[6]. Thus, the proposed diagnostic method using UWB radar can be considered incomplete. Therefore, we are currently working on the development of a microwave tomography method for early breast cancer screening by reconstructing the dielectric property distributions of the examined breast [7]-[9].

In microwave tomography, the electromagnetic waves are efficiently incident on the object. The influence of surrounding structures is avoided by immersing the antenna and imaged object into a lossy matching fluid [10]. In order to achieve accurate image reconstruction, it is necessary to obtain diverse observations data. Several methods must be considered. The amount of observation data can be increased by increasing the number of antennas; however, the antennas must be arranged at a certain distance from each other. Thus, the scale of the apparatus increases and the computational cost becomes substantial. Furthermore, the signal-noise ratio (SNR) is degraded by this method, which creates difficulties in the correct reconstruction of the image. A method using multiple frequencies has been proposed [11]. In general, biological tissue is a medium with frequency dependence, and its behavior is modeled using the Debye approximation with several parameters. In this method, the number of unknown parameters is increased along with the number of frequencies; thus, the reconstruction becomes difficult and the images are reproduced with low accuracy.

Furthermore, the multiple-polarization method has been examined as a means to obtain a variety of observations data. The changes in the polarization of the scattered waves are influenced by the composition and shape of the biological tissue. For example, a dual-polarized MWI system that can simultaneously collect both TE and TM polarizations using scattering probes has been built to obtain 2-D tomographic images [12]. The polarized data are inverted using contrast source inversion (CSI) algorithm to reconstruct the electrical properties. However, there are some issues regarding the imaging quality of the system. In [13], the impact of polarization on image reconstruction was evaluated with a shielded array of patch antennas using truncated singular-value decomposition (TSVD) analysis. The multi-polarization layout was observed to enhance the imaging performance with a higher truncation index compared to uniform array configurations. Although the effectiveness of employing multi-polarization to reconstruct images have been established, to our knowledge, the physical considerations related to antenna arrangement in order to achieve sufficient image reconstruction have not yet been investigated. This information can be used as a viable parameter in sensor design.

In this paper, we present a compact-sized imaging sensor using the multi-polarization method for breast cancer detection. Based on our previous paper [7], we confirmed the image reconstruction by solving the inverse problem using the Newton-Kantorovich method, as explained in [14]. Because of the extensive calculations of the inverse matrix performed to compute the Jacobian, this method involves problems, such as low resolution and high computational cost. Therefore, we employ the Distorted Born Iterative Method (DBIM), described in [15] [16], which does not require inverse matrix calculation for the computation of the Jacobian. We use a different antenna arrangement from [7] [9] to show the impact of polarizations on 3-D image reconstruction at various antenna locations. We also clarify the impact on physical measurements using the analysis of the correlation coefficient of the received data from adjacent antennas.

This paper is organized as follows. The imaging algorithm approach employed for image reconstruction is described in Section 2. Section 3 presents our proposed compact-sized imaging sensor and breast model used for numerical simulation. Further in Section 3, the imaging results are presented and discussed. Finally, Section 4 discusses the conclusion of our paper.

2. Imaging Algorithm

In tomography, the breast model is expressed as a cube and discretized as voxels, and the dielectric properties are estimated in each voxel. The total number of voxels is *K*, and the dielectric properties, which consist of the relative permittivity and conductivity distribution, are represented by contrast. The calculated data group, Y_{mn}, which is based on the estimated model is compared with the measured data group, X_{mn}, which is based on the actual model ($m = 1,\cdots,N; n = 1,\cdots,N$, where *N* is the total number of antennas, *m* is the number of transmitters, and *n* is the number of receivers). The contrast distribution of the breast model is iteratively updated by the DBIM until $X_{mn} = Y_{mn}$ is reached. The image reconstruction is accomplished by minimizing the norm of the difference between measured and calculated scattering data.

2.1. Inverse Scattering Problem and DBIM

The governing equation for the scattering field at an observation point *r* for a given frequency is

$$E^s(r) = \omega^2 \mu \int_V \bar{G}^b(r \mid r') E^t(r') \left[\chi(r') - \chi^b(r')\right] \mathrm{d}r'. \tag{1}$$

Here, r' is an arbitrarily position vector in the imaging regionV, $\chi(r')$ is the unknown contrast, $\chi^b(r')$ is the contrast of the background, $E^s(r)$ is the scattered electric field at the observation points, $E^t(r')$ is the total field, G^b is the dyadic Green's function, ω is the angular frequency, and μ is the magnetic permeability.

Equation (1) is a nonlinear integral equation; therefore, we employ the DBIM to solve the nonlinear problem. DBIM is equivalent to the Gauss-Newton method that is used in nonlinear least-squares optimization problems to obtain an approximate solution for the contrast. In the iterative DBIM, we start with an initial guess of the contrast in the imaging region. At each iteration, we determine the contrast perturbations based on the approximate solution of the inverse problem, and then update the contrast. The updated contrastis referred to as the new contrast of the background; the total field and the scattering field at the observation points are recalculated based on the new contrast. Then, the calculated scattering field is compared with the measured scattering field to minimize the residue error until convergence is reached.

2.2. Linearization of the Scattering Equation

In this paper, the contrast function, χ is represented by Equation (2), where ε_0 is the vacuum permittivity, ε^b is the relative permittivity of the background, σ and ε_r are the conductivity and relative permittivity of the imaging object, respectively. The conductivity, σ and relative permittivity, ε_r are the unknown parameters that will be measured.

$$\chi = \sigma + \mathrm{j}\omega\varepsilon_0\left(\varepsilon_r - \varepsilon^b\right) \tag{2}$$

As we mentioned earlier, Equation (1) is a nonlinear equation. In order to linearize Equation (1) in the Born approximation, the incident field is adopted instead of total field. In this paper, we use the recalculated total field E^t obtained at each iteration with respect to the unknown contrast function to linearize the nonlinear equation. Equation (2) is applied to Equation (1) and is discretized, thus resulting in a linear relationship between the scattered field and the contrast function. Because the two unknown parameters in the contrast function are real numbers, the equation is divided into a real and an imaginary part to form linear equations of a real numbers, as expressed by Equation (3).

$$\begin{bmatrix} \Re\left\{\frac{\partial\chi}{\partial\sigma}B\right\} & \Re\left\{\frac{\partial\chi}{\partial\varepsilon_r}B\right\} \\ \Im\left\{\frac{\partial\chi}{\partial\sigma}B\right\} & \Im\left\{\frac{\partial\chi}{\partial\varepsilon_r}B\right\} \end{bmatrix} \begin{bmatrix} \Delta\sigma \\ \Delta\varepsilon_r \end{bmatrix} = \begin{bmatrix} \Re\left\{\Delta E^S\right\} \\ \Im\left\{\Delta E^S\right\} \end{bmatrix} \tag{3}$$

$$B = \omega^2 \mu\varepsilon_0 \bar{G}^b(r \mid r') E^t(r \mid r')$$

$$\Delta\sigma = \sigma - \sigma^b \qquad \Delta\varepsilon_r = \varepsilon_r - \varepsilon_r^b$$

ΔE^s is the difference between the measured scattered field, which is based on the actual model, and the calculated scattered field, which is based on the current contrast distribution. $\Re\{\}$ and $\Im\{\}$ denote the real and imaginary parts of the complex number, respectively. From Equation (3), a linear equation of the form $\boldsymbol{Ax} = \boldsymbol{b}$ is obtained. Here, vector $\boldsymbol{x}$ is the perturbation used to update the contrast; meanwhile, the residual scattering field between the measured data X_{mn} and the calculated data from the current contrast Y_{mn} at the observation points is placed in vector $\boldsymbol{b}$. Furthermore, the sensitivity matrix, known as the Jacobian, that shows the changes in the calculated data at the current contrast distribution is equivalent to matrix $\boldsymbol{A}$. For a system with K unknowns and N^2 measurements, the unknown contrast $\boldsymbol{x}$ is $K\times1$, the residual scattered fieldb is $N^2\times1$, and the Jacobian matrix $\boldsymbol{A}$ is configured as $N^2\times K$.

In the linear model of Equation (3), an approximation of the contrast perturbations, $\Delta\boldsymbol{x}=\{\Delta\sigma,\Delta\varepsilon_r\}$ is determined by Equation (4).

$$\Delta\boldsymbol{x}_k=\left[\boldsymbol{A}^{+}\boldsymbol{A}\right]^{-1}\boldsymbol{A}^{+}\Delta\boldsymbol{b}_k \tag{4}$$

Here, k is the iteration number, and + denotes the conjugate transpose of a matrix. Because the inversion in Equation (4) is ill-posed, conjugate gradient (CG) regularization is applied at each iteration to estimate the contrast perturbations and to update and generate a new contrast. Then, the imaging region is reconstructed using the new contrast, and this process is repeated until the residual norm is smaller.

3. Numerical Simulation

In the inverse scattering problem, a large amount of diverse observation data is required to achieve accurate image reconstruction with high resolution. In addition, the measurement error increases if the SNR decreases; thus, it cannot reconstruct the image accurately. Therefore, it is necessary to minimize the analysis region in order to reduce the computational cost. For this reason, the implementation of a compact sensor that involves a small distance between the antenna and the breast is preferred. In this case, large number of antennas must be arranged in a small and limited space. Various observation data can be obtained even in a small space by changing the plane of polarization.

3.1. Imaging Sensor and Breast Model

Figure 1 shows the imaging sensor with dimensions 48 × 96 × 48 mm (width × length × height). The imaging region is discretized into 8-mm resolution. For simplicity, a point source is used for the antennas, and arranged in a 4 × 2 configuration on each of the four side-panels of the sensor. **Figure 1(a)** shows the position of the antenna. One of the antennas is used for transmitting, and all the elements including transmitter are used for receiving data. The lines in **Figure 1** represent the polarization direction of the antenna, where the y-axis indicates vertical polarization, and the x- or z-axis horizontal polarization. The antenna arrangements for each side are identical, and the four side-panels are parallel to either the xy- or zy-plane. In this paper, we investigated three different configurations, as shown in **Figure 1**, to examine the effectiveness of polarization in breast cancer detection. **Figure 1(a)** illustrates the vertical polarization, **Figure 1(b)** the horizontal polarization, and **Figure 1(c)** vertical and horizontal polarization (hereafter referred to as multi-polarization). Here, we did not arrange antennas on the top surface of the panel to clearly show the effect of polarization.

We assumed that the imaging sensor is constructed from resin, which has properties similar to those of adipose (fatty) tissue. A hemispheric volume is providedinside the sensor to accommodate the breast model,

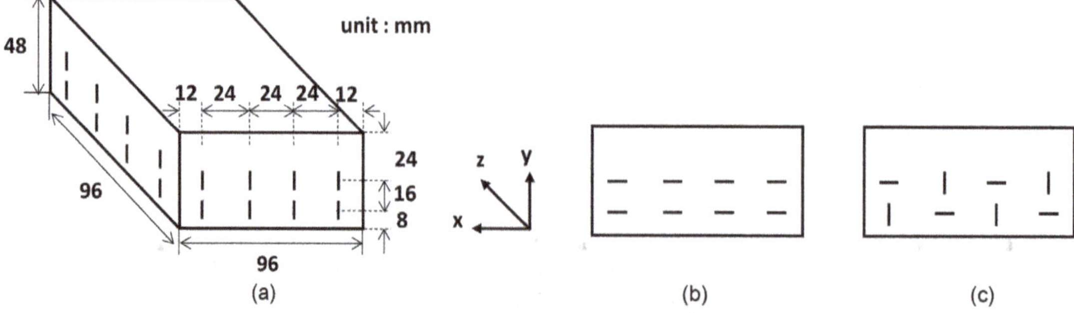

Figure 1. Imaging sensor with dimensions and various polarizations. (a) Vertical polarization; (b) Horizontal polarization; (c) Multi-polarization.

similar to the imaging sensor with fixed suction proposed in [4]. The simple breast model shown in **Figure 2** consists of adipose tissues, fibro-glandular tissues, and a tumor. The breast model is a hemisphere with a radius of 48 mm, and the tumor has a radius of 4 mm. The analysis region is limited to the hemispheric space containing the breast model; the chest wall under the breast was also modeled. The analysis region is discretized into 8-mm voxels, thus it composed of 538 voxels. We characterized the dielectric properties of the breast model in each voxel, as shown in **Table 1**.

3.2. Numerical Results and Discussion

The total field within the scattering object was calculated according to the method of moment (MOM), as described in [14]. In previous research investigating MWI for biomedical applications, operating frequencies of 1.0 - 2.3 GHz [17] and 2.45 GHz [18] were used. Here, we used a single frequency of 2.5 GHz.

3.2.1. Image Reconstruction of Breast Models

In Model 1, the analysis region consists of the chest wall, adipose tissue, and a tumor. The distributions of the two unknown parameters, *i.e.*, the relative permittivity and conductivity, are shown in **Figure 3**. The parameter-setting model of Model 1 is shown in **Figure 3(a)**, where the chest wall has been omitted. **Figures 3(b)-(d)**

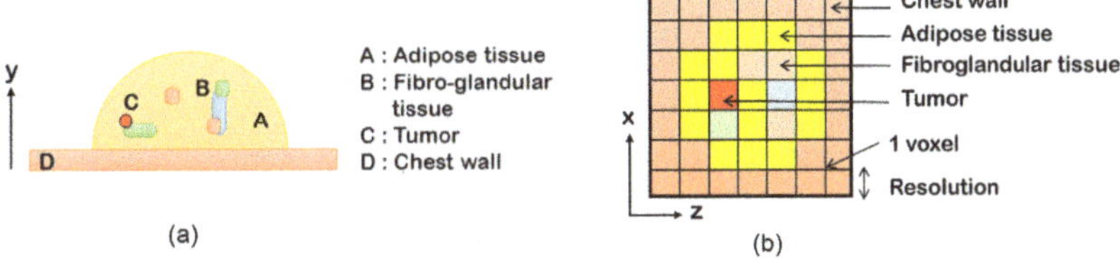

Figure 2. A simple breast model. (a) Configuration of breast model; (b) Allocation of dielectric properties.

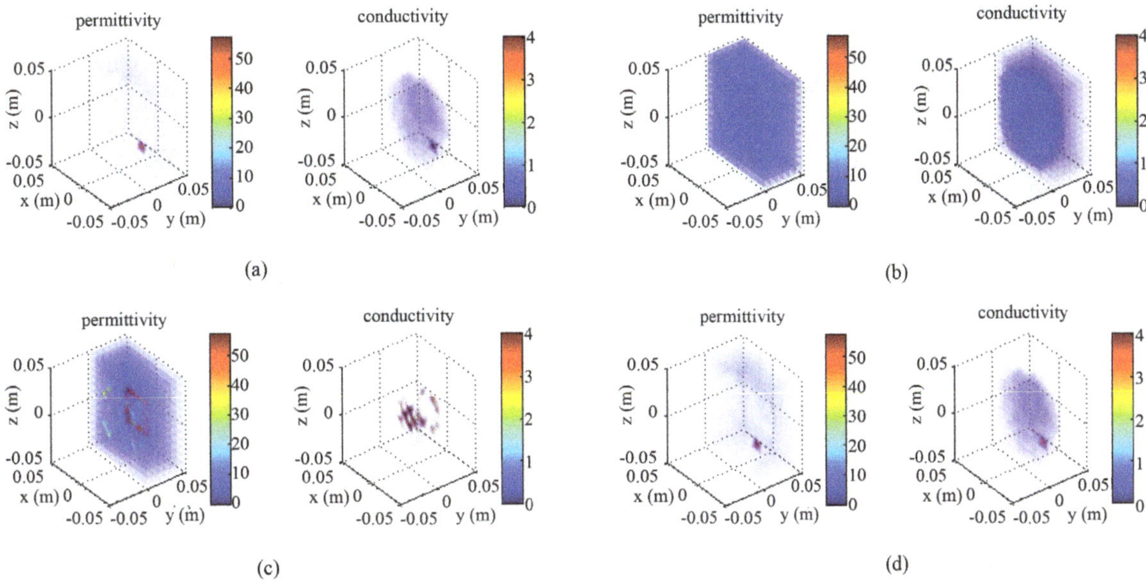

Figure 3. Setting model and reconstructed images of Model 1. (a) Setting model; (b) Vertical polarization; (c) Horizontal polarization; (d) Multi-polarization.

Table 1. Dielectric properties of breast model.

Parameter	Background	Chest wall	Adipose tissue	Fibro-glandular tissue	Tumor
Relative permittivity, ε_r	8.2	57	7	25 - 40	52
Conductivity, σ [S/m]	0.15	2	0.4	1.0 - 2.2	4

show the results of the 3-D image reconstruction after 100 iterations, using vertical, horizontal, and multi-polarization, respectively, for transmitting and receiving data. From the results, we cannot estimate both the relative permittivity and the conductivity of the tumor using single polarizations, *i.e.*, vertical and horizontal polarization. In contrast, **Figure 3(d)** clearly indicates the presence of the tumor and the reconstruction is successfully for both parameters using multi-polarization.

Figures 4(a)-(b) show the setting and reconstruction value of the relative permittivity and conductivity through the tumor voxel. The x-axis indicates the x-coordinate of the voxel, and the y-axis indicates either the relative permittivity or conductivity. The setting values of the relative permittivity and conductivity of the voxel corresponding to the tumor are 52 and 4 [S/m], respectively. The reconstructed parameters of the tumor for the different polarizations are summarized in **Table 2**. We observed that the dielectric properties of the tumor are lower than the expected setting values when using single polarizations. However, the dielectric properties of the tumor are improved and accurately estimated using multi-polarization.

For Model 2, the analysis region consists of the chest wall, adipose tissue, fibro-glandular tissue, and a tumor. Here, 10% of the volume ratio of the breast is occupied by fibro-glandular tissues that are distributed randomly. **Figure 5(a)** is the setting model of Model 2 and the tumor is marked by an arrow for clarity. **Figures 5(b)-(d)**

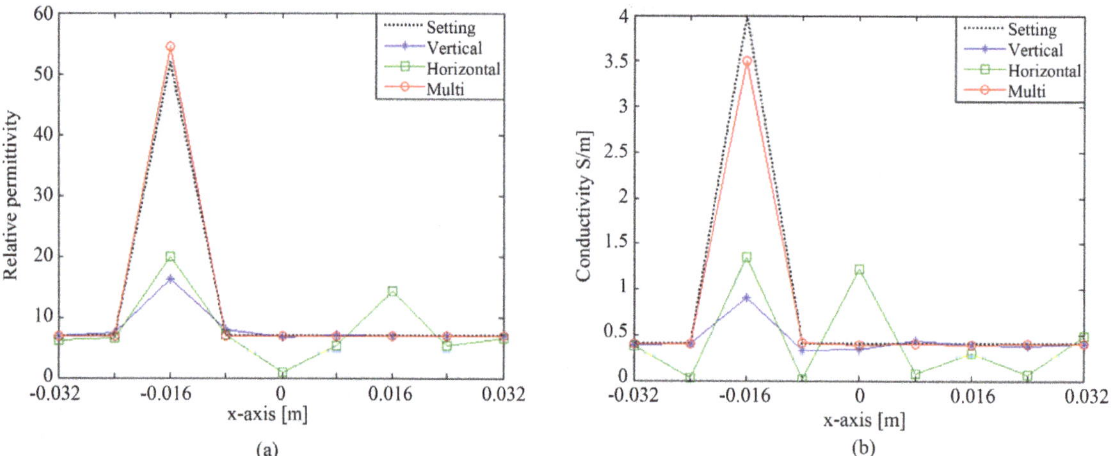

Figure 4. Dielectric property distributions of Model 1 in the cross section including the tumor voxel. (a) Relative permittivity; (b) Conductivity.

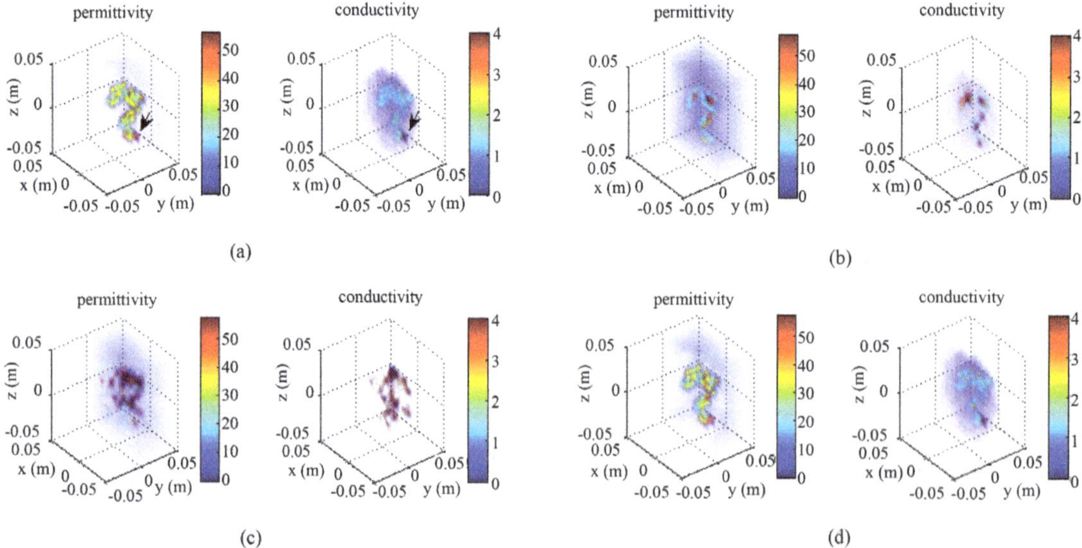

Figure 5. Setting model and reconstructed images of Model 2. (a) Setting model; (b) Vertical polarization; (c) Horizontal polarization; (d) Multi-polarization.

Table 2. Reconstructed values of the dielectric properties of the tumor for different polarizations.

Parameter	Setting value	Model 1			Model 2		
		VP[a]	HP[b]	MP[c]	VP	HP	MP
Relative permittivity, ε_r	52	16.24	19.93	54.48	24.26	73.38	48.98
Conductivity, σ [S/m]	4	0.91	1.36	3.50	2.25	0	3.86

[a]Vertical polarization. [b]Horizontal polarization. [c]Multi-polarization.

show the results of the 3-D reconstructed images after 230 iterations, using vertical, horizontal, and multi-polarization, respectively. As shown in **Figure 5(b)**, the reconstruction of the tumor cannot be performed correctly when vertical polarization antennas are used. Moreover, the image reconstruction is insufficient when using horizontal polarization. However, the results in **Figure 5(d)** indicate that the image reconstruction of the breast is accurate when using multi-polarization antennas for transmitting and receiving.

Figures 6(a)-(b) show the setting and reconstruction values of the dielectric properties of the tumor voxel in the x-axis direction. **Figure 6(a)** demonstrates that the relative permittivity of the tumor is relatively high when using horizontal polarization and low when using vertical polarization. Nevertheless, the reconstruction value obtained using multi-polarization is similar to the setting value. In addition, the conductivity of the tumor is improved and accurately estimated when using multi-polarization. **Table 2** shows the reconstructed relative permittivity and conductivity values of the tumor when using different polarizations in Model 2.

In Model 2, the tumor is adjacent to fibro-glandular tissues and the contrast difference between them is small. The above findings confirm that we can detect the tumor accurately with the presence of fibro-glandular tissues by employing multi-polarization. Overall, the results from both breast models indicate that vertical and horizontal polarization provide different information of the image reconstruction and dielectric properties. The multi-polarization array configuration consistently performed better than single polarizations.

3.2.2. Correlation Coefficient of Adjacent Antennas

In this section, we investigated the impact on image reconstruction by the analysis of the correlation coefficient of the received data between adjacent antennas in our imaging sensor. First, we examined the positions of the antennas shown in **Figure 1(a)**, with the upper antenna at 24 mm and the lower antenna at 8 mm (Position 1). **Table 3** summarizes the correlation coefficients for different polarizations. At this position and for Model 1, we observed that single polarizations achieve correlation coefficients close to 1 and the correlation coefficient of the multi-polarization result is 0.1555, relatively low compared with the single polarizations. Model 2 also shows a similar tendency. Next, we changed the position of the upper antenna to 32 mm (Position 2), and finally the position of the lower antenna to 16 mm (Position 3). The results for the two new positionsare tabulated in **Table 3**. We observed that the results show a similar trend to that exhibited at Position 1 for both breast models.

Overall, the correlation coefficient of the receiver pair was significantly reduced when using multi-polarization compared to single polarizations. At Position 1, adequate image reconstruction was obtained using multi-polarization owing to the low correlation coefficient, as shown in **Figures 3-6**. Furthermore, the correlation coefficient of the multi-polarization results at Position 3 is higher than that at Position 1 for both models. **Figures 7(a)-(c)** show the x-axis projection of the setting and reconstruction values of the dielectric properties of the tumor voxel for Model 1 obtained using multi-polarization. These results demonstrate that positions with low correlation coefficients reconstruct the dielectric properties sufficiently. Therefore, we can conclude that a low correlation coefficient is a viable condition for successful image reconstruction.

3.2.3. Condition Number of the Inverse Matrix

The condition number of a matrix characterizes the solution sensitivity of a linear equation system to errors in the data and it indicates the accuracy of the results from of the matrix inversion. We examined the condition number of the inverse matrix for the two breast models using different polarizations; these are summarized in **Table 4**. For Model 2, when horizontal polarization is used, the condition number is slightly decreased from 12.07 to 11.76 at the second iteration. Then, the condition number exhibits unstable changes with increasing iteration number and obtains a significantly high value of 3.08×10^3 at the 230th iteration. This suggests that the

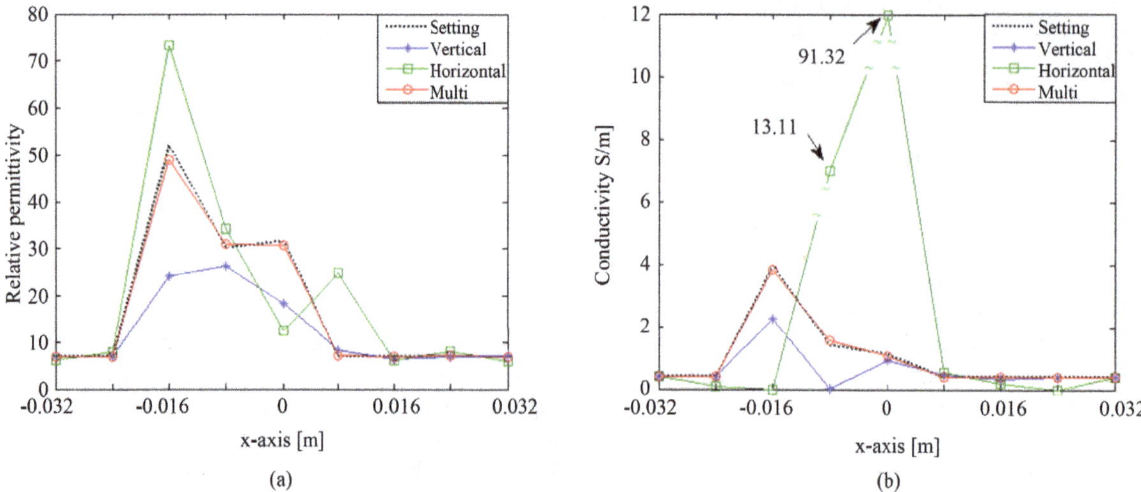

Figure 6. Dielectric property distributions of Model 2 in the cross section including the tumor voxel. (a) Relative permittivity; (b) Conductivity.

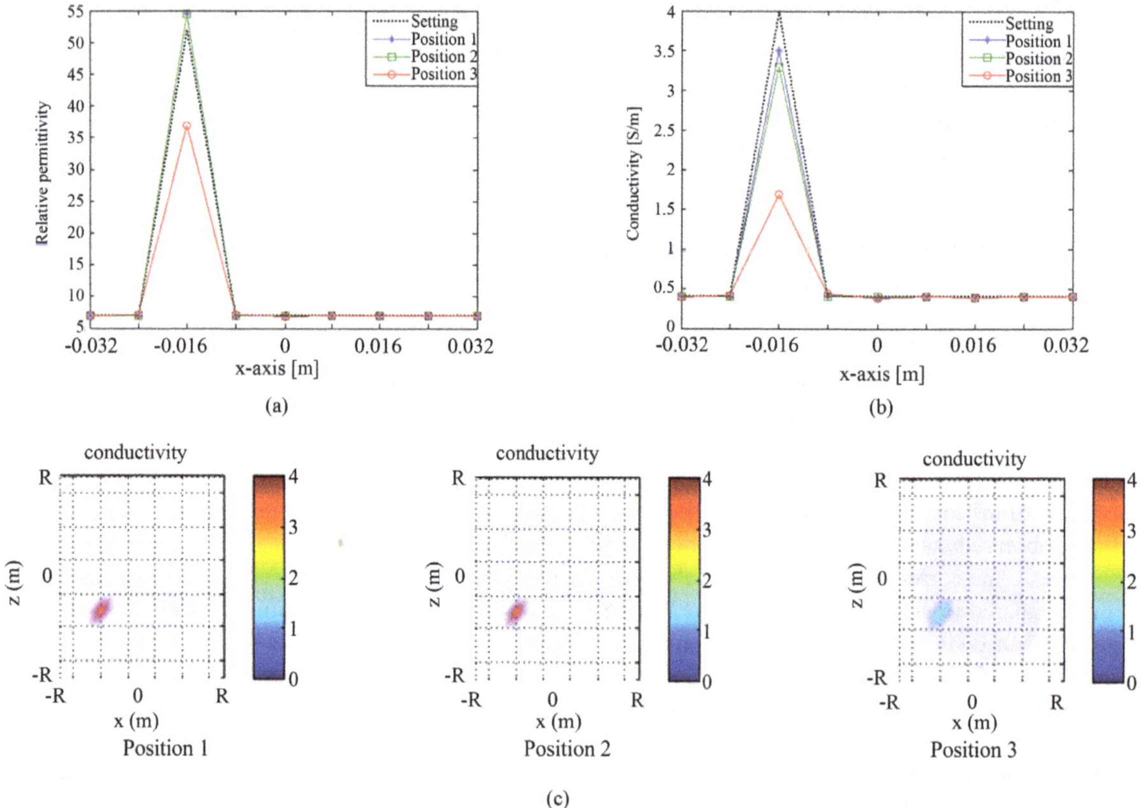

Figure 7. Dielectric property distributions of Model 1 between different positions of the antenna obtained using multi-polarization with $R = 0.048$ m. (a) Relative permittivity; (b) Conductivity; (c) Reconstructed conductivity for different positions of the antennas.

reconstruction of the dielectric property distributions may not be performed correctly owing to the ill-posed problem of the inverse matrix. In contrast, when vertical polarization is used, the condition number decreases to 11.78 at the second iteration. Subsequently, it increases with increasing iteration number. Although an increase occurred in the condition number, it was small compared to that observed for the horizontal polarizations.

Lastly, when multi-polarization is applied to the imaging sensor, the condition number slightly increases to 12.16 at the second iteration. Subsequently, the condition number varies moderately with the iteration number and reaches 13.80 at the 230th iteration. Model 1 showed a similar tendency. From these results, we conclude that the ill-posed problem does not occur when multi-polarization is applied. Thus, the images and dielectric properties of the breast can be reconstructed. **Figure 8** shows the changes in the condition number as a function of iteration number for Model 1. The figure demonstrates that multi-polarization is an effective method for image reconstruction nowing to the small and smooth variations in the condition number.

4. Conclusion

We have confirmed the effectiveness of applying multi-polarization to transmit and receive antennas to determine the dielectric property distributions of a simple breast model. For the imaging algorithm, this is accomplished using the MOM and DBIM in the inverse scattering problem. The numerical simulation results demonstrated that the ill-posed problem can be avoided because of the improvement of the condition number by multi-polarization. Furthermore, the correlation coefficient of multi-polarization is relatively low compared to those

Table 3. The correlation coefficient for different positions of the antennas.

Position	Position of antenna (upper, lower) mm	Model 1			Model 2		
		VP	HP	MP	VP	HP	MP
1	24, 8	0.9536	0.9253	0.1555	0.9469	0.9036	0.1598
2	32, 8	0.8422	0.7605	0.1655	0.8068	0.6748	0.1618
3	32, 16	0.9311	0.8475	0.2044	0.9020	0.7435	0.2092

Table 4. The condition number of inverse matrix in breast models for different polarizations.

Iteration number	Model 1			Model 2		
	VP	HP	MP	VP	HP	MP
1	12.07	12.07	12.07	12.07	12.07	12.07
2	11.63	11.67	11.87	11.78	11.76	12.16
50	23.29	726.87	14.91	22.04	637.02	16.03
100	32.25	1.36×10^3	13.85	69.37	1.66×10^3	15.33
150	-	-	-	100.88	2.64×10^3	14.35
230	-	-	-	98.19	3.08×10^3	13.80

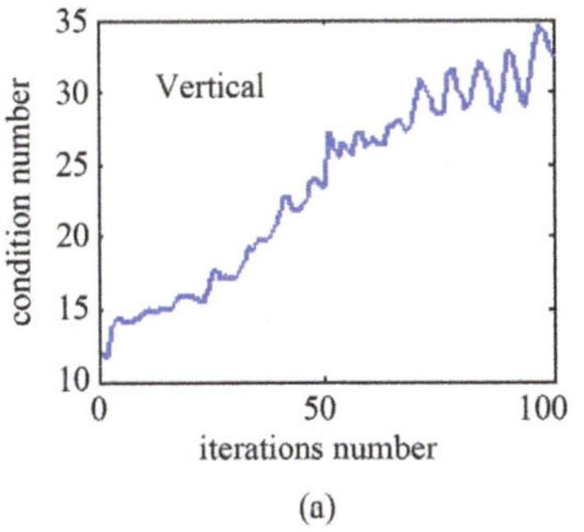

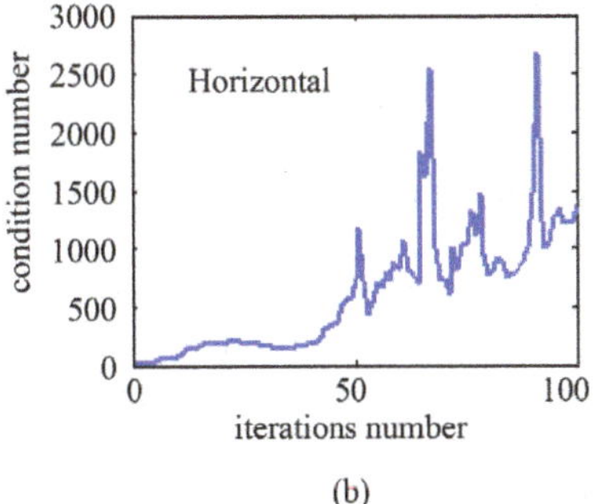

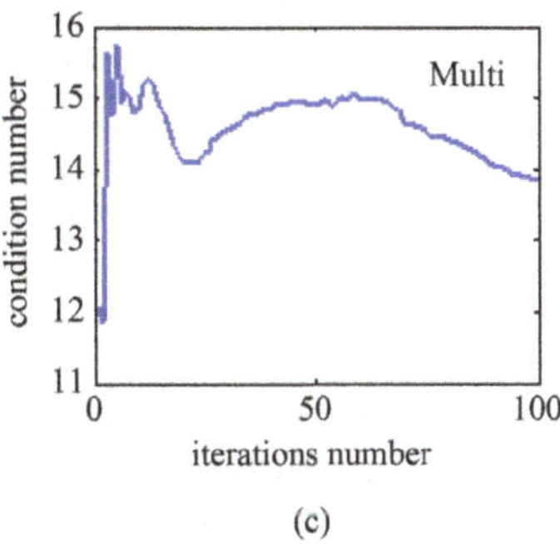

Figure 8. Condition number of Model 1 using vertical, horizontal, and multi-polarization of the sensor.

corresponding to single polarizations. For this reason, the correlation coefficient may represent a viable parameter for image reconstruction in microwave tomography aimed at breast cancer detection.

References

[1] Saika, K. and Sobue, T. (2009) Epidemiology of Breast Cancer in Japan and the US. *Japan Medical Association Journal*, **52**, 39-44.

[2] Nikolova, N.K. (2011) Microwave Imaging for Breast Cancer. *IEEE Microwave Magazine*, **12**, 78-94. http://dx.doi.org/10.1109/MMM.2011.942702

[3] Lazebnik, M., Popovic, D., McCartney, L., Watkins, C.B., Lindstrom, M.J., Harter, J., Sewall, S., Ogilvie, T., Magliocco, A., Breslin, T.M., Temple, W., Mew, D., Booske, J.H., Okonniewsk, M. and Hagness, S.C. (2007) A Large Scale Study of the Ultra-Wideband Microwave Dielectric Properties of Normal, Benign, and Malignant Breast Tissues Obtained from Cancer Surgeries. *Physics in Medicine and Biology*, **52**, 6093-6115. http://dx.doi.org/10.1088/0031-9155/52/20/002

[4] Kuwahara, Y., Miura, S., Nishina, Y., Mukumoto, K., Ogura, H. and Sakahara, H. (2013) Clinical Setup of Microwave Mammography. *IEICE Transactions Communication*, **E96-B**, 2553-2562. http://dx.doi.org/10.1587/transcom.E96.B.2553

[5] Kuwahara, Y., Ogura, H. and Sakahara, H. (2013) Microwave Mammography—Considerations on Clinical Test. *MWE* 2013 *Microwave Workshop Digest*, 50-53.

[6] Kuwahara, Y., Ogura, H. and Sakahara, H. (2013) Microwave Mammography—Considerations on Clinical Test. *Proceedings of the* 26*th Annual International Conference of the IEEE EMBS*, San Francisco, 1-5 September 2004, 2758-2761.

[7] Kuwahara, Y. (2013) Microwave Mammography Technology (Japanese). *RF World*, **25**, 8-19.

[8] Ozawa, N. and Kuwahara, Y. (2014) Considerations of Antennas for Microwave Mammography. *Proceedings of Thailand-Japan Microwave* (*TJMW*), TH3-4.

[9] Mohamed, L. and Kuwahara, Y. (2014) Distortion Born Iterative Method in Microwave Tomography—A Numerical Study of 3D Non-Debye and Debye Model. *Proceedings of Thailand-Japan Microwave* (*TJMW*), TH4-1.

[10] Meaney, P.M., Fanning, M.W., Li, D., Poplack, S.P. and Paulsen, K.D. (2000) A Clinical Prototype for Active Microwave Imaging of the Breast. *IEEE Transactions on Microwave Theory and Techniques*, **48**, 1841-1853. http://dx.doi.org/10.1109/22.883861

[11] Fang, Q., Meaney, P.M. and Paulsen, K.D. (2004) Microwave Image Reconstruction of Tissue Property Dispersion Characteristics Utilizing Multiple Frequency Information. *IEEE Transactions on Microwave Theory and Techniques*, **52**, 1866-1875. http://dx.doi.org/10.1109/TMTT.2004.832014

[12] Ostadrahimi, M., Zakaria, A., LoVetri, J. and Shafai, L. (2013) A Near-Field Dual Polarized (TE-TM) Microwave Imaging System. *IEEE Transactions on Microwave Theory and Techniques*, **61**, 1376-1384. http://dx.doi.org/10.1109/TMTT.2012.2237181

[13] Mays, R.O., Behdad, N. and Hagness, S.C. (2015) A TSVD Analysis of the Impact of Polarization on Microwave Breast Imaging Using an Enclosed Array of Miniaturized Patch Antennas. *IEEE Antennas and Wireless Propagation Letters*, **14**, 418-421. http://dx.doi.org/10.1109/LAWP.2014.2365755

[14] Joachimowicz, N., Pichot, C. and Hugonin, J.P. (1991) Inverse Scattering: An Iterative Numerical Method for Electromagnetic Imaging. *IEEE Transactions on Antennas and Propagation*, **39**, 1742-1752. http://dx.doi.org/10.1109/8.121595

[15] Shea, J.D., Kosmas, P., Veen, V. and Hagness, S.C. (2010) Contrast-Enhanced Microwave Imaging of Breast Tumors: A Computational Study Using 3D Realistic Numerical Phantoms. *Inverse Problem*, **26**, 1-22. http://dx.doi.org/10.1088/0266-5611/26/7/074009

[16] Shea, J.D., Kosmas, P., Hagness, S.C. and Veen, B.D.V. (2010) Three-Dimensional Microwave Imaging of Realistic Numerical Breast Phantoms via a Multiple Frequency Inverse Scattering Technique. *Medical Physics*, **37**, 4210-4226. http://dx.doi.org/10.1118/1.3443569

[17] Semenov, S., Kellam, J., Sizov, Y., Nazarov, A., Williams, T., Nair, B., Pavlovsky, A., Posukh, V. and Quinn, M. (2011) Microwave Tomography of Extremities: 1. Dedicated 2-D System and Physiological Signatures. *Physics in Medicine and Biology*, **56**, 2005-2017. http://dx.doi.org/10.1088/0031-9155/56/7/006

[18] Franchois, A., Joisel, A., Pichot, C. and Bolomey, J. (1998) Quantitative Microwave Imaging with a 2.45-GHz Planar Microwave Camera. *IEEE Transactions on Medical Imaging*, **17**, 550-561. http://dx.doi.org/10.1109/42.730400

14

Heating of the Electrons and the Rectified Current at the Contacts That Are in an Alternating Electromagnetic Field

G. Gulyamov, M. G. Dadamirzaev

Namangan Engineering Pedagogical Institute, Namangan, Uzbekistan
Email: gulyamov1949@mail.ru

Abstract

The paper deals with the heating of electrons and current rectification in contact, which is located in an alternating electromagnetic field. It was found that the electrical component of the microwave (UHF) waves inside the p-n-junction was curved. This leads to the perpendicular component of the electric field of the microwave wave. This component modulates the height of the potential barrier with the frequency of the microwave. In the p-n-junction, straightening microwave current occurs. It is shown that the rectifying contact in the microwave electromagnetic field is always an electromotive force. This is due to carrier heating and straightening microwave current. It is shown that electron heating and straightening of the microwave power will lead to higher ideality factor of the diode.

Keywords

Hot Electrons, The Microwave Field, The Open Circuit Voltage, Short Circuit Current, CVC, p-n-Junction

1. Introduction

Experimental studies of the heating of the charge carriers by the strong electric field in inhomogeneous semiconductors began with thermopower measurements of hot charge carriers occurring at the p-n-junction. In [1], it used an ultrahigh frequency (UHF)-electric field to warm up the charge carriers in the measurement of the thermopower of hot electrons at UT p-n-junction. In this case, the electric field vector is oriented along a concentration gradient, so that the ends of the sample are oriented along the longitudinal thermopower of hot charge carriers, a signal rectification of alternating current. In order to avoid straightening AC p-n-junction, in the measurement of the thermopower of hot carriers, the vector of the microwave electric field is directed perpendicular to the concentration gradient [2].

[3] studied the current and the emf arising in asymmetrical p-n-junctions under the influence of a strong microwave field. It revealed that, in asymmetric p-n-junction, a strong microwave field for the analysis of voltages and currents must be considered as the heating of electrons and holes.

In [4], the results of theoretical and experimental studies of the emergence of negative differential resistance in the diode structures are based on the p-n-junction in a strong microwave field.

Under the influence of the electromagnetic wave, it increases the average energy of the carriers in the contact area with the metal-semiconductor interface. As a result, the potential barrier of electrons is heated by flowing current [2]. On the other hand, due to changes in the barrier height at the contact, alternating current generated by the electric field of the wave is rectified. Due to the fact that the directions of the currents of hot carriers and rectified currents are identical, establishing the true mechanism of the generated EMF diode is an important issue.

The aim of this work is to study electron heating and current rectification to contact in an alternating electromagnetic field.

2. Influence of Electron Heating and Current Rectification on Track in a Strong Microwave Field

We estimate the current arising under the influence of electromagnetic waves on a thin Schottky diode. If the wave period longer than the transit time of electrons through the barrier, on the basis of the theory of electron diode current passing through the barrier is given by:

$$j = j_s\left[\exp\left(\frac{\mathrm{e}\varphi_0}{kT} - \frac{\mathrm{e}\varphi + \mathrm{e}U_B\cos(\omega t)}{kT_e(t)}\right) - 1\right] \tag{1}$$

where j_s is the saturation current; φ_0 is the height of the potential barrier in the absence of an electromagnetic wave; $\varphi = \varphi_0 - U$; U is the resulting voltage across the diode; $U_B = -\int_0^d E_B \mathrm{d}x$ is the AC voltage of the incident wave, created by the barrier diode; T is the temperature of the lattice; k is the Boltzmann constant; $T_e(t)$ is the temperature of the electron gas in contact with the region; E_B is the electric field of the wave; e is the charge of the electron.

The average value of the diode current is determined by the following integral:

$$\frac{\bar{j}}{j_s} = \int_0^{2\pi}\left\{\left(\frac{T_e}{T}\right)^{\frac{1}{2}} \exp\frac{\mathrm{e}}{k}\left[\frac{\varphi_0}{T} - \frac{\varphi_0 + U(t) + U_B\cos(\omega t)}{T_e(\omega t)}\right] - 1\right\}\frac{\mathrm{d}(\omega t)}{2\pi}. \tag{2}$$

The values of the electron temperature are determined by the wave field strength.

At low wave power can not take into account the heating of the electrons, while the constant current arises only because of rectification:

$$\frac{\bar{j}}{j_s} + 1 = \exp\left(-\frac{\mathrm{e}U_0}{kT}\right)\int_0^{2\pi}\exp\left(-\frac{\mathrm{e}U_B\cos(\omega t)}{kT}\right)\frac{\mathrm{d}(\omega t)}{2\pi}. \tag{3}$$

Here U_0 is the DC offset applied to the barrier.

Detected voltage in the diode is given by:

$$U = \frac{kT}{\mathrm{e}}\left\{\ln\left(1 + \frac{\bar{j}}{j_s}\right) - \ln\left[\int_0^{2\pi}\exp\left(-\frac{\mathrm{e}U_B}{kT}\cos(\omega t)\right)\frac{\mathrm{d}(\omega t)}{2\pi}\right]\right\}. \tag{4}$$

It follows that the open circuit voltage at $j = 0$:

$$U_{oc} = \frac{kT}{e}\ln\left[\int_0^{2\pi}\exp\left(\frac{eU_B}{kT}\cos(\omega t)\right)\frac{d(\omega t)}{2\pi}\right]. \tag{5}$$

Short-circuit current is the following:

$$\frac{j}{j_s}+1 = \frac{1}{2\pi}\int_0^{2\pi}\exp\left(-\frac{eU_B}{kT}\cos(\omega t)\right)d(\omega t). \tag{6}$$

Equation (5) defines the CVC diode situated in the field of the electromagnetic wave when the current occurs only through rectification of alternating current.

At high power microwave energy when $T_e \neq T$, CVC diode has the following form:

$$\frac{\bar{j}}{j_s}+1 = \left(\frac{T_e}{T}\right)^{\frac{1}{2}}\exp\left[\frac{e}{k}\left(\frac{\varphi_0}{T}-\frac{\varphi_0}{T_e}-\frac{U}{T_e}\right)\right]\int_0^{2\pi}\exp\left(-\frac{eU_B\cos(\omega t)}{kT_e}\right)\frac{d(\omega t)}{2\pi}. \tag{7}$$

Using Formula (7) we will investigate the effect of electron heating in the CVC p-n-junction in a strong microwave field. CVC p-n-junction temperature change of the electrons is shown in **Figure 1**.

This shows that the inclusion of straightening and heating of electrons increases in the strong currents of the diode microwave field. The degree of increase of the current is determined by the following factor:

$$\int_0^{2\pi}\exp\left(-\frac{eU_B\cos(\omega t)}{kT_e}\right)\frac{d(\omega t)}{2\pi} \tag{8}$$

when $\left|\frac{eU_B}{kT_e}\right| \gg 1$ this value increases rapidly and strongly affects the CVC diode.

Now consider the short-circuit current of the Formula (7) (for $U = 0$) we have the following expression for the short-circuit current:

$$\frac{\bar{j}_{sc}}{j_s}+1 = \left(\frac{T_e}{T}\right)^{\frac{1}{2}}\exp\left[\frac{e}{k}\left(\frac{\varphi_0}{T}-\frac{\varphi_0}{T_e}\right)\right]\int_0^{2\pi}\exp\left(-\frac{eU_B\cos(\omega t)}{kT_e}\right)\frac{d(\omega t)}{2\pi}. \tag{9}$$

This shows that the inclusion of rectification increases the short circuit current by a factor of (8).

C using the Formula (9) is plotted as the short circuit current of the electron temperature (**Figure 2**).

Figure 2 that the short circuit current is strongly dependent on the temperature of the electrons. With the increase of the electron temperature increases rapidly short-circuit current.

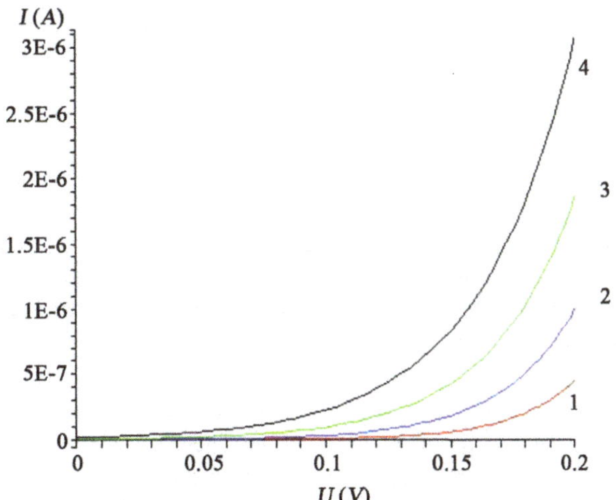

Figure 1. CVC p-n-junction at different electron temperature. 1—T_e = 300 K, 2—T_e = 350 K, 3—T_e = 400 K, 4—T_e = 450 K.

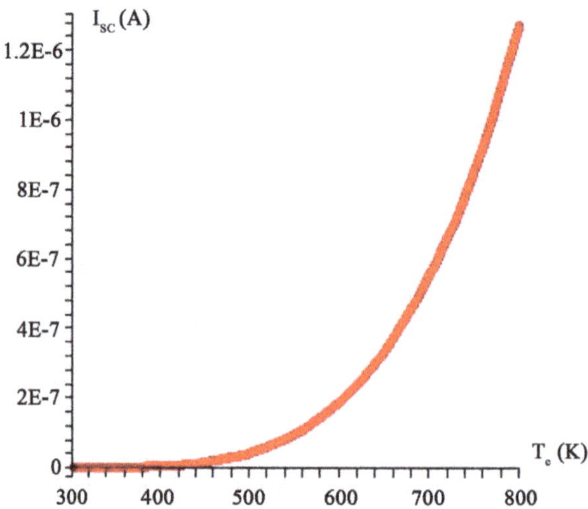

Figure 2. Dependence of short circuit current on the electron temperature according to the Formula (9).

For the open-circuit voltage of hot electrons from the Formula (7) (for $j = 0$) we have the following expression:

$$U_{oc} = -\varphi_0\left(\frac{T_e}{T}-1\right)-\frac{kT_e}{\mathrm{e}}\ln\left[\left(\frac{T_e}{T}\right)^{\frac{1}{2}}\int_0^{2\pi}\exp\left(-\frac{\mathrm{e}U_B\cos(\omega t)}{kT_e}\right)\frac{\mathrm{d}(\omega t)}{2\pi}\right]. \tag{10}$$

The first term of this formula gives emf, heated by the electron in the potential barrier [2], and the second part—the voltage generated due to rectification. However, it should be noted that the second portion (10) differs from the rectifying electromotive force (5), when the electrons are not warmed up. If you do not take into account current rectification caused by electromagnetic wave, the first term of the Formula (10) corresponds to the EMF of the heated electrons produced in the potential barrier φ_0 [2].

Formula (10) allows to analyze graphically the dependence of $f(U_{oc}, T_e)$. **Figure 3** shows the dependence of the open circuit voltage of the electron temperature $f(U_{oc}, T_e)$. This shows that in the p-n-junction due to the heating of electrons increases the open circuit voltage.

Analyze the second part of the Formula (10) is associated with straightening:

$$U_{\text{rect}} = \frac{kT_e}{\mathrm{e}}\ln\left[\theta^{\frac{1}{2}}\int_0^{2\pi}\exp\left(-\frac{Y_B}{\theta}\cos(2\pi z)\right)\mathrm{d}z\right] \tag{11}$$

where $\theta = \frac{T_e}{T}$ is the dimensionless electron temperature; $Y_B = \frac{\mathrm{e}U_B}{kT_e}$ is the dimensionless potential wave; $z = \omega t$.

This formula shows that the dependence of $Y_B U_{\text{rect}}$ differs from linearity. The linear dependence of the detected voltage of the electric field of the wave, is often regarded as the result of rectification of the AC signal. From the analysis of (11) we can conclude that this conclusion is not always justified. This is due to the fact that the deviation from the linear dependence of the rectified voltage on the intensity of the microwave field is due to the heating of the electrons.

In [5] in order to eliminate the current rectification in the microwave electromagnetic field of the p-n-junction plane is set parallel to the electric field component of the wave. There is assumed that the detection signal occurs only due to heating of charge carriers. The temperature value of the electron gas, calculated from the experimental data on the basis of this assumption turned several times greater than the expected [5]. To explain these anomalously high values of the thermoelectric power of hot carriers in [5] was introduced ideality factor, and it was believed that he was connected with the process of recombination in the p-n-junction.

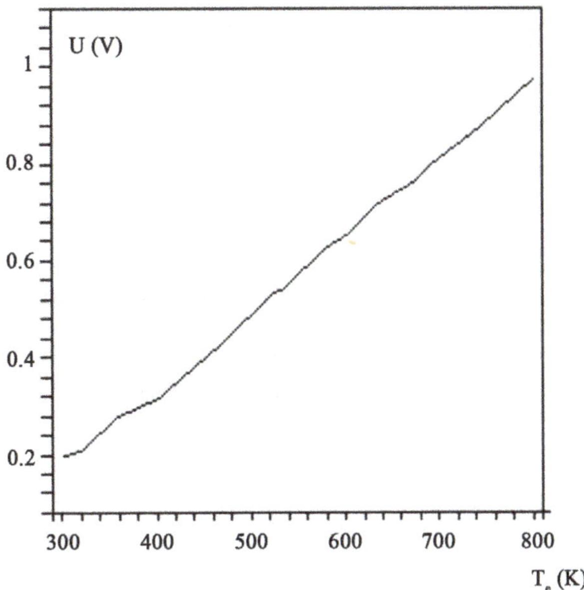

Figure 3. The dependence of the open circuit voltage of the electron temperature.

The emergence of large values of EMF [5], in our opinion may be due to the following reasons yet.

When the location of the p-n-junction in a microwave electromagnetic field is a redistribution of the field E_B waves in the space inside the sample. The result is distorted parallel to the plane of the E_B p-n-junction. This leads to the inevitable emergence of the electric field component perpendicular to the plane of the p-n-transition, which is not uniformly distributed in the p-n-junction. By turning the p-n-junction relative to E_B can be reduced to a minimum average value of the perpendicular component $E_\perp$, but it is impossible to completely eliminate. The parallel component of the electric field to the p-n-junction is not rectify the high frequency electric current, and is involved only in the heating media. A perpendicular component is at the p-n-junction rectification microwave current [6] [7]. Thus, in the rectifying contact in the electromagnetic field occurs as a heating carriers and straightening of the microwave power. Consequently, there is excessive DC component and generate additional EMF.

Therefore, a strong microwave field in the calculation of EMF occurring at the p-n-junction in addition to the heating of electrons must be considered and straightening of the microwave power.

The expression for the effective electron temperature, calculated from Formulas (7)-(10) has the form:

$$T_e^* = T_e\left[1+\frac{kT}{e\varphi_0}\ln\int_0^{2\pi}\exp\left(-\frac{eU_\perp\cos(\omega t)}{kT_e}\right)\frac{d(\omega t)}{2\pi}\right]. \quad (12)$$

Here $U_\perp = -\int_0^d E_\perp dx$, $E_\perp$—perpendicular component of the wave field to the surface of the p-n-junction.

You can enter the ideality factor m in the following form:

$$T_e^* = mT_e \quad (13)$$

$$m = \left[1+\frac{kT}{e\varphi_0}\ln\int_0^{2\pi}\exp\left(-\frac{eU_{B\perp}\cos(\omega t)}{kT_e}\right)\frac{d(\omega t)}{2\pi}\right]. \quad (14)$$

The second term in the square brackets is due to the more effective temperature of the electrons due to rectification. When $U_\perp = 0$ we have $T_e^* = T_e$, effective temperature coincides with the actual temperature of the electrons. In this case there is no extension. When $U_\perp \neq 0$ always have $T_e^* > T_e$, straightening as it effectively increases the temperature T_e^*. With the increasing power of the electromagnetic wave incident on the diode and the difference between T_e^* and T_e increases (see Equation (12)) T_e^* strong increase was also observed in the experiment [5]. The authors of these studies neglect the distortion of the microwave field, the increase associated with the increase T_e^* nonideality factor *m* with increasing electric field strength. However, there is no quantitative analysis. If we consider the rectification, we can quantitatively explain the relationship between m and the field strength. Effect of carrier heating charges on nonideality factor CVC p-n-junction is discussed in [8] [9].

3. Conclusion

The rectifying junction in the microwave electromagnetic field is always an electromotive force, which is due to electron heating and straightening microwave current. This is due to the distortion of the electric component of the microwave inside the p-n-junction. Distortion direction of the electric component of the microwave will cause the perpendicular component of the electric field that causes rectification of the microwave current through the p-n-junction. Heating of the electrons and the microwave current rectification increases the ideality factor of the diode.

References

[1] Pozhela, J.K. and Repshas, K.K. (1968) Thermoelectric Force of Hot Carriers. *Physica Status Solidi*, **27**, 757-762. http://dx.doi.org/10.1002/pssb.19680270233

[2] Veynger, A.I., Paritsky, L.G., Hakobyan, E.A. and Dadamirzaev, G. (1975) Thermopower Hot Carriers on the p-n-Transition. *Semiconductors-Leningrad*, **9**, 216-224.

[3] Dadamirzaev, M.G. (2013) Heating of Electrons and Holes in Asymmetric p-n-Junction, Located in a Microwave Field. *Physical Surface Engineering-Ukraine*, **11**, 191-193.

[4] Usanov, D.A., Skripal, A.V., Ugryumova, N.V., Wenig, S.B. and Orlov, V.E. (2000) The Occurrence of Negative Differential Resistance Mode and Switch to the Tunnel Diode by an External Microwave Signal. *Semiconductors-St. Petersburg*, **34**, 567-571.

[5] Ablyazimova, N.A., Veynger, A.I. and Nutrition, V. (1988) Elektricheskie Properties of Silicon p-n-Junctions in Strong Microwave Fields. *Semiconductors-St. Petersburg*, **22**, 2001-2007.

[6] Gulyamov, G., Dadamirzaev, M.G. and Boydedaev, S.R. (2000) Kinetics of the Establishment of the Thermopower of Hot Carriers in the p-n-Junction, Taking into Account the Heating of the Lattice. *Semiconductors-St. Petersburg*, **34**, 266-269.

[7] Gulyamov, G., Dadamirzaev, M.G. and Boydedaev, S.R. (2000) Emf of Hot Carriers Due to the Modulation of the Surface Potential in a Strong Microwave Field. *Semiconductors-St. Petersburg*, **34**, 572-575.

[8] Shamirzaev, S.H., Gulyamov, G., Dadamirzaev, M.G. and Gulyamov, A.G. (2009) Ideality Factor of the Current-Voltage Characteristics of p-n-Junctions in the Strong Field of the Microwave. *Fizika i Tekhnika Poluprovodnikov-St. Petersburg*, **43**, 53-57.

[9] Shamirzaev, S.H., Guliamov, G., Dadamirzaev, M.G. and Guliamov, A.G. (2009) Electromotive Force at Rectifying Barrier in Microwave Electromagnetic Field. *Uzbek Journal of Physiks Tashkent*, **11**, 122-127.

15

Microwave-Hydrothermal Synthesis of Ferric Oxide Doped with Cobalt

Eman Alzahrani, Abeer Sharfalddin, Mohamad Alamodi

Chemistry Department, Faculty of Science, Taif University, Taif, Kingdom of Saudi Arabia
Email: em-s-z@hotmail.com, sharfalddin.aa@hotmail.com, dr_alamoudi@yahoo.com

Abstract

Ferric oxides have drawn significant interest due to their unique properties, relatively low cost, and due to their potential applications in different fields. In this work, cobalt (Co) doped iron oxide (Fe_2O_3) powders, with crystalline size 36.97 nm were successfully prepared using a microwave-hydrothermal process for the first time and characterised using different techniques. The morphology of the samples was characterised by scanning electron microscopy (SEM), transmission electron microscopy (TEM), energy dispersive analysis of X-ray spectroscopy (EDAX), Fourier transform infrared (FT-IR) spectroscopy and ultraviolet-visible (UV-Vis) spectroscopy. The images show monodispersed particles with a sharp-edged square morphology. It was found that the average size was about 33.3 nm for Fe_2O_3 and 36.97 nm for Co-Fe_2O_3. The Co atomic percentage dopants were approximately 5.73%. The nanosized synthesised materials in this study may find an application in the areas of removal of toxic metal and dyes research.

Keywords

Nanostructures, Ferric Oxide, Doping, Cobalt, Characterisation

1. Introduction

Iron is found in nature in different chemical compounds. Normally, iron has eight electrons on its valence shield, and because of oxygen's electronegativity it can form bivalent and trivalent combinations. These have many applications in different fields such as drug delivery systems [1], cancer treatment [2], magnetic resonance imaging [3], rechargeable lithium batteries, catalysis gas sensors and biosensors [4].

Iron oxides nanocrystals have attracted increasing attention for their outstanding new properties such as their biocompatibility, catalytic activity and low toxicity. In addition, they can be easily separated and removed from a solution by simply using an external magnet. There are three different forms of Iron oxide; mainly FeO, Fe_2O_3 and Fe_3O_4. Fe_2O_3 is the most common oxide of iron and it has four crystallographic phases; namely α-Fe_2O_3 (hematite), β-Fe_2O_3, γ-Fe_2O_3 (maghemite) and ε-Fe_2O_3 [5] [6].

Much effort has been devoted to preparing nanoparticles and many methods have been reported for fabricating metal oxides such as forced hydrolysis [7], combustion [8], anhydrous solvent [9], so-gel [10], wet chemical synthesis [11], microwave-hydrothermal synthesis [12] and spray pyrolysis [13]. Among them, the microwave-hydrothermal process for fabrication of nanoparticles is a new technique. It is a combination of hydrothermal and microwave processes. There are many advantages for using this method such as savings in energy and time and the low temperature requirements for the synthesis of anaphase materials.

Doping of transition metal ions into Fe_2O_3 can improve the properties of nanocrystalline materials by narrowing the energy-band gap and inhibiting electron-hole recombination [4]. So far there are no reports for using the microwave-hydrothermal method for the preparation of Co-Fe_2O_3 nanoparticles; therefore this technique was used in this study to fabricate Co-doped Fe_2O_3 nanopowders. The physical properties of the prepared nanoparticles were then studied.

2. Experiment

2.1. Chemicals and Materials

Nonahydrate ferric nitrate ($Fe(NO_3)_3 \cdot 9H_2O$), hexhydrate cobalt nitrate ($Co(NO_3)_2 \cdot 6H_2O$), and absolute ethanol (C_2H_5OH) were purchased from Sigma-Aldrich (Poole, UK). Hexamethylenetetramine (HMT) was purchased from Fisher Scientific (Loughborough, UK) Distilled water was employed for preparing all the solutions.

2.2. Instruments

The microwave digestion system was sourced from CEM Corporation (North Carolina, USA), and the centrifuge from Hettich (Kirchlengern, Germany). A Scholar 171 magnetic stirrer plate was sourced from Corning stirrer (Tewksbury, USA). The oven came from F.LLI GALLI Company (Milano, Italy). The transmission electron microscopy (TEM) was from JEOL Ltd. (Welwyn Garden City, UK), and the scanning electron microscope (SEM) and energy dispersive analysis of X-ray spectroscopy (EDAX) equipment were a Cambridge S360 from Cambridge Instruments (Cambridge, UK). The FT-IR spectra were PerkinElmer RX FTIR ×2 with diamond ATR, and DRIFT attachment from PerkinElmer (Buckinghamshire, UK). The UV-Vis analysis was collected by the Shimadzu UV-2550 spectrophotometer double beam (Nakagyo-Japan).

2.3. Preparation of Co-Fe_2O_3 Nanoparticles

The nanoparticles were prepared by the microwave-hydrothermal method using a typical procedure described in previous work [14] with some modification: 4.05 g of $Fe(NO_3)_3 \cdot 9H_2O$ and 1.2 g of hexamethylenetetramine (HMT) were dissolved in 30 mL of distilled water and ethanol mixture (1:1, v/v) and stirred vigorously (1100 rpm) until dissolved. Then, 2.32 g of $Co(NO_3)_2 \cdot 6H_2O$ was added to the mixture with constant stirring. After 30 minutes, the mixture was transferred into a Teflon-lined stainless-steel autoclave that was placed in the microwave at 160°C for 90 minutes. The mixture was left to cool down to room temperature. The resulting precipitate was collected by centrifugation for 10 minutes. The Co-Fe_2O_3 nanoparticles were washed with distilled water and ethanol. Finally, the prepared nanoparticles were dried in an oven at 60°C for 24 hours. Undoped Fe_2O_3 was also prepared using the same procedure without adding $Co(NO_3)_2 \cdot 6H_2O$.

2.4. Characterisation of the Fabricated Materials

The surface morphology of the prepared nanoparticles was characterised using scanning electron microscopy (SEM), and transmission electron microscopy (TEM). In addition, the compositional analysis was studied using energy dispersive analysis of X-ray spectroscopy (EDAX). The FT-IR spectra were collected in the attenuated total reflectance (ATR) mode in the range of 500 - 4000 cm^{-1}. For UV-Vis absorption measurements, the powder samples were dispersed in deionised water with a fixed concentration (5 $mg \cdot 4\ mL^{-1}$) [15]. The nanomolar suspensions were prepared by milling in order to minimise the reflection of light [16].

3. Results and Discussion

3.1. Preparation of the Co-Fe_2O_3 Nanoparticles

Iron oxide nanoparticles (Fe_2O_3) have attracted intensive attention because they are common in nature, and are consequently eco-friendly and inexpensive [6]. In this study, a microwave-hydrothermal method was utilised to fabricate nanosized materials to decrease energy consumption, decrease preparation time from days to minutes, and simplify procedures.

Due to the competition between electron-hole pair recombination, metal doping is the perfect modification method to prevent recombination and charge carrier trapping. In literature, Mg is the most studied P-dopant for Fe_2O_3, besides Ca and Ti, which can be used for P-doping Fe_2O_3 [17]. In this work, a new doping method using Co will be discussed. Cobalt is one of the transition metal ions and it can result in higher photocatalytic activity compared with undoped Fe_2O_3. Moreover, it can improve optical activity by narrowing the energy-band gap and extend absorption to the visible region [18].

In this study, Fe $(NO_3)_3{\cdot}9H_2O$ and Co $(NO_3)_2{\cdot}6H_2O$ were used as iron and cobalt sources, respectively. Hexamethylenetetramine (HMT) was used as a molecular building block for self-assembled molecular crystals [19] [20]. **Figure 1** shows an image of the fabricated Co-Fe_2O_3 powders.

3.2. Characterisation of the Fabricated Materials

3.2.1. SEM Analysis

SEM analysis was used to study the surface morphology of the prepared nanoparticles. **Figure 2** represents the SEM images of Fe_2O_3 and Co-Fe_2O_3 nanoparticles in different magnifications. They demonstrate that the grain size is homogeneous, polygonal and agglomerates. By comparing the micrographs, it was found that no significant morphological differences can be viewed. The agglomeration is ascribed to the removal of nanostructure-stabilising ions by washing with water [21]. Moreover, SEM analysis gives only the average grain size of the samples, which simply represents the fact that each grain is formed by aggregation of a number of nanocrystals.

3.2.2. EDAX Analysis

Energy dispersive X-spectroscopy (EDAX) was used to identify elements that exist in the prepared nanosised powders. **Figure 3** shows the EDAX patterns and compositions of Fe_2O_3 and Co-Fe_2O_3. The results confirm that all the elements appear at their corresponding keV values. It was observed that there was a new peak in Co-Fe_2O_3, representing Co, **Figure 3(b)**, which confirms doping of iron oxide (Fe_2O_3) with cobalt. **Table 1** shows the atomic percentages of the nanoparticle elements, which were iron, oxygen and cobalt. It was found that the Co atomic dopant percentage was nearly 5.73%.

3.2.3. TEM Analysis

The TEM morphologies and microstructures of the prepared Fe_2O_3 and Co-Fe_2O_3 powders are shown in **Figure 4**. Clearly, they were composed of uniformly dispersed particles, which indicates that high disparity and uni-

Figure 1. Image of the fabricated Co-Fe_2O_3 powder.

Figure 2. SEM images of (A) (C) and (E) Fe_2O_3 and (B, D, and F) Co-Fe_2O_3 using different magnifications.

Table 1. The atomic percentages for Fe_2O_3 and Co-Fe_2O_3.

Sample	Atom%			Total
	Fe %	O %	Co %	
Fe_2O_3	36.02	63.98	0	100.00
Co-Fe_2O_3	29.58	64.69	5.73	100.00

formity are achieved using this route. In addition, it was found that Co-Fe_2O_3 maintains the cubic particle structure of the undoped Fe_2O_3. The monodispersed particles had a sharp-edged square morphology with an average size range of about 33.3 nm for Fe_2O_3 and 36.97 nm for Co-Fe_2O_3. It was concluded that the average size of nanoparticles was increased slightly due to Co doping in the Fe_2O_3 lattice.

3.2.4. FT-IR Analysis

Figure 5 shows the FT-IR spectra of Fe_2O_3 and Co-Fe_2O_3. The bands cantered at 3335 cm^{-1} and 1574 cm^{-1} are ascribed to the O-H bonding stretching and bending vibrational modes, respectively [22]. It suggests the presence of very small amount of free and adsorbed water on the surface of the samples. In addition, a peak at around 563 cm^{-1} is ascribed to the stretching between iron and oxygen in Fe_2O_3. The absorption band located at 523 cm^{-1} in the Co-Fe_2O_3 nanoparticles samples was attributed to metal dopant-oxygen stretching modes. Similar observations have been documented in literature [23] [24]. The absorption bands around 1030 cm^{-1} and 1110 cm^{-1} are caused by the vibration of crystalline Fe-O modes, which are characteristic of Fe_2O_3 [25]. On doping,

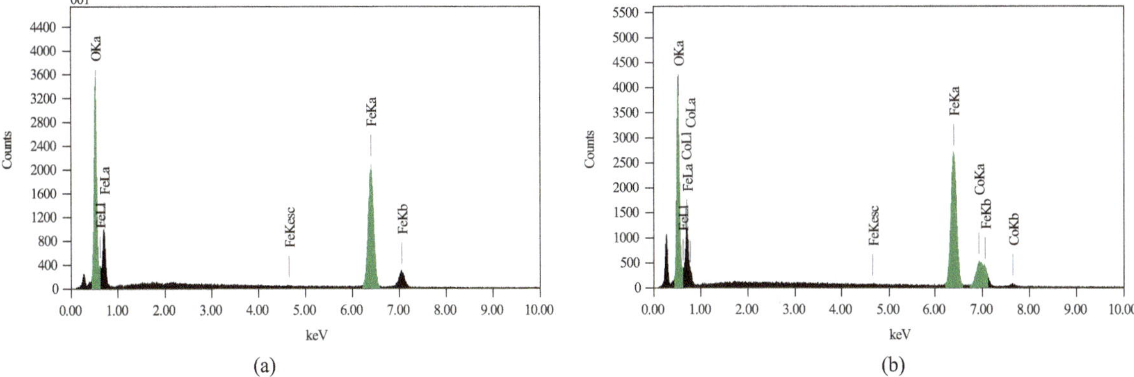

Figure 3. EDAX spectra of (a) Fe_2O_3 and (b) $Co-Fe_2O_3$.

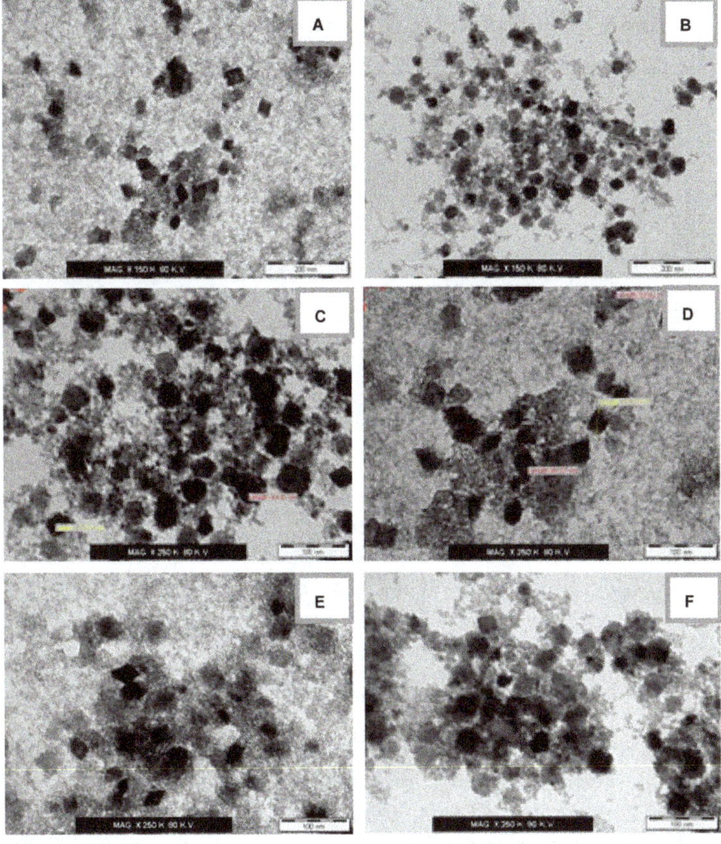

Figure 4. TEM micrographs of Fe_2O_3 (A) (C) and (E) and $Co-Fe_2O_3$ (B) (D)and (F).

the band at 537 cm^{-1} shifts toward a lower frequency suggesting the possible formation of a Co-O-Fe bond. The decrease in the intensity of bands suggests the possible interaction of dopants with surface hydroxyl groups of Fe_2O_3 [26].

3.2.5. UV-Vis Spectroscopy

It has been believed that these narrow band gap values are beneficial for the efficient utilisation of visible light for photocatalysis. The UV-Vis spectrophotometer was used to investigate the absorption regions of Fe_2O_3 and $Co-Fe_2O_3$, as can be seen in **Figure 6**. The energy band gaps (E) were calculated using the following equation [16]:

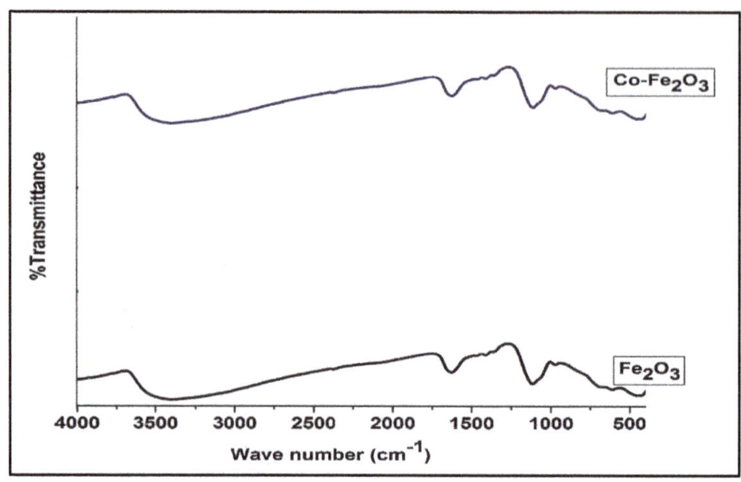

Figure 5. FT-IR spectra of Fe_2O_3 and $Co\text{-}Fe_2O_3$.

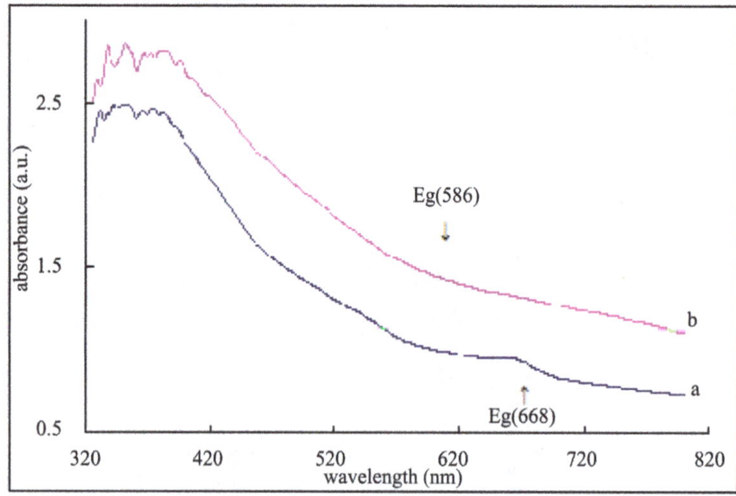

Figure 6. The UV-Vis spectra for (a) Fe_2O_3 and (b) $Co\text{-}Fe_2O_3$.

$$E = h\frac{C}{\lambda}$$

As can be seen in **Figure 6**, there was a decrease in the spectra at the absorption edge of ≈668 nm, and 586 nm for Fe_2O_3 and $Co\text{-}Fe_2O_3$ respectively (indicated by an arrow) [16]. The energy band gaps were calculated to be 1.86 eV for Fe_2O_3 and 2.01 eV for $Co\text{-}Fe_2O_3$. The reported values of the indirect band gap of Fe_2O_3 are in the range of 1.38 - 2.09 eV, the results are in good agreement with previous reports [27]. Moreover, it was observed that the absorption edge is extended towards the visible region for $Co\text{-}Fe_2O_3$ compared with Fe_2O_3. This phenomenon arises due to the transfer of charge from the dopant (Co^{2+}) to the conduction or valence band of Fe_2O_3, which enhances visible light absorption and promotes the photocatalytic activity [27].

4. Conclusion

Nanocrystaline transition metal doped Fe_2O_3 powders were successfully fabricated using the microwave-hydrothermal method. The properties of the fabricated materials were investigated using different techniques, and the morphology of the fabricated nanoparticles was analysed by SEM analysis and TEM analysis. TEM images showed the $Co\text{-}Fe_2O_3$ powders formed 36.97 nm crystals. Moreover, the product was characterised using an EDAX analysis that confirmed doping of iron oxide with cobalt. Work is currently in progress to use nanosized fabricated materials in this work in water purification and the removal of pollutants from wastewater.

References

[1] Jain, T.K., Morales, M.A., Sahoo, S.K., Leslie-Pelecky, D.L. and Labhasetwar, V. (2005) Iron Oxide Nanoparticles for Sustained Delivery of Anticancer Agents. *Molecular Pharmaceutics*, **2**, 194-205. http://dx.doi.org/10.1021/mp0500014

[2] Jordan, A., Scholz, R., Wust, P., Fähling, H. and Felix, R. (1999) Magnetic Fluid Hyperthermia (MFH): Cancer Treatment with AC Magnetic Field Induced Excitation of Biocompatible Superparamagnetic Nanoparticles. *Journal of Magnetism and Magnetic Materials*, **201**, 413-419. http://dx.doi.org/10.1016/S0304-8853(99)00088-8

[3] Chertok, B., Moffat, B.A., David, A.E., Yu, F.Q., Bergemann, C., Ross, B.D. and Yang, V.C. (2008) Iron Oxide Nanoparticles as a Drug Delivery Vehicle for MRI Monitored Magnetic Targeting of Brain Tumors. *Biomaterials*, **29**, 487-496. http://dx.doi.org/10.1016/j.biomaterials.2007.08.050

[4] Suresh, R., Prabu, R., Vijayaraj, A., Giribabu, K., Stephen, A. and Narayanan, V. (2012) Facile Synthesis of Cobalt Doped Hematite Nanospheres: Magnetic and Their Electrochemical Sensing Properties. *Materials Chemistry and Physics*, **134**, 590-596. http://dx.doi.org/10.1016/j.matchemphys.2012.03.034

[5] Zboril, R., Mashlan, M. and Petridis, D. (2002) Iron (III) Oxides from Thermal Processes Synthesis, Structural and Magnetic Properties, Mössbauer Spectroscopy Characterization, and Applications. *Chemistry of Materials*, **14**, 969-982. http://dx.doi.org/10.1021/cm0111074

[6] Chirita, M., *et al.* (2009) Fe_2O_3—Nanoparticles, Physical Properties and Their Photochemical and Photoelectrochemical Applications. *Chemical Bulletin of Politehnica*, University of Timisoara, Romania, **54**, http://www.chim.upt.ro/buletin_chimie/index.htm

[7] Chen, M., *et al.* (2008) Preparation of Akaganeite Nanorods and Their Transformation to Sphere Shape Hematite. *Journal of Nanoscience and Nanotechnology*, **8**, 3942-3948.

[8] Hiremath, V.A. and Venkataraman, A. (2003) Dielectric, Electrical and Infrared Studies of γ-Fe_2O_3 Prepared by Combustion Method. *Bulletin of Materials Science*, **26**, 391-396. http://dx.doi.org/10.1007/BF02711182

[9] Hyeon, T., Lee, S.S., Park, J., Chung, Y., and Na, H.B. (2001) Synthesis of Highly Crystalline and Monodisperse Maghemite Nanocrystallites without a Size-Selection Process. *Journal of the American Chemical Society*, **123**, 12798-12801. http://dx.doi.org/10.1021/ja016812s

[10] Huang, J.Y., *et al.* (2010) *In Situ* Observation of the Electrochemical Lithiation of a Single SnO_2 Nanowire Electrode. Science, **330**, 1515-1520. http://dx.doi.org/10.1126/science.1195628

[11] Chakrabarti, S., Mandal, S. and Chaudhuri, S. (2005) Cobalt Doped γ-Fe_2O_3 Nanoparticles: Synthesis and Magnetic Properties. *Nanotechnology*, **16**, 506-511. http://dx.doi.org/10.1088/0957-4484/16/4/029

[12] Katsuki, H. and Komarneni, S. (2011) Low Temperature Synthesis of Nano-Sized $BaTiO_3$ Powders by the Microwave-Assisted Process. *Journal of the Ceramic Society of Japan*, **119**, 525-527. http://dx.doi.org/10.2109/jcersj2.119.525

[13] Katsuki, H. and Komarneni, S. (2001) Microwave-Hydrothermal Synthesis of Monodispersed Nanophase α-Fe_2O_3. *Journal of the American Ceramic Society*, **84**, 2313-2317. http://dx.doi.org/10.1111/j.1151-2916.2001.tb01007.x

[14] Sun, P., Wang, C., Zhou, X., Cheng, P.F., Shimanoe, K., Lu, G.Y. and Yamazoe, N. (2014) Cu-Doped α-Fe_2O_3 Hierarchical Microcubes: Synthesis and Gas Sensing Properties. *Sensors and Actuators B*: *Chemical*, **193**, 616-622. http://dx.doi.org/10.1016/j.snb.2013.12.015

[15] Morales, A.E., Mora, E.S. and Pal, U. (2007) Use of Diffuse Reflectance Spectroscopy for Optical Characterization of Un-Supported Nanostructures. *Revista Mexicana de Fisica Supplement*, **53**, 18-22.

[16] D'Souza, L.P., Shree, S. and Balakrishna, G.R. (2013) Bifunctional Titania Float for Metal Ion Reduction and Organics Degradation, via Sunlight. *Industrial & Engineering Chemistry Research*, **52**, 16162-16168. http://dx.doi.org/10.1021/ie402592k

[17] Pozan, G.S., Isleyen, M. and Gokcen, S. (2013) Transition Metal Coated TiO_2 Nanoparticles: Synthesis, Characterization and Their Photocatalytic Activity. *Applied Catalysis B*: *Environmental*, **140**, 537-545. http://dx.doi.org/10.1016/j.apcatb.2013.04.040

[18] Liu, J., Liang, C.H., Xu, G.P., Tian, Z.F., Shao, G.S. and Zhang, L.D. (2013) Ge-Doped Hematite Nanosheets with Tunable Doping Level, Structure and Improved Photoelectrochemical Performance. *Nano Energy*, **2**, 328-336. http://dx.doi.org/10.1016/j.nanoen.2012.10.007

[19] Merkle, R.C. (2000) Molecular Building Blocks and Development Strategies for Molecular Nanotechnology. *Nanotechnology*, **11**, 89-99. http://dx.doi.org/10.1088/0957-4484/11/2/309

[20] Habibi, M.H. and Habibi, A.H. (2013) Effect of the Thermal Treatment Conditions on the Formation of Zinc Ferrite Nanocomposite, $ZnFe_2O_4$, by Sol-Gel Method. *Journal of Thermal Analysis and Calorimetry*, **113**, 843-847. http://dx.doi.org/10.1007/s10973-012-2830-4

[21] Zhou, K.W., Wu, X.H., Wu, W.W., Xie, J., Tang, S.Q. and Liao, S. (2013) Nanocrystalline $LaFeO_3$ Preparation and Thermal Process of Precursor. *Advanced Powder Technology*, **24**, 359-367. http://dx.doi.org/10.1016/j.apt.2012.08.009

[22] Linxu, S. and Ping, L. (2010) Preliminary Exploration on Water Pollution from Non-Point Source in Xiang Xi River. *Proceedings of the 4th International Conference on Bioinformatics and Biomedical Engineering* (*iCBBE*), Chengdu, 18-20 June 2010, 1-5.

[23] Wu, J.-M., Zhang, T.-W., Zeng, Y.-W., Hayakawa, S., Tsuru, K. and Osaka, A. (2005) Large-Scale Preparation of Ordered Titania Nanorods with Enhanced Photocatalytic Activity. *Langmuir*, **21**, 6995-7002. http://dx.doi.org/10.1021/la0500272

[24] Pfitzner, A., Dankesreiter, S., Eisenhofer, A. and Cherevatskaya, M. (2013) Heterogeneous Semiconductor Photocatalysis. In: König, B., Ed., *Chemical Photocatalysis*, de Gruyter, Berlin, 211-246. http://dx.doi.org/10.1515/9783110269246.211

[25] Pal, B. and Sharon, M. (2000) Preparation of Iron Oxide Thin Film by Metal Organic Deposition from Fe (III)-Acetylacetonate: A Study of Photocatalytic Properties. *Thin Solid Films*, **379**, 83-88. http://dx.doi.org/10.1016/S0040-6090(00)01547-9

[26] Devi, L.G., Kottam, N., Murthy, B.N. and Kumar, S.G. (2010) Enhanced Photocatalytic Activity of Transition Metal Ions Mn^{2+}, Ni^{2+} and Zn^{2+} Doped Polycrystalline Titania for the Degradation of Aniline Blue under UV/Solar Light. *Journal of Molecular Catalysis A*: *Chemical*, **328**, 44-52. http://dx.doi.org/10.1016/j.molcata.2010.05.021

[27] Satheesh, R., Vignesh, K., Suganthi, A. and Rajarajan, M. (2014) Visible Light Responsive Photocatalytic Applications of Transition Metal (M = Cu, Ni and Co) Doped α-Fe_2O_3 Nanoparticles. *Journal of Environmental Chemistry Engineering*, **2**, 1956-1968. http://dx.doi.org/10.1016/j.jece.2014.08.016

16

Microwave Assisted Synthesis of Guar Gum Grafted Acrylic Acid/Nanoclay Superabsorbent Composites and Its Use in Crystal Violet Dye Absorption

S. B. Shruthi[1], Chandan Bhat[1], S. P. Bhaskar[1], G. Preethi[1], R. R. N. Sailaja[2*]

[1]Chemical Engineering Department, Dayananda Sagar College of Engineering, Kumaraswamy Layout, Bengaluru, India

[2]Resource Efficient Process Technology Application, The Energy and Resources Institute, Bengaluru, India

Email: [*]sailajab@teri.res.in

Abstract

Natural gums can be tailored and used for the removal of toxic dyes like crystal violet via grafting techniques. However, grafting via microwave irradiation showed both higher yield and fast reaction kinetics as compared to conventional grafting. Silane modified nanoclay has been used to prepare acrylic acid grafted guar gum nanocomposites via microwave irradiation technique. The grafting was confirmed via infra-red spectroscopy while XRD diffractograms suggested exfoliation of modified nanoclay in guar gum grafted acrylic acid. The reaction kinetic parameters have been optimized. The effect of nanoclay on swelling characteristics has been examined. The sensitivity of pH on swelling capabilities has also been assessed. The efficiency of the superabsorbent nanocomposite on the absorption of crystal violet dye has been studied. The superabsorbent nanocomposite loaded with 1.75% modified nanocaly was found to be optimal concentration for the removal of crystal violet dye.

Keywords

Microwave, Guar Gum, Superabsorbent, Nanocomposites, Dye Absorption

1. Introduction

Superabsorbents are three dimensional networks with a capacity to absorb and retain large quantities of water.

[*]Corresponding author.

The water absorbency is facilitated by various functional groups such as amine, hydroxyl, amide and carboxyl groups. The loosely crosslinked networks are made up of homopolymers or grafted copolymers which can hold water [1]. Superabsorbents are used for a wide variety of applications. Further, depleting petrochemical resources prompts us to look for benign and sustainable alternatives. Thus, plant and animals based renewable resources such as starch, cellulose, chitosan various gums etc. have been widely studied. Although biopolymeric materials are abundantly available, sustainable, environment friendly, they have inferior properties as compared to synthetic polymers. Hence, they have to be modified for various applications. Thus, grafted polysaccharides have been synthesized and used for various drug delivery applications [1]-[4].

Grafted chitosan has been covalently immobilized on enzyme for biosensor applications [5]. Gums such as Xanthan gum, Arabic gum, Acacia gum etc. are abundantly available. Most of these gums are commonly used as thickening, stabilizing, binding and emulsifying agents. They are mainly classified based on their ionic nature and origin. Guar gum is a plant based hydrocolloid and an edible polysaccharide and is extracted from cyamopsistetragonoloba [6]. It mainly consists of galactomannans with (1, 4)-linked-B-D-mannopyranose units [7]. It is a high molecular weight carbohydrate polymer which swells in cold water. Hence, it is widely used as a thickener emulsifier in food, cosmetics and pharmaceutical industries. Gum based materials modified by grafting crosslinking irradiation are attractive for agricultural applications as reviewed by Guiclherme *et al.* [8]. An improved moisture retention capacity for acrylic-acid-aniline grafted gum hydrogels has been synthesized by Sharma *et al.* [9]. Similar observations that have been made by Sharma *et al.* for grafted gum exhibited both water retention as well as dye removal properties [10]. Grafted gums with electrical conductivity applications have also been synthesized [11]. Similar usage of guar gum for sensor based applications has been developed by Dharela *et al.* [12]. Various researchers have also used gums for drug delivery applications such as for antibiotics. Interpenetrating networks for controlled release of amoxicillin trihydrateexibhited Fickian diffusion characteristics [2] [3]. Guar gum showed potential for tablet coating, carriers for drug release as reviewed by Prabaharan [13]. Thus, as seen in all these applications, modification via grafting leads to development of products which can be tailored to get tunable properties.

Microwave assisted grafting is a convenient approach with rapid reaction rates and higher yields. Singh *et al.* [14] [15] synthesized and characterized methylated starch and cassia marginata gum grafted acrylonitrile via microwave irradiation. Modified gum showed enhanced thermal stability as observed by Singh *et al.* [16]. Review articles by Singh *et al.* [17] and Sosnik *et al.* [18] highlight the importance of microwave irradiation for polysaccharide modification although upscaling to produce large quantities is still being developed. Similar studies on the modification of guar gum by grafting sodium acrylate and montmorillonite have been carried out by Wang *et al.* to study the swelling characteristics and pH sensitivity [19]. Recently studies on superabsorbent composites in combination with clay nanocomposites led to improvement in swelling and adsorption characteristics as observed by Kabiri *et al.* [20]. However, studies on grafted guar gum loaded with nanofiller superabsorbent nanocomposites for dye removal applications are very few. Yan *et al.* [21] used carbon nanotubes along with modified guar gum for the removal of neutral red and methylene blue dyes. Hybrid nanocomposites comprised of Xanthan gum loaded with nanosilica particles led to enhanced absorption ability for the removal of Cango red dye [22]. Similar observations have also been made for guar gum/nanosilica composite for the removal of reactive blue and Congo red dyes [20]. In this study, crystal violet dye has been chosen as it is hazardous in nature. Hence, in this study guar gum along with silane modified nanoclay has been grafted with acrylic acid via microwave irradiation technique. The various kinetic parameters have been optimized. The synthesized super absorbent nanocomposite has been investigated for crystal violet dye removal.

2. Experimental

2.1. Materials

Nanoclay (Montmorillonite clay surface modified with 15 - 35 wt% octadecylamine and 0.5 - 5 wt% aminopropyltriethoxysilane) was purchased from Sigma Aldrich, USA. Guar gum with molecular weight 4.22×10^6, Ammonium persulphate (Extrapure), NN'-Methylene bis-acrylamide has been procured from SD Fine-Chem Limited, Mumbai. Methyl Violet (10 B, practical grade) was obtained from Himedia Laboratories Pvt. Ltd. Mumbai. All the other solvents and chemicals of analytical grade were purchased from SD Fine-Chem Limited, Mumbai.

2.2. Synthesis of Control (0% NC) Sample

1 g of guar gum is dissolved in 100 ml of distilled water. 6 ml of acrylic acid is added to the guar gum solution

and stirred for 10 minutes. 0.09 g of ammonium persulphate (APS) initiator and 0.09 g of methylene bisacrylamide (MBA) cross linker was added to the mixture. Then it was subjected to microwave irradiation using locally fabricated Microwave reactor (Enerzi Microwave Systems, India) at different temperatures and varied reaction times. To carry out the reaction power of 800 W was supplied. The reaction mixture of acrylic acid grafted guar gum was cooled to room temperature. In order to remove unreacted acrylic acid and homopolymer, the mixture was neutralized using 1N NaOH till the pH reaches 7. This was followed by precipitation using acetone. The precipitate obtained was washed several times with acetone and kept for drying at 80°C for 5 - 6 hours.

2.3. Synthesis of Superabsorbent/Nanoclay Composites

The procedure for synthesis of superabsorbent with nanoclay is similar to that of control sample with sonication. Varied quantities of nanoclay was added to guar gum solution and sonicated for 30 minutes. After sonication, acrylic acid, MBA and APS were added. Similar process of microwave irradiation was carried out and the dried composite of superabsorbent with nanoclay was obtained. Amount of nanoclay was varied from 0 to 5 wt%.

2.4. Optimization of Reaction Parameters

The reaction parameters *i.e.*, monomer concentration, initiator dosage, cross-linker concentration, nanoclay concentration, time of reaction and temperature of the reaction has been optimized to obtain the maximum grafting percentage. The effects of these reaction parameters on the percentage swelling were investigated.

2.5. Swelling Kinetics

0.1 ± 0.01 g of powdered superabsorbent was taken in a test tube and 10 ml of prepared buffer solution was added. For every interval of 10 minutes the solution was decanted using filter paper. The weight of swollen sample was noted. The procedure was repeated till 120 minutes. Equilibrium swelling of sample was calculated using Equation (1).

$$\textbf{Equilibrium Swelling in } (g/g) = \frac{(W_s - W_d)}{W_d} \quad (1)$$

where, W_s = weight of swollen sample (g);
W_d = weight of dried sample (g).

2.6. Dye Removal Studies

Dye adsorption was carried out by immersing the 0.05 ± 0.01 g of grafted guar gum into 20 ml of dye solution. All adsorption experiments were examined through a batch method. To study the adsorption kinetics, at specified time intervals, the amount of adsorbed crystal violet dye was evaluated using a UV spectrometer at λ_{max}= 590 nm [23]. Adsorption capacity (q_t, mg/g) is calculated using following Equation 2 given below. Similar procedure was carried out for the composite with optimal concentration of nanoclay. Standard dye solution of crystal violet dye was prepared for five different concentrations namely 10 mg/l, 20 mg/l, 30 mg/l, 40 mg/l and 50 mg/l. The absorbance of dye solution was measured using UV spectrometer at a wavelength of 590 nm.

$$\textbf{Adsorption capacity, } q_t = \frac{(C_o - C_t)}{m} \times V \quad (2)$$

where,
C_o is the initial dye concentration (mg/l);
C_t is the remaining dye concentrations in the solution at time t (mg/l);
V is the volume of dye solution used (l);
m is the weight of superabsorbent (mg).

2.7. Fourier Transform Infrared Spectroscopy

The Fourier transform infrared spectroscopy (FTIR) analysis of pure guar gum, pure nanoclay and guar gum grafted acrylic acid were carried out using FTIR spectrophotometer (Perkin-Elmer spectrum 1000) between 300 and 4000 cm^{-1}. The samples were coated on a potassium bromide (KBr) plate and dried in a vacuum oven at 120°C before it was tested.

2.8. X-Ray Diffraction Studies

X-ray diffraction (XRD) measurements for the composites have been performed using advanced diffractometer (analytical, XPERT-PRO) equipped with Cu-Kα radiation source (X = 0.154 nm). The diffraction data were collected in the range of $2\theta = 3° - 60°$ using a fixed time mode with a step interval of 0.05°.

2.9. Thermogravimetric Analysis (TGA)

The TGA of guar gum and acrylic acid grafted guar gum nanocomposite were carried out by using Perkin-Elmer Pyris Diamond 6000 analyzer in an atmosphere of nitrogen. The sample was subjected to a heating rate of 10°C /min in a heating range of 20°C - 900°C.

2.10. Scanning Electron Microscope (SEM)

The morphological characterization of the specimen was carried out using a scanning electron microscope (SEM) (JEOL, JSM-840A microscope). The specimens were gold sputtered prior to microscopy.

2.11. Transmission Electron Microscopy (TEM)

Transmission electron microscopy (TEM) for nanocomposites has been performed using a JEOL, Model 782, operating at 200 kV. TEM specimens were prepared by dispersing the composite powders in methanol by ultra-sonication. A drop of the suspension was put on a TEM support grid (300 mesh copper grid coated with carbon). After drying in air, the composite powder remained attachedto the grid and was viewed under the transmission electron microscope.

3. Results and Discussion

3.1. Fourier Transform Infrared Spectroscopy

Figure 1 shows FTIR spectrograms of neat guar gum and acrylic acid guar gum. Neat guar gum (curve a) shows characteristic bands at 3430 and 2945 cm^{-1} for –O–H stretching vibration along with –C–H stretching. The bands at 1468 and 1098 cm^{-1} corresponds to –C–H and –O–H bending vibrations [24]. A new band at 1584 cm^{-1} can be seen for guar gum grafted acrylic acid (curve b) indicating formation of ester group and the presence of carboxylic anion [25].

3.2. Morphology

Figure 2(a) and **Figure 2(b)** show the SEM micrographs of grafted guar gum (without nanoclay) and guar gum-g-acrylic acid loaded with 1.75% nanoclay respectively. The grafted guar gum (without nanoclay) shows a smooth but dense surface. The nanocomposite exhibits a coarse dimpled surface indicating dispersion of nanoclay in the matrix. Similar observation has been made by Wang *et al.* [19]. Rough surfaces lead to enhancement of overall surface area and hence increased the adsorption sites [26] [27]. TEM images of 1.75% NC loaded guar gum are shown in **Figure 2(b)** and **Figure 2(c)**. TEM images show clusters of various sizes. This agglomeration has been caused during TEM grid preparation due to drying induced clustering [28] [29]. Hence, the presence of uniformly spread both large and small aggregates can be seen. Similar observations have been made by Liu *et al.* [27] and Giri *et al.* [28]. Due to intercalation of the nanoclay particles, the interlayer space increased due to which, the water absorbency improved [30]-[32].

3.3. XRD Analysis

XRD diffractograms for the nanocomposite is shown in **Figure 3**. In the case of guar gum grafted acrylic acid (without *i.e.* 0% nanoclay); the peaks appear at 2θ values of 5.2° and 19.77° respectively. The nanoclay used has crystalline peaks at 2θ values of 7.8°, 19.73°, 24.17°, 26.63° and 34.98°. For the nanocomposite loaded with 1.75% NC, the peaks both in the lower as well as higher range are not to be seen. This indicates that the silane modified

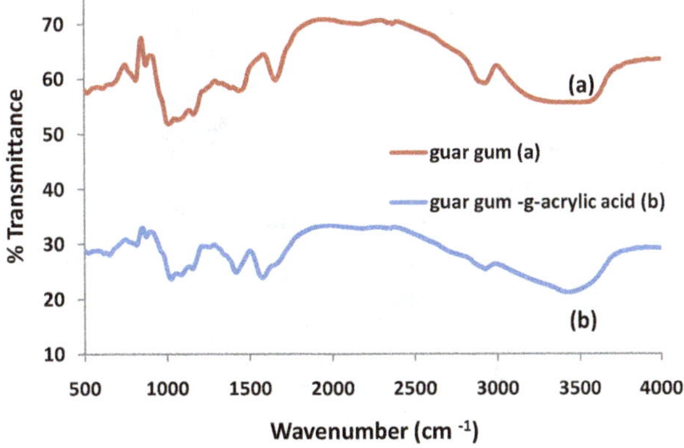

Figure 1. FTIR spectrograms of neat guar gum and acrylic acid guar gum.

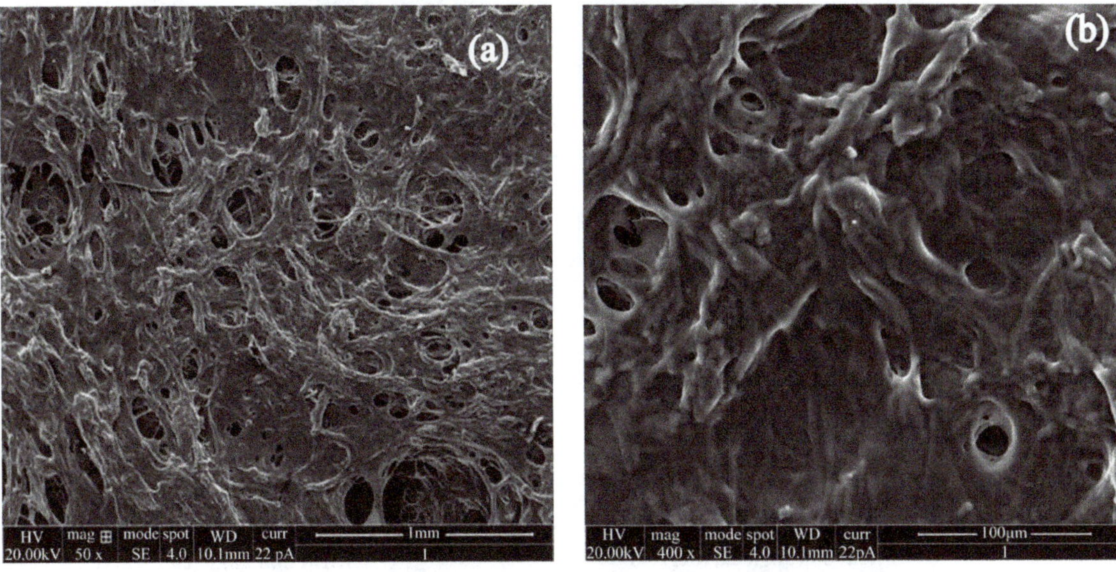

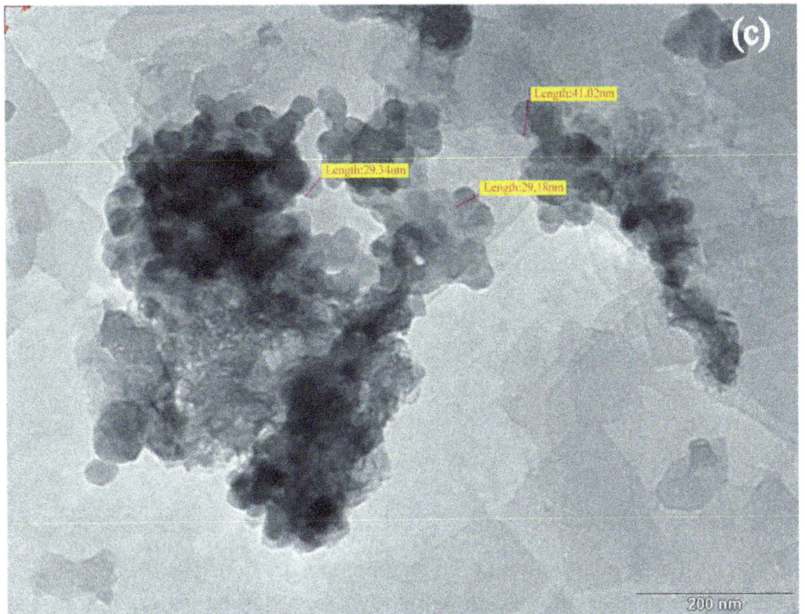

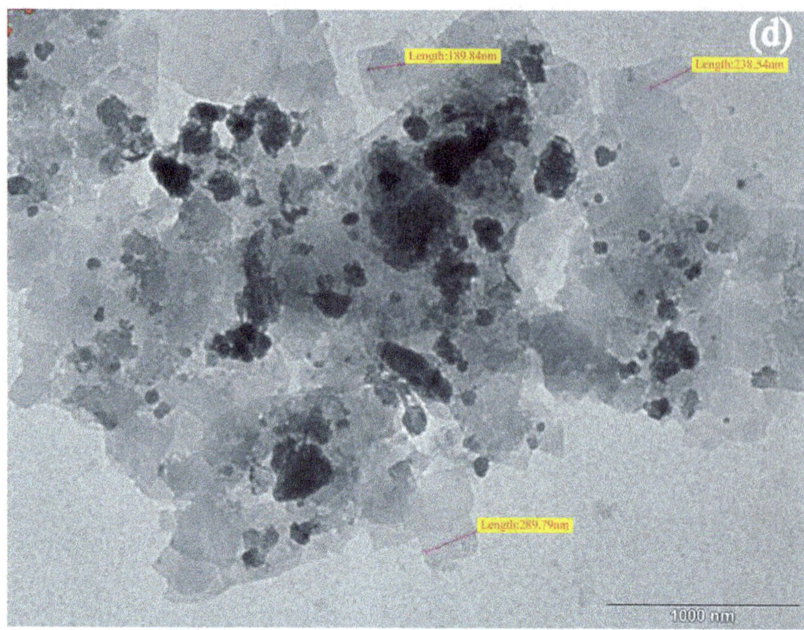

Figure 2. SEM micrographs of (a) without *i.e.* 0% NC and (b) with 1.75% NC and TEM images of composite loaded with 1.75% NC ((c) and (d)).

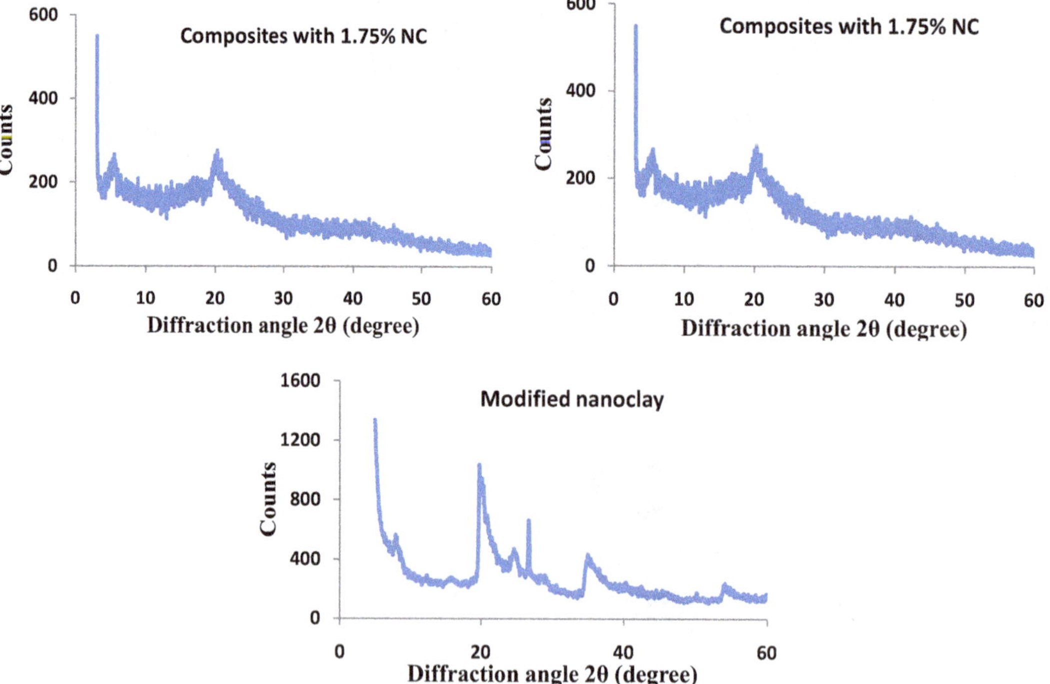

Figure 3. XRD diffractograms of nanocomposites.

nanoclay has exfoliated in grafted guar gum. The amine groups of silane modified nanoclay can interact effectively with the carboxyl and unsubstituted hydroxyl groups of grafted guar gum.

3.4. Thermogravimetric Analysis

The TGA thermograms of neat guar gum and the nanocomposite loaded with 1.75% NC is shown in **Figure 4**. The initial mass loss for neat guar gum is due to removal of absorbed water. The first zone of weight loss of

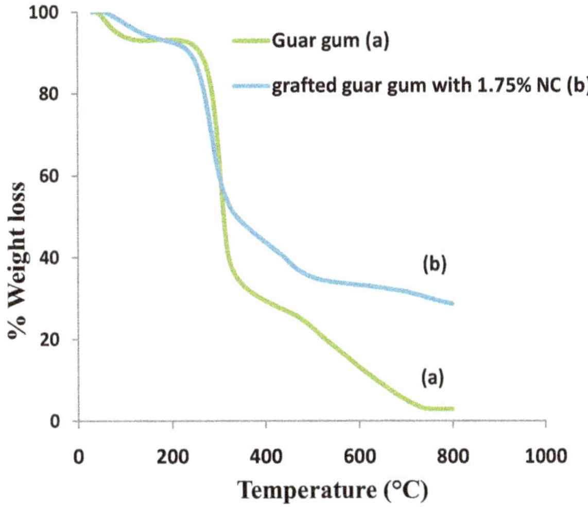

Figure 4. TGA thermograms of neat guar gum (a) and grafted guar gum loaded with 1.75% NC (b).

10% occurs at 260.3˚C due to the onset of decomposition of guar gum. The second major zone with 63% weight loss starts at 328.4˚C due to cleavage of guar gum backbone [20]. For the nanocomposite loaded with 1.75% NC, 63% weight loss occurs at 436˚C indicating enhanced thermal stability due to the presence of ceramic nanoparticles. The initial decrease in decomposition temperature for grafted guar gum is due to anhydride decomposition which forms due to dehydration of acrylic acid [33] [34].

3.5. Optimization of Reaction Parameters

The influence of monomer (acrylic acid) concentration has been studied by synthesizing series of composites with varied quantities of acrylic acid. Volume of acrylic acid has been varied from 2 ml to 8 ml, **Figure 5(a)** indicates that addition of 6 ml of acrylic acid gives maximum grafting percentage of 9%, and thereafter it reduces. Co-polymerization between acrylic acid and guar gum increases the grafting percentage thereafter it decreases due to formation of homopolymer of acrylic acid. Similar observation has been reported by Bardajee *et al.* [35] for salep grafted sodium acrylate/alumina based superabsorbent.

Initiator has greater influence on rate of polymerization as it involves in initiating free radicals. **Figure 5(b)** depicts that continuous increase in grafting percentage upto 25% is observed till an optimum weight of 90 mg which indicates that initiation of free radicals increases the polymerization which in turn increases the grafting percentage. At higher levels of initiator concentration, grafting is unfavorable and this behavior is attributed to the formation of excess number of free radical sites leading to shorter polymer chains. Yu *et al.* [36] analyzed that in case of carboxymethyl chitosan/vinyl monomer superabsorbent with Azobis (isobutyl amide hydrochloride) as initiator, the mean kinetic chain length decreased with increase in initiator concentration.

Concentration of crosslinker has considerable effects on water absorbency. In **Figure 5(c)** it is observed that highest Equilibrium swelling of 6.96 g/g is achieved at 90 mg of MBA, thereafter it is decreases. This behavior indicates that initially crosslinker forms a continuous three-dimensional network which facilitates water to penetrate into polymer network space. Higher crosslinker concentration leads to increased number of crosslink points and crosslink density, due to which network space for water retention is reduced. Similar observation has been reported by Wang *et al.* [37] in case of guar gum/attalpulgite based superabsorbent with methylene bisacrylamide as crosslinker.

Reaction time is the deciding factor for efficiency of reaction. Series of nanocomposites at time interval of 0 to 6 minutes has been synthesized. From **Figure 5(d)**, it can be seen that the maximum grafting percentage of 14% has been observed at reaction time of 4 minutes thereafter it decreases gradually. This behavior is attributed to increase of co-polymerization at initial intervals of reaction time which favors increase in grafting percentage. However, at higher reaction times (after 4 minutes) formation of short chain polymers is favorable which decreases the grafting percentage. Similar observations are reported in the study carried out by Likitha *et al.* [38].

Figure 5(e) shows the effect of temperature on grafting percentage. From figure it can be seen that as the

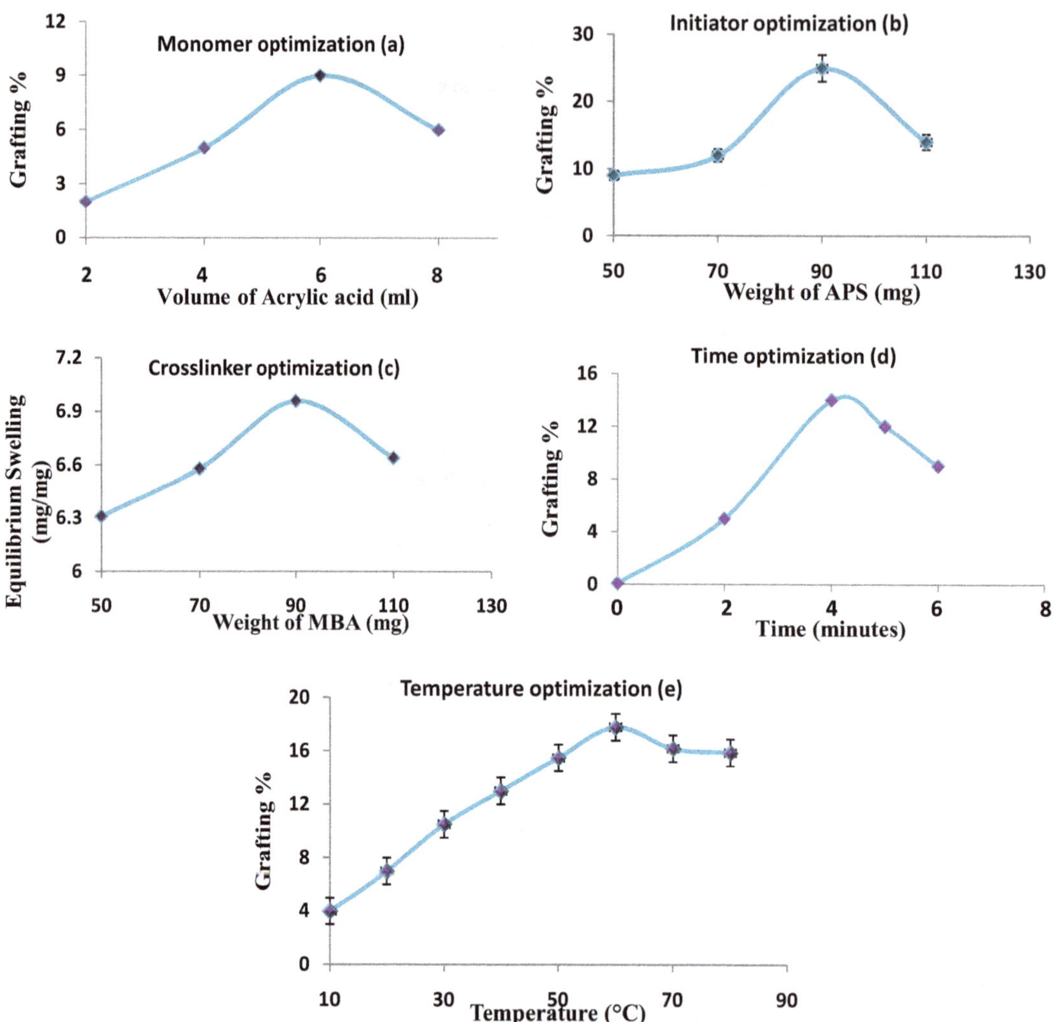

Figure 5. Variation of grafting percentage with (a) monomer concentration, (b) initiator concentration, (c) crosslinker concentration, (d) reaction time and (e) reaction temperature.

temperature increased grafting percentage also increased till 60°C then it has decreased slightly. This behavior indicates that a temperature of 60°C is desirable for radical formation thereafter excess radical's forms short chain polymers leading to decreased grafting percentage.

3.6. Effect of Nanoclay on Water Absorbency Characteristics (Optimization of Nanoclay Percentage)

The water absorbency increases with increased nanoclay content upto 1.75% (**Figure 6**). Further nanoclay loading led to a decrease in equilibrium swelling value. The initial increase in absorbency may be due to inhibition of entanglement of grafted polymer chains due to the addition of rigid nanoclay particles. Further, the hydrophilic groups *i.e.* –OH and –COOH weakened the hydrogen bonding interactions which allow penetration of water. At higher nanoclay contents, the voids would be filled and increase the hydrophobicity leading to reduction in swelling properties. Similar observations on polysaccharide based superabsorbent nanocomposites have been made by Zhang *et al.* [39] and Wang *et al.* [32].

3.7. Effect of PH Medium on Equilibrium Swelling Kinetics of Superabsorbent Nanocomposite

Figures 7(a)-(c) show the swelling behavior of guar gum-g-acrylic acid in various pH buffer solutions. The

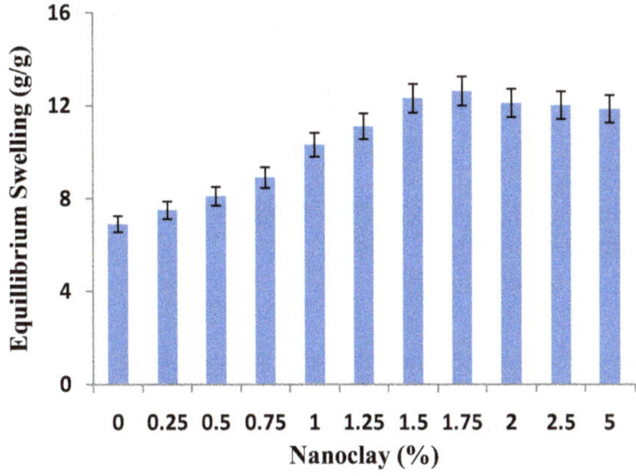

Figure 6. Effect of nanoclay on swelling characteristics.

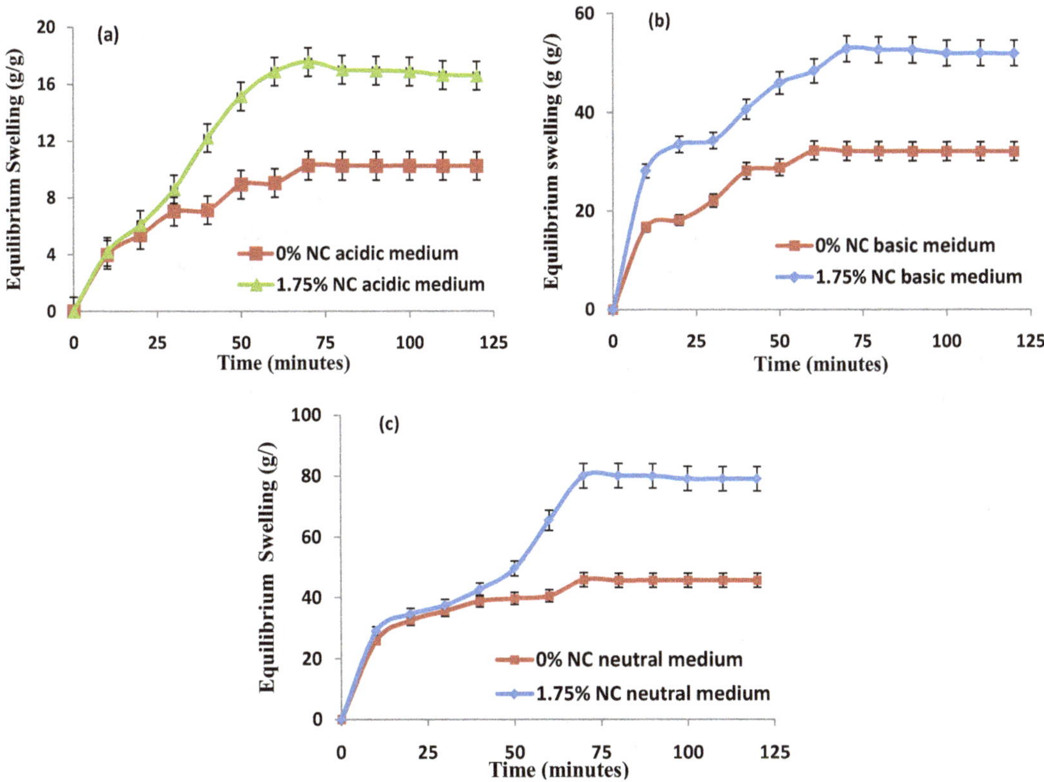

Figure 7. Swelling characteristics in (a) acidic (b) basic and (c) neutral medium.

swelling is lowest in acidic medium and highest in neutral medium. The hydrogen bonding interactions between –OH and –COOH led to increase in physical crosslinking and this in turn hindered swelling of the superabsorbent. As the pH increased beyond the acidic media range, the hydrogen bonding interactions are weakened leading to an increase in negatively charged –COO groups causing repulsion of polymer chains. Hence, this allows diffusion of water molecules in the network leading to an increase in swelling properties [40]. The addition of 1.75% NC showed a similar trend in various pH mediums although the magnitude is higher. The nanoclay enhanced the holding capacity of water due to increased surface area as reported by Wang *et al.* [37]

3.8. Swelling Kinetics

The swelling behavior with time is indicated in **Figures 7(a)-(c)**, where the rate of swelling is high till 70 minutes and thereafter equilibrium is attained. The swelling kinetics of composites in three different buffer solutions can be evaluated using Schott's second order swelling kinetic model.

$$\frac{t}{ES} = \frac{1}{Ki} + \left(\frac{1}{E}\right)t \quad (3)$$

where

E_S is the equilibrium swelling (g/g) at a given time t (minutes);
K_i is the initial swelling rate constant ($g{\cdot}g^{-1}{\cdot}min^{-1}$);
E_i is the theoretical equilibrium swelling (g/g).

The t/ES versus t plots of experimental data showed straight lines with linear correlation coefficient (R^2) > 0.95, indicating that Schott's swelling theoretical model is suitable for evaluating the kinetic swelling behavior of the composites. By fitting the experimental data with equation (3), the swelling kinetic parameters like the initial swelling rate constant (K_i) and the theoretical equilibrium water absorption (E), were calculated from the slope and intercept of the lines shown in **Figure 8** and are listed in **Table 1**. Value of Ki for neutral buffer > basic buffer > acidic buffer indicates that the order of swelling capability in neutral medium is higher than that of basic and acidic medium.

3.9. Dye Removal Studies

3.9.1. Effect of Nanoclay on Dye Removal

Nanoclay has the positive effect on adsorption of dye solutions. **Figure 9** indicates that for superabsorbent loaded with 1.75% NC shows a higher dye adsorption capacity (of 25.2% than the control sample) for an initial dye concentration of 30 mg/l. The reason for this behavior is similar as explained earlier, nanoclay has active –OH group on its surface which increases the porous surface area and thus dye adsorption capacity increases [41].

3.9.2. Effect of Contact Time on Adsorption Capacity

Contact time between superabsorbent composites and dye solutions is an important factor on adsorption capacity. **Figure 10** indicates variation of adsorption capacity with contact time for different dye concentrations (10 mg/l,

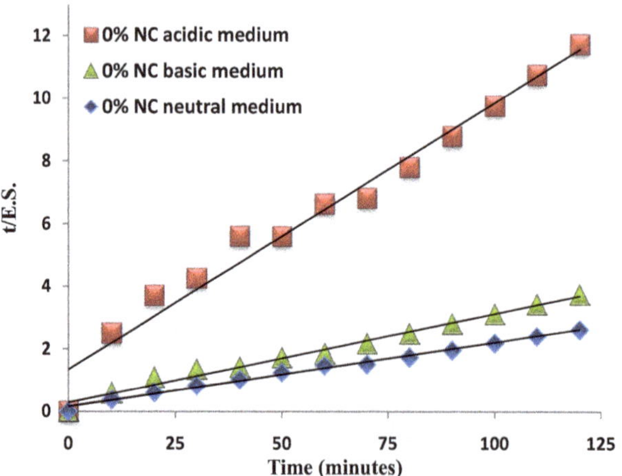

Figure 8. Schott's model for pseudo second order at 3 different pH mediums.

Table 1. Variation of swelling kinetic parameter of Schott's kinetic model in different buffer solutions.

Buffer solution	K_i	E	$<R^2>$
Acidic	0.741	11.764	0.972
Basic	3.508	35.714	0.984
Neutral	6.25	50	0.992

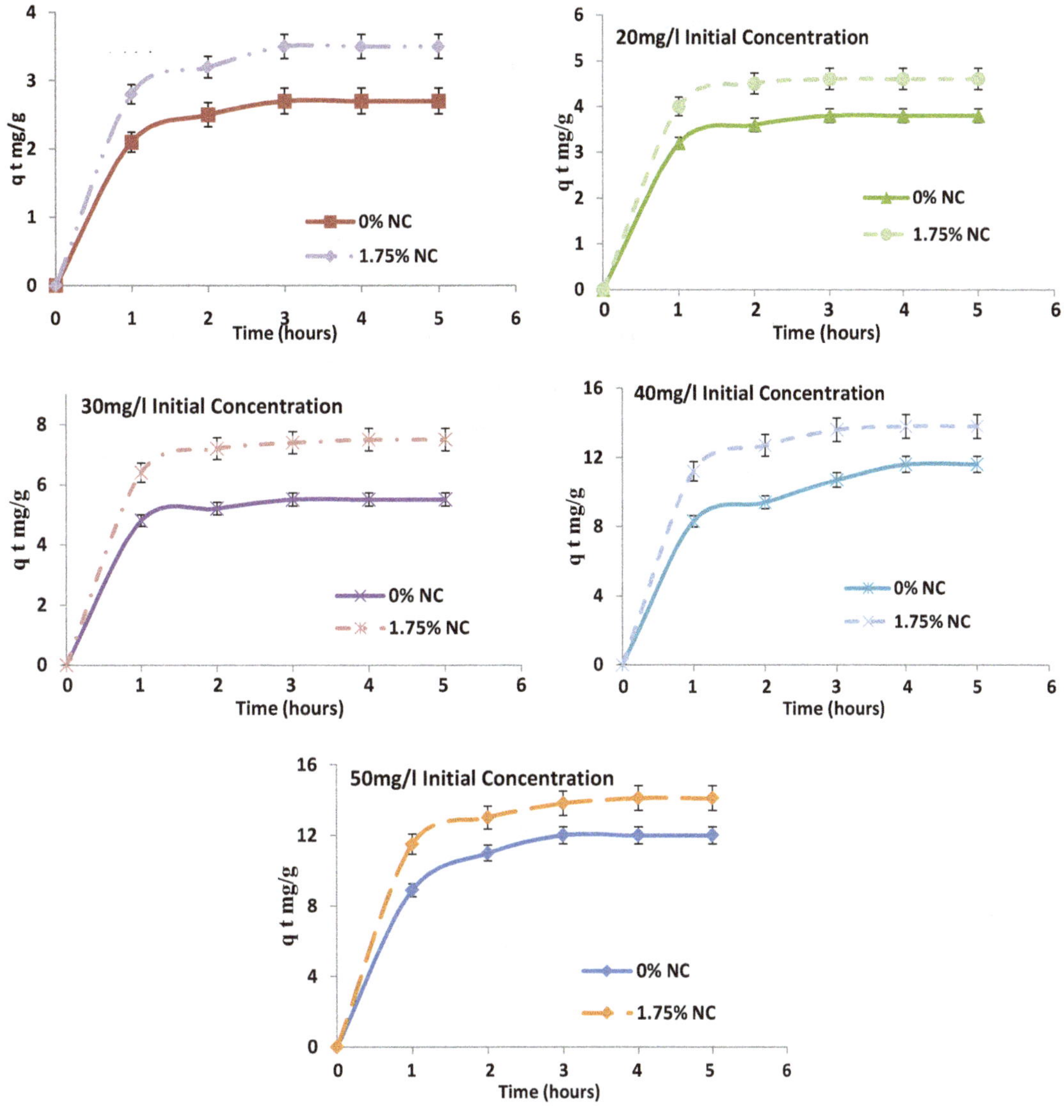

q t mg/g
Time (hours)
0% NC
1.75% NC
20mg/l Initial Concentration
q t mg/g
Time (hours)
0% NC
1.75% NC
30mg/l Initial Concentration
q t mg/g
Time (hours)
0% NC
1.75% NC
40mg/l Initial Concentration
q t mg/g
Time (hours)
0% NC
1.75% NC
50mg/l Initial Concentration
q t mg/g
Time (hours)
0% NC
1.75% NC

Figure 9. Effect on dye absorption capacity for guar gum based superabsorbent with and without nanoclay for varied initial concentration of crystal violet dye.

q t mg/g
Time (hours)
0% NC 10mg/l
0% NC 20mg/l
0% NC 30mg/l
0% NC 40mg/l
0% NC 50mg/l

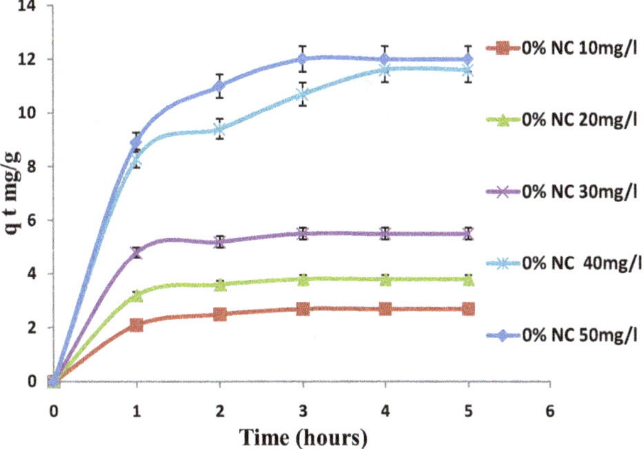

Figure 10. Variation of dye adsorption capacity of superabsorbent (without nanoclay) for different initial dye concentration w.r.t contact time.

Table 2. Variation of dye adsorption kinetic parameters using pseudo second order model.

Dye concentration (mg/l)	K_i (g mg^{-1} $hour^{-1}$)	q (mg/g)	<R^2>
10	18.52	2.78	0.996
20	40	3.87	0.998
30	66.66	5.62	0.998
40	41.66	12.19	0.987
50	71.43	12.50	0.994

20 mg/l, 30 mg/l, 40 mg/l and 50 mg/l). It is observed that dye adsorption capacity has increased till 3 hours of contact time, beyond which adsorption capacity reaches equilibrium. Also from the figure it can be observed that adsorption capacity will be higher at higher dye adsorption concentration. The reason for this behavior is adsorption capacity mainly depends on surface area of superabsorbent composites, at initial interval of time dye adsorption will be higher due to large surface area. Similar analyses have been reported by Wang *et al.* in case of methyl violet removal [41].

3.9.3. Dye Adsorption Kinetics

In order to investigate the dye adsorption mechanism pseudo second order model fits the best to experimental data like concentration of dye and adsorption capacity. The equation for pseudo-second order reaction is as follows,

$$\frac{t}{q_t} = \frac{1}{K_i} + \left(\frac{1}{q}\right)t \tag{4}$$

where

q_t is the dye adsorption capacity (mg/g) at a given time t (hours);
k_i is the initial dye adsorption rate constant (g·mg^{-1}·$hour^{-1}$);
q is the theoretical dye adsorption capacity (mg/g).

Plot of t/q_t vs t for different dye concentrations gives straight line with linear correlation (R^2) > 0.98, indicates that Schott's pseudo second order model fits best for the obtained experimental data. The values of initial dye adsorption rate constant (K_i) and theoretical dye adsorption capacity (q) can be determined using intercept and slope, and are tabulated in **Table 2**. Initial dye adsorption rate constant K_i values is in the order 50 mg/l > 40 mg/l > 30 mg/l > 20 mg/l > 10 mg/l which indicates that higher the initial dye concentration higher is the dye adsorption rate.

4. Conclusion

Superabsorbent nanocomposites based on acrylic acid grafted guar gum and silane modified nanoclay of varied quantitates have been synthesized via microwave irradiation. The reaction parameters for the grafting have been optimized. The addition of modified nanoclay enhanced swelling properties in various pH mediums ranging from acidic, neutral and alkaline buffers. The optimal nanoclay content has been found to be 1.75% w.r.t maximum swelling. Silane groups on nanoclay interacted effectively with the carboxylic and hydroxyl groups of grafted guar gum leading to exfoliation of nanoclay as seen in XRD diffractograms. The swelling kinetics was found to obey Schott's model with <R^2> values above 0.97. The crystal violet dye absorption increases with increases in initial dye concentration. The dye absorption was found to follow pseudo second order kinetics with <R^2> values above 0.99. Addition of 1.75% nanoclay enhanced the dye absorption capability as compared to the control sample without nanoclay. The synthesized grafted natural gum superabsorbent nanocomposites can be used for the removal of toxic dyes from various contaminated water bodies. It can also be used to improve water retention and effective use of water resources in agriculture.

References

[1] Tiwari, A. and Prabaharan, M. (2010) An Amphiphilic Nanocarrier Based on Guar Gum-Graft-Poly(epsilon-caprolactone) for Potential Drug-Delivery Applications. *Journal of Biomaterials Science*, **21**, 937-949.

http://dx.doi.org/10.1163/156856209X452278

[2] Sharma, K., Kumar, V., Kaith, B.S., Som, S., Kumar, V., Pandey, A., Kalia, S. and Swart, H.C. (2015) Synthesis of Biodegradable Gum Ghatti Based Poly(methacrylic acid-aniline) Conducting IPN Hydrogel for Controlled Release of Amoxicillin Trihydrate. *Industrial & Engineering Chemistry Research*, **54**, 1982-1991. http://dx.doi.org/10.1021/ie5044743

[3] Sharma, K., Kumar, V., Kaith, B.S., Som, S., Kumar, V., Pandey, A., Kalia, S. and Swart, H.C. (2015) Evaluation of a Conducting Interpenetrating Network Based on Gum Ghatti-G-Poly(acrylic acid-aniline) as a Colon-Specific Delivery System for Amoxicillin Trihydrate and Paracetamol. *New Journal of Chemistry*, **39**, 3021-3034. http://dx.doi.org/10.1039/C4NJ01982B

[4] Jassal, M. (2011) Design and Production Techniques for Hygiene Textiles. In: McCarthy, B.J., Ed., *Textiles for Hygiene and Infection Control*, Woodhead Publishing Ltd., Cambridge, 3-13.

[5] Benedetti, M. (2011) Biodegradable Hygiene Products, In: McCarthy, B.J., Ed., *Textiles for Hygiene and Infection Control*, Woodhead Publishing Ltd., Cambridge, 68-84.

[6] Singh, V.K., Banerjee, I., Agarwal, T., Pramanik, K., Bhattacharya, M.K. and Pal, K. (2014) Guar Gum and Sesame Oil Based Novel Bigels for Controlled Drug Delivery. *Colloids and Surfaces B: Biointerfaces*, **123**, 582-592. http://dx.doi.org/10.1016/j.colsurfb.2014.09.056

[7] Finley, J,W., Soto-Vaca, A., Heimbach, J., Rao, T., Juneja, L.R., Slavin, J. and Fahey, G.C. (2013) Safety Assessment and Caloric Value of Partially Hydrolyzed Guar Gum. *Journal of Agricultural and Food Chemistry*, **61**, 1756-1771. http://dx.doi.org/10.1021/jf304910k

[8] Guilherme, M.R., Aouada, F.A., Fajardo, A.R., Martins, A.F., Paulino, A.T., Davi, M.F.T., Rubira, A.F. and Muniz, E.C. (2015) Superabsorbent Hydrogels Based on Polysaccharides for Application in Agriculture as Soil Conditioner and Nutrient Carrier: A Review. *European Polymer Journal*, **72**, 365-385. http://dx.doi.org/10.1016/j.eurpolymj.2015.04.017

[9] Sharma, K., Kaith, B.S., Kumar, V., Kalia, S., Kumar, V., Som, S. and Swart, H.C. (2014) Gum Ghatti Based Novel Electrically Conductive Biomaterials: A Study of Conductivity and Surface Morphology. *eXPRESS Polymer Letters*, **8**, 267-281. http://dx.doi.org/10.3144/expresspolymlett.2014.30

[10] Sharma, K., Kaith, B.S., Kumar, V., Kalia, S., Kumar, V., Som, S. and Swart, H.C. (2015) Synthesis, Characterization and Water Retention Study of Biodegradable Gum Ghatti-Poly(acrylic acid-aniline) Hydrogels. *Polymer Degradation and Stability*, **111**, 20-31. http://dx.doi.org/10.1016/j.polymdegradstab.2014.10.012

[11] Sharma, K., Kaith, B.S., Kumar, V., Kalia, S., Kumar, V. and Swart, H.C. (2014) Water Retention and Dye Adsorption Behavior of Gg-cl-poly(acrylic acid-aniline) Based Conductive Hydrogels. *Geoderma*, **232-234**, 45-55. http://dx.doi.org/10.1016/j.geoderma.2014.04.035

[12] Dharela, R., Raj, L. and Chauhan, G.S. (2012) Synthesis, Characterization, and Swelling Studies of Guar Gum-Based pH, Temperature, and Salt Responsive Hydrogels. *Journal of Applied Polymer Science*, **126**, E259-E264. http://dx.doi.org/10.1002/app.36983

[13] Prabaharan, M. (2011) Prospective of Guar Gum and Its Derivatives as Controlled Drug Delivery Systems. *International Journal of Biological Macromolecules*, **49**, 117-124. http://dx.doi.org/10.1016/j.ijbiomac.2011.04.022

[14] Singh, V., Kumari, P.L., Tiwari, A. and Sharma, A.K. (2007) Alumina Supported Synthesis of *Cassia marginata* Gum-g-poly(acrylonitrile) under Microwave Irradiation. *Polymers for Advanced Technologies*, **18**, 379-385. http://dx.doi.org/10.1002/pat.899

[15] Singh, V. and Tiwari, A. (2008) Microwave-Accelerated Methylation of Starch. *Carbohydrate Research*, **43**, 151-154. http://dx.doi.org/10.1016/j.carres.2007.09.006

[16] Singh, V., Srivastava, A. and Tiwari, A. (2009) Structural Elucidation, Modification and Characterization of Seed Gum from *Cassia javahikai* Seeds: A Non-Traditional Source of Industrial Gums. *International Journal of Biological Macromolecules*, **45**, 293-297. http://dx.doi.org/10.1016/j.ijbiomac.2009.06.007

[17] Singha, V., Kumara, P. and Sanghi, R. (2012) Use of Microwave Irradiation in the Grafting Modification of Polysaccharides—A Review. *Progress in Polymer Science*, **37**, 340-364. http://dx.doi.org/10.1016/j.progpolymsci.2011.07.005

[18] Sosnika, A., Gotelli, G. and Abraham, G.A. (2011) Microwave-Assisted Polymer Synthesis (MAPS) as a Tool in Biomaterials Science: How New and How Powerful. *Progress in Polymer Science*, **36**, 1050-1078. http://dx.doi.org/10.1016/j.progpolymsci.2010.12.001

[19] Wang, W. and Wang, A. (2009) Synthesis and Swelling Properties of Guar Gum-g-poly(sodium acrylate)/Na-Montmorillonite Superabsorbent Nanocomposite. *Journal of Composite Materials*, **23**, 2805-2819. http://dx.doi.org/10.1177/0021998309345319

[20] Kabiri, K., Omidian, H., Zohuriaan-Mehr, M.J. and Doroudiani, S. (2011) Superabsorbent Hydrogel Composites and Nanocomposites: A Review. *Polymer Composites*, **32**, 277-289. http://dx.doi.org/10.1002/pc.21046

[21] Yan, L., Chang, P.R., Zheng, P. and Ma, X. (2012) Characterization of Magnetic Guar Gum-Grafted Carbon Nanotubes and the Adsorption of the Dyes. *Carbohydrate Polymers*, **87**, 1919-1924. http://dx.doi.org/10.1016/j.carbpol.2011.09.086

[22] Ghorai, S., Sarkar, A.K., Panda, A.B. and Pal, S. (2013) Effective Removal of Congo Red Dye from Aqueous Solution Using Modified Xanthan Gum/Silica Hybrid Nanocomposite as Adsorbent. *Bioresource Technology*, **144**, 485-491. http://dx.doi.org/10.1016/j.biortech.2013.06.108

[23] Mahdavinia, G.H. and Zhalebaghy, R. (2012) Removal Kinetic of Cationic Dye Using Poly(sodium acrylate)-Carrageenan/Na-Montmorillonite Nanocomposite Superabsorbents. *Journal of Materials and Environmental Science*, **3**, 895-906.

[24] Taunk, K. and Behari, K. (2006) Graft Copolymerization of Acrylic Acid onto Guar Gum. *Journal of Applied Polymer Science*, **77**, 39-44. http://dx.doi.org/10.1002/(SICI)1097-4628(20000705)77:1<39::AID-APP6>3.0.CO;2-Z

[25] Reddy, T.T., Reddy, N.S. and Tammishetti, S. (2003) Synthesis and Characterization of Guar Gum-Graft-Polyacrylonitrile. *Polymers for Advanced Technologies*, **14**, 663-668. http://dx.doi.org/10.1002/pat.372

[26] Kumar, A., Aerry, S., Saxena, A., De, A. and Mozumdar, S. (2012) Copper Nanoparticulates in Guar-Gum: A Recyclable Catalytic System for the Huisgen [3 + 2]-Cycloaddition of Azides and Alkynes without Additives under Ambient Conditions. *Green Chemistry*, **14**, 1298-1301. http://dx.doi.org/10.1039/c2gc35070j

[27] Liu, Y., Chen, H., Zhang, J.P. and Wang, A.Q. (2013) Effect of Number of Grindings of Attapulgite on Enhanced Swelling Properties of the Superabsorbent Nanocomposites. *Journal of Composite Materials*, **47**, 969-978. http://dx.doi.org/10.1177/0021998312443398

[28] Giri, A., Bhunia, T., Mishra, S.R., Goswami, L., Panda, A., Pal, S. and Bandyopadhyay, A. (2013) Acrylic Acid Grafted Guargum-Nanosilica Membranes for Transdermal Diclofenac Delivery. *Carbohydrate Polymers*, **91**, 492-501. http://dx.doi.org/10.1016/j.carbpol.2012.08.035

[29] Amuda, O.S., Olayiwola, A.O., Alade, A.O., Farombi, A.G. and Adebisi, S.A. (2014) Adsorption of Methylene Blue from Aqueous Solution Using Steam-Activated Carbon Produced from *Lantana camara* Stem. *Journal of Environmental Protection*, **5**, 1352-1363. http://dx.doi.org/10.4236/jep.2014.513129

[30] Bhanu, P. and Chauhan, S. (2011) Hybrid Nanomaterials. Wiley, Hoboken.

[31] Sharma, K., Kaith, B.S., Kumar, V., Kumar, V., Som, S., Kalia, S. and Swart, H.C. (2013) Synthesis and Properties of Poly(acrylamide-aniline)-Grafted Gum Ghatti Based Nanospikes. *RSC Advances*, **3**, 25830-25839. http://dx.doi.org/10.1039/c3ra44809f

[32] Wang, W., Zhang, J. and Wang, A. (2009) Preparation and Swelling Properties of Superabsorbent Nanocomposites Based on Natural Guar Gum and Organo-Vermiculite. *Applied Clay Science*, **46**, 21-26. http://dx.doi.org/10.1016/j.clay.2009.07.001

[33] Singha, A.S. and Rana, A.K. (2011) Kinetics Study on Acrylic Acid (AAc) Graft Copolymerized Cannabis Indica Fibre. *Iranian Polymer Journal*, **20**, 913-919.

[34] Kalia, S., Kumar, A. and Kaith, B.S. (2011) Sunn Hemp Cellulose Graft Copolymers Polyhydroxybutyrate Composites: Morphological and Mechanical Studies. *Advanced Materials Letters*, **2**, 17-25. http://dx.doi.org/10.5185/amlett.2010.6130

[35] Bardajee, G.R., Pourjavadi, A., Soleyman, R., Pourjavadi, A., Soleyman, R. and Ghavami, S. (2011) Salep-g-Poly(sodium acrylate)/Alumina as an Environmental-Sensitive Biopolymer Superabsorbent Composite: Synthesis and Investigation of Its Swelling Behavior. *Advances in Polymer Technology*, **31**, 41-51. http://dx.doi.org/10.1002/adv.20233

[36] Chen, Y., Liu, Y.-F., Tan, H.-M. and Jiang, J.-X. (2008) Synthesis and Characterization of a Novel Superabsorbent Polymer of *N,O*-Carboxymethyl Chitosan Graft Copolymerized with Vinyl Monomers. *Carbohydrate Polymers*, **72**, 287-292.

[37] Wang, Y., Wang, W., Shi, X. and Wang, A. (2013) A Superabsorbent Nanocomposite Based on Sodium Alginate and Illite/Smectite Mixed-Layer Clay. *Journal of Applied Polymer Science*, **130**, 161-167. http://dx.doi.org/10.1002/app.39141

[38] Likhitha, M., Sailaja, R.R.N., Priyambika, V.S. and Ravibabu, M.V. (2014) Microwave Assisted Synthesis of Guar Gum Grafted Sodium Acrylate/Cloisite Superabsorbent Nanocomposites: Reaction Parameters and Swelling Characteristics. *International Journal of Biological Macromolecules*, **65**, 500-508. http://dx.doi.org/10.1016/j.ijbiomac.2014.02.008

[39] Zhang, Y., Gu, Q., Yin, J., Wang, Z. and He, P. (2014) Effect of Organic Montmorillonite Type on the Swelling Behavior of Superabsorbent Nanocomposites. *Advances in Polymer Technology*, **33**, Article ID: 21400.

[40] Wang, W. and Wang, A. (2010) Preparation, Swelling and Water-Retention Properties of Crosslinked Superabsorbent Hydrogels Based on Guar Gum. *Advanced Materials Research*, **96**, 177-182.

http://dx.doi.org/10.4028/www.scientific.net/AMR.96.177

[41] Wang, Y., Zeng, L., Ren, X., Song, H. and Wang, A. (2010) Removal of Methyl Violet from Aqueous Solutions Using Poly(acrylic acid-co-acrylamide)/Attapulgite Composite. *Journal of Environmental Sciences*, **1**, 7-14. http://dx.doi.org/10.1016/S1001-0742(09)60068-1

17

Catalyst and Solvent-Free Microwave Assisted Expeditious Synthesis of 3-Indolyl-3-hydroxy Oxindoles and Unsymmetrical 3,3-Di(indolyl)indolin-2-ones

Prasanna K. Vuram[1], C. Kabilan[1], Anju Chadha[1,2*]

[1]Laboratory of Bioorganic Chemistry, Department of Biotechnology, Indian Institute of Technology Madras, Chennai, India
[2]National Centre for Catalysis Research, Indian Institute of Technology Madras, Chennai, India
Email: *anjuc@iitm.ac.in

Abstract

A simple and efficient method for the synthesis of 3-indolyl-3-hydroxy oxindoles and unsymmetrical 3,3-di(indolyl)indolin-2-ones using microwave irradiation without catalyst and solvent is described. A series of 3-indolyl-3-hydroxy oxindoles and unsymmetrical 3,3-di(indolyl)indolin-2-ones have been synthesized in very short reaction times of 5 and 10 minutes and in yields ranging from 31% to 98% and from 53% to 78% respectively. This method offers a significant advantage over the conventional methods in terms of simplicity and shorter reaction time. To the best of our knowledge compounds N-allyl-3-hydroxy-3-(1-methyl-indol-3-yl)indolin-2-one (6c), N-allyl-3-hydroxy-3-(5-methoxy-indol-3-yl)indolin-2-one (8c), N-benzyl-3-hydroxy-3-(1-methyl-indol-3-yl) indolin-2-one (10c), N-propargyl-3-hydroxy-3-(1-methyl-indol-3-yl)indolin-2-one (13c), N-propargyl-3-hydroxy-3-(5-methoxy-indol-3-yl)indolin-2-one (14c), 3-(5-methoxy-1*H*-indol-3-yl)-3-(1*H*-indol-3-yl)indolin-2-one (1e), 3-1-methyl(5-methoxy-1*H*-indol-3-yl)-3-(1*H*-indol-3-yl)indolin-2-one (2e), 3-1-allyl(5-methoxy-1*H*-indol-3-yl)-3-(1*H*-indol-3-yl)indolin-2-one (3e), 3-1-benzyl(5-methoxy-1*H*-indol-3-yl)-3-(1*H*-indol-3-yl)indolin-2-one (4e) and 3-1-(prop-2-ynyl)(5-methoxy-1*H*-indol-3-yl)-3(1*H*-indol-3-yl)indolin-2-one (5e) are reported here for the first time. All the compounds are characterized by IR, ^{1}H, ^{13}C NMR and HRMS.

*Corresponding author.

Keywords

Catalyst-Free, Isatin, Indole, Microwave, Solvent-Free

1. Introduction

Indoles and their derivative oxindoles are a privileged class of molecules in synthetic as well as biological chemistry. The hybrid molecules of indole and oxindole are also important due to their prevalent biological activities. For instance, the spermicidal activity of 3,3-bis(5-methoxy-1*H*-indol-3-yl)indolin-2-one and 3,3-bis(3-carboxymethyl-1*H*-indol-2-yl)indolin-2-one is higher than that of the standard spermicide "Nonoxynol-9" (N-9) [1]; di(indolyl)indolin-2-one derivatives showed strong and selective cytotoxicity against cancer cells [2]. These molecules are also known to possess antibacterial, anti-inflammatory and antiprotozoal activities [3]. In addition to these manifold biological applications, very recently 3-indolyl-3-hydroxy oxindoles (hybrid molecule of isatin and indole) were used as precursors in the synthesis of natural products (+)-gliocladin [4] and (+)-folicanthine [5] and in the synthesis of the heterocyclic analogue of BINAP [6]. The synthesis of 3-indolyl-3-hydroxy oxindoles involves a Friedel-Crafts type of electrophilic substitution between the electron-rich third position of the indole and electron deficient carbonyl group of the isatin [7]. This reaction usually results in the formation of symmetrical 3,3'-diindolyl oxindoles in a single step. Controlling the reaction at monosubstituted 3-indolyl-3-hydroxy oxindole stage is quite challenging.

The synthesis of symmetrical 3,3-di(indolyl)indolin-2-ones has been reported [8]-[22], while in the case of unsymmetrical 3,3-di(indolyl)indolin-2-ones, so far only three reports are available in the literature to the best of our knowledge. The synthesis involves a stepwise process. Wang and Ji reported an ultrasound irradiation method in the presence of ceric ammonium nitrate (CAN) for 1 to 5 h [23]; Moghadam and co-workers reported the reaction in the ionic-liquid N,N,N,N-tetramethylguanidinium trifluoroacetate (TMGT) for 1 h [24] and Nikpassand and co-workers reported it with a montmorillonite in 30 - 35 minutes [25].

In the case of 3-indolyl-3-hydroxy oxindoles, Kumar and co-workers reported the synthesis of various 3-indolyl-3-hydroxy oxindoles by supramolecular catalysis (β-Cyclodextrin) in 45 - 190 minutes [7]; Shanthi and co-workers reported a time of 60 - 120 minutes using K_2CO_3 as a catalyst [26]; Meshram and co-workers reported a time of 15 minutes in the presence of Triton B [27]; Hosseini and Tavakolian reported a reaction time of 2.5 h using ZnO nanorods in aqueous medium [28]. EtOH and water medium (60:40) in the presence of Lewis acid ($FeCl_3 \cdot 6H_2O$) and ultrasonic irradiation in 5 - 25 minutes was reported by Khorshidi and Tabatabaeian [29] while Makarem and co-workers reported an electrochemical method using EtOH/Propanol (60 - 240 minutes) [30]. Srihari and Murthy reported a heterogeneous catalyst (Kaolin/KOH) in the presence of MeOH as a solvent in 138 - 470 minutes [31]. Using various ionic-liquids, Moghadam and co-workers reported a reaction time of 10 - 20 minutes [24]. Jing Deng and coworkers reported an enantioselective version by using cupreine [32]. Nadine and coworkers also reported the reaction using chiral scandium(III) and indium(III) pybox complexes [33]. Recently, Pravathaneni Sai Prathima and co-workers reported the synthesis in aqueous medium using a base catalyst-diethanolamine [34]. The focus of our lab is to develop "green" methods for organic synthesis. In addition to using biocatalysts for organic transformations, we have now used microwave irradiation for organic synthesis as reported in the present study. Use of solvent-free and catalyst-free reaction conditions is an attractive proposition as seen in numerous microwave-assisted reactions. To the best of our knowledge, this is the first report for catalyst and solvent-free synthesis of 3-indolyl-3-hydroxy oxindoles and unsymmetrical 3,3-di(indolyl)indolin-2-ones under microwave irradiation in 5 and 10 minutes respectively.

2. Results and Discussion

Optimization of reaction conditions for the synthesis of 3-indolyl-3-hydroxy oxindoles, was carried out using a isatin (**1a**) [1 equivalent] and an indole (**1b**) [1.2 equivalents]. Using a microwave oven temperature at 100˚C and irradiation for 5 minutes, gave only the symmetrical 3,3-di(indolyl)indolin-2-ones as product (**Scheme 1**). Decreasing the time of irradiation to 3 minutes and then to 1 minute also resulted in the formation of the symmetrical 3,3-di(indolyl)indolin-2-one product and in 1 minute, 3-hydroxy-3-(1*H*-indol-3-yl)indolin-2-one (**1c**) and unreacted starting materials isatin (**1a**) and indole (**1b**), were also detected. Then we envisioned that 100˚C was not a suitable temperature for controlling the reaction at the mono substituted hydroxyl stage. The reaction was therefore carried out at reduced temperatures: at 90˚C and 80˚C for 5 minutes, a mixture of 3-hydroxy-3-

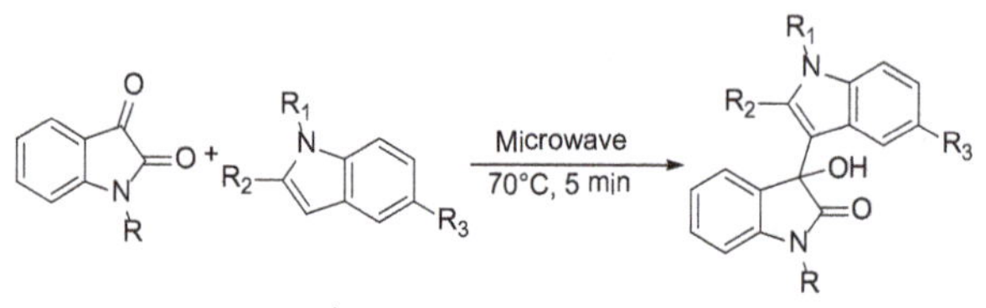

Microwave
70°C, 5 min

a b c

Scheme 1. Synthesis of various 3-indolyl-3-hydroxy oxindoles.

(1*H*-indol-3-yl)indolin-2-one (**1c**), symmetrical 3,3-di(indolyl)indolin-2-one in addition to unreacted starting material were observed. At 70°C for five minutes, only 3-hydroxy-3-(1*H*-indol-3-yl)indolin-2-one (**1c**) and unreacted starting materials were detected. But an increase in the reaction time from the sixth minute onwards, resulted in the formation of symmetrical 3,3-di(indolyl)indolin-2-one, even at 70°C. Based on this observation, 70°C and 5 minutes were optimized as the reaction conditions for synthesizing 3-hydroxy-3-(1*H*-indol-3-yl) indolin-2-one (**1c**), but the isolated yield was only 31%. Further increasing the amount of one of the starting materials *i.e.* three and five equivalents of indole (**1b**), showed no improvement in the yield of 3-hydroxy-3-(1*H*-indol-3-yl)indolin-2-one (**1c**). Also with N-allyl, and N-benzyl isatins, the yields of N-allyl-3-hydroxy-3-(1*H*-indol-3-yl)indolin-2-one (**5c**) and N-benzyl-3-hydroxy-3-(1*H*-indol-3-yl)indolin-2-one (**9c**) were only 34% and 40% respectively. Based on the nucleophilicity index of the indole ring [35], N-methyl indole, 2-methyl indole and 5-methoxy indole were selected [these are electron-rich at the third position]. In addition, 5-nitro indole which is electron deficient at the third position was also selected. The reaction between 5-methoxy indole and simple isatin, gave a yield of 64% for 5-Methoxy-3-hydroxy-3-(1*H*-indol-3-yl)indolin-2-one (**3c**). Interestingly excellent yields were obtained for products synthesized from the reactions between N-allyl, N-benzyl, N-propargyl and N-methyl isatins and 5-methoxy indole which is electron rich at the third position. Thus, N-allyl-3-hydroxy-3-(5-methoxy-indol-3-yl)indolin-2-one (**8c**), N-benzyl-3-hydroxy-3-(5-methoxy-indol-3-yl) indolin-2-one (**12c**), N-propargyl-3-hydroxy-3-(5-methoxy-indol-3-yl)indolin-2-one (**14c**) and N-methyl-3-hydroxy-3-(5-methoxy-indol-3-yl)indolin-2-one (**15c**) gave yields of 97%, 98%, 89% and 96% respectively. 76% yield was obtained in the reaction of 2-methyl indole with simple isatin to give 3-hydroxy-3-(2-methyl-1*H*-indole-3-yl)indolin-2-one (**2c**), while N-allyl isatin and N-benzyl isatin gave N-allyl-3-hydroxy-3-(2-methyl-indol-3-yl)indolin-2-one (**7c**) and N-benzyl-3-hydroxy-3-(2-methyl-indol-3-yl)indolin-2-one (**11c**) with 90% and 92% yields respectively. Moderate yields were obtained in the case of N-methyl indole with various isatins. The yields were 56% for N-allyl-3-hydroxy-3-(1-methyl-indol-3-yl)indolin-2-one (**6c**) 56% for N-benzyl-3-hydroxy-3-(1-methyl-indol-3-yl)indolin-2-one (**10c**) and 58% for N-propargyl-3-hydroxy-3-(1-methyl-indol-3-yl) indolin-2-one (**13c**). No product was obtained in the case of electron withdrawing nitro group present at fifth position of the indole ring with simple isatin even after a prolonged reaction time of 10 minutes and at a temperature of 100°C (**Table 1**).

The proposed reaction mechanism for the synthesis of 3-hydroxy-3-(1*H*-indol-3-yl)indolin-2-one (**1c**) is shown in **Scheme 2**.

Entry 8 (**Table 1**) was selected as a representative example for comparing reaction rates in various solvents such as acetonitrile (ACN), 1,4-dioxane, water (H_2O), ethanol (EtOH) and 1,2-dichloroethane (1,2-DCE). Moderate yield was observed only in the case of ACN (53%) in five minutes which improved to 96% in a reaction time of 15 minutes. Comparatively less yields were observed in the case of H_2O (46%), EtOH (46%), 1,2-DCE (36%) and 1,4-dioxane (15%).

Synthesis of Unsymmetrical 3,3-Di(indolyl)indolin-2-ones

Formation of symmetrical diindolyl was observed as a major product at 100°C in the synthesis of 3-hydroxy-3-(1*H*-indol-3-yl)indolin-2-one (**1c**). The same conditions were adopted for synthesizing unsymmetrical 3,3-di (indolyl)indolin-2-ones. For optimization, 5-methoxy-3-hydroxy-3-(1*H*-indol-3-yl)indolin-2-one (**1d**) [1 equivalent] and indole [1.2 equivalents] were selected. At a microwave oven temperature of 150°C and irradiation for 5 minutes, 53% product 3-(5-methoxy-1*H*-indol-3-yl)-3-(1*H*-indol-3-yl)indolin-2-one (**1e**) was formed. At a temperature higher than 150°C and a reaction time of 10 minutes, no significant increase in the yield was observed. When the same conditions were applied to the reaction [*i.e.* 150°C for 10 minutes] using various N-substituted isatins, good yields were obtained in all the cases (**Table 2**, **Scheme 3**).

All the compounds were characterized by IR, ^1H, ^{13}C NMR and HRMS. To the best of our knowledge com-

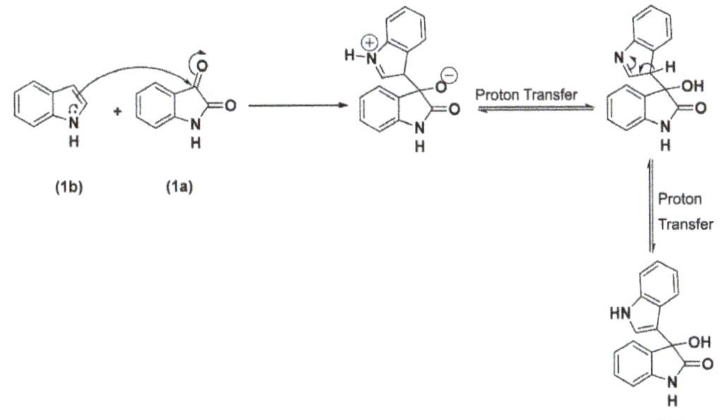

Scheme 2. Proposed mechanism for the formation of 3-indolyl-3-hydroxy oxindole.

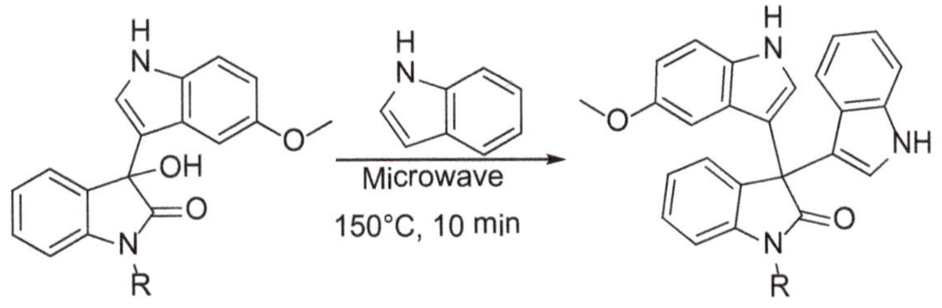

Scheme 3. Synthesis of various unsymmetrical 3,3-di(indolyl)indolin-2-ones.

Table 1. Synthesis of various 3-indolyl-3-hydroxy oxindoles.

SI. No.	Isatin **a**	Indole **b**	Product **c**	Yield %[a]
1	R = H	R_1 = H, R_2 = H, R_3= H	1c	31
2	R = H	R_1 = H, R_2 = Me, R_3 = H	2c	76
3	R = H	R_1 = H, R_2 = H, R_3 = OMe	3c	64
4	R = H	R_1 = H, R_2 = H, R_3 = NO_2	4c	00
5	R = allyl	R_1 = H, R_2 = H, R_3= H	5c	34
6	R = allyl	R_1 = Me, R_2 = H, R_3= H	6c	56
7	R = allyl	R_1 =H, R_2 = Me, R_3 =H	7c	90
8	R = allyl	R_1 = H, R_2 = H, R_3 = OMe	8c	97
9	R = benzyl	R_1 = H, R_2 = H, R_3= H	9c	40
10	R = benzyl	R_1 = H, R_2 = H, R_3= H	10c	56
11	R = benzyl	R_1 =H, R_2 = Me, R_3 =H	11c	92
12	R = benzyl	R_1 = H, R_2 = H, R_3 = OMe	12c	98
13	R = propargyl	R_1 = Me, R_2 = H, R_3= H	13c	58
14	R = propargyl	R_1 = H, R_2 = H, R_3 = OMe	14c	89
15	R = propargyl	R_1 = H, R_2 = H, R_3 = OMe	15c	96

[a] Isolated yield.

Table 2. Synthesis of various unsymmetrical 3,3-di(indolyl)indolin-2-ones.

SI.No.	Substrate **(R)d**	Product **e**	Yield %[a]
1	H	1e	53
2	methyl	2e	72
3	allyl	3e	78
4	benzyl	4e	77
5	propargyl	5e	77

[a] Isolated yield.

pounds N-Allyl-3-hydroxy-3-(1-methyl-indol-3-yl)indolin-2-one **(6c)**, N-Allyl-3-hydroxy-3-(5-methoxy-indol-3-yl)indolin-2-one (**8c**), N-Benzyl-3-hydroxy-3-(1-methyl-indol-3-yl)indolin-2-one (**10c**), N-Propargyl-3-hydroxy-3-(1-methyl-indol-3-yl)indolin-2-one (**13c**), N-propargyl-3-hydroxy-3-(5-methoxy-indol-3-yl)indolin-2-one (**14c**); 3-(5-methoxy-1*H*-indol-3-yl)-3-(1*H*-indol-3-yl)indolin-2-one(**1e**), 3-1-methyl(5-methoxy-1*H*-in-dol-3-yl)-3-(1*H*-indol-3-yl)indolin-2-one (**2e**), 3-1-allyl(5-methoxy-1*H*-indol-3-yl)-3-(1*H*-indol-3-yl)indolin-2-one (**3e**), 3-1-benzyl(5-methoxy-1*H*-indol-3-yl)-3-(1*H*-indol-3-yl)indolin-2-one (**4e**) and 3-1-(prop-2-ynyl)(5-methoxy-1*H*-indol-3-yl)-3-(1*H*-indol-3-yl)indolin-2-one (**5e**) are reported here for the first time.

3. Material and Methods

Chemicals were obtained from Spectrochem, and used without further purification. All known products were identified by comparison of their physical and spectral data with those of authentic samples. ^{1}H and ^{13}C NMR spectra were recorded on a Bruker AV-400 spectrometer operating at 400 and 100 MHz, respectively. The spectra were calibrated on the solvent residual peak (DMSO-d6: δ = 2.50 ppm for ^{1}H and δ = 39.52 ppm for ^{13}C; $CDCl_3$: δ = 7.26 ppm for ^{1}H and δ = 77.0 ppm for ^{13}C). Analytical thin layer chromatography (TLC) was performed on Kieselgel 60 F254 aluminum sheets (Merck 1.05554). Column chromatography was performed on silica gel (100 - 200 mesh). FT-IR spectra were recorded on a Nicolet-6700. High-resolution mass spectra were recorded with Thermo Scientific-Orbitrap Elite (Electro spray Ionization). Anton Paar microwave synthesizer was used at 600 rpm and temperature was kept constant at 70°C and 150°C.

3.1. General Procedure for the Synthesis of 3-Indolyl-3-hydroxy Oxindoles (1c to 15c)

A mixture of the isatin (0.6797 mmol, 100 mg) and indole (0.8156 mmol, 96 mg) was added to a microwave-oven reaction vial then irradiated for 5 minutes at 70°C. After completion of the reaction (as indicated by TLC) the residue was washed with 2 mL of DCM and mixed with silica gel and then evaporated the solvent under reduced pressure, and the mixture was purified by column chromatography using EtOAc in hexanes. All the known products have spectral and physical data consistent with those reported in literatures.

3.1.1. 3-Hydroxy-3-(1*H*-indol-3-yl)indolin-2-one (1c)

White solid; mp: 294 - 296 (lit[26]: 294°C - 296°C). ^{1}H NMR (400 MHz, DMSO-d_6): δ 10.98 (s, 1H), 10.34 (s, 1H), 7.35 (t, J = 9.6 Hz, 2 H), 7.27 - 7.24 (m, 2H), 7.09 (d, J = 2.4 Hz, 1H), 7.03 - 7.02 (m, 1H), 6.98 - 6.86 (m, 4H), 6.36 (s, 1H). ^{13}C NMR (100 MHz, DMSO-d_6): δ 178.5, 141.7, 136.8, 133.5, 129.1, 124.9, 124.8, 123.5, 121.7, 121.1, 120.3, 118.5, 115.5, 111.5, 109.6, 74.9. IR (KBr): υ 3264, 1711, 1617, 1549, 1467, 1424, 1335, 1226, 1184, 1107, 1071, 938, 910, 745, 686, 655 cm^{-1}. HRMS: Calcd for $C_{16}H_{12}O_2N_2$ $[M + Na]^+$ 287.0899, found 287.0795.

3.1.2. 3-Hydroxy-3-(2-methyl-1*H*-indol-3-yl)indolin-2-one (2c)

White solid; mp: 178°C - 180°C (lit[26]: 176°C - 178°C). ^{1}H NMR (400 MHz, DMSO-d_6): δ 10.86 (s, 1H), 10.33 (s, 1H), 7.25 - 7.17 (m, 3H), 6.96 - 6.88 (m, 4H), 6.74 - 6.70 (m, 1H), 6.26 (s, 1H), 2.34 (s, 3H). ^{13}C NMR (100 MHz, DMSO-d_6): δ 178.7, 141.6, 134.8, 134.1, 133.4, 129.0, 126.6, 124.9, 121.7, 119.8, 119.2, 118.2, 110.2, 109.6, 109.4, 75.8, 13.3. IR (KBr): υ 3348, 3229, 3034, 1693, 1611, 1487, 1454, 1431, 1374, 1345, 1291, 1239, 1211, 1163, 1024, 996, 954, 927, 893,853, 832, 737, 696 cm^{-1}. HRMS: Calcd for $C_{17}H_{14}O_2N_2$ $[M + Na]^+$ 301.0948; found 301.0953.

3.1.3. 5-Methoxy-3-hydroxy-3-(1*H*-Indol-3-yl)indolin-2-one (3c)

White solid; mp: 196°C - 198°C (lit[26]: 196°C - 198°C). ^{1}H NMR (400 MHz, DMSO-d_6): δ 10.84 (s, 1H), 10.33 (s, 1H), 7.28 - 7.23 (m, 3H), 7.02 (d, J = 2.8 Hz, 1H), 6.98 (t, J = 7.6 Hz, 1H), 6.92 (d, J = 7.26 Hz, 1H), 6.86 (d, J = 2 Hz, 6.71 (dd, J = 2.4 Hz, J = 8.8 Hz, 1 H), 6.34 (s, 1H), 5.76 (s, 1H), 3.62 (s, 3H). ^{13}C NMR (100 MHz, DMSO-d$_6$): δ 178.5, 152.7, 141.7, 133.4, 132.0, 129.0, 125.4, 124.8, 124.2, 121.7, 115.0, 112.0, 110.9, 109.6, 102.7, 75.0, 55.2. IR (KBr): υ 3318, 3166, 1701, 1614, 1581, 1465, 1433, 1344, 1304, 1235, 1207, 1173, 1039, 927, 837, 779, 752, 650 cm^{-1}. HRMS: Calcd for $C_{17}H_{14}O_3N_2$ $[M + Na]^+$ 317.0897; found 317.0902.

3.1.4. N-allyl-3-hydroxy-3-(1*H*-indol-3-yl)indolin-2-one (5c)

White solid; mp: 148°C - 150°C. ^{1}H NMR (400 MHz, DMSO-d_6): δ 11.01 (s, 1H), 7.35 - 7.33 (m, 4H), 7.06 - 7.04 (m, 4H), 6.89 - 6.86 (m, 1H), 6.49 (s, 1H), 5.89 - 5.82 (m, 1H), 5.24 - 5.16 (m, 2H), 4.38-4.27 (m, 2H). ^{13}C NMR (100 MHz, DMSO-d$_6$): δ 176.4, 142.2, 136.8, 132.7, 132.0, 129.0, 124.9, 124.5, 123.6, 122.3, 121.1, 120.4, 118.5, 117.0, 115.2, 111.5, 109.1, 74.6, 41.4. IR (KBr): υ 3219, 3056, 1688, 1603, 1485, 1459, 1435, 1365, 1332, 1294, 1241, 1223, 1182, 1104, 981, 921, 905, 826, 735, 682 cm^{-1}. HRMS: Calcd for $C_{19}H_{16}O_2N_2$ $[M + Na]^+$ 327.1109; found 327.1110.

3.1.5. N-allyl-3-hydroxy-3-(1-methyl-indol-3-yl)indolin-2-one (6c)

White solid; mp: 136°C - 138°C. ^{1}H NMR (400 MHz, CDCl$_3$): δ 7.67 (d, J = 8 Hz, 1H), 7.52 (d, J = 7.2 Hz, 1H), 7.33 - 7.19 (m, 3H), 7.09 - 7.05 (m, 2H), 6.97 (s, 1H), 6.90 (d, J = 7.6 Hz, 1H), 5.91 - 5.81 (m, 1H), 5.30 - 5.22 (m, 2H), 4.44 (dd, J = 4.8 Hz, J = 16 Hz, 1H), 4.28 (dd, J = 4.8 Hz, 16.4 Hz, 1H), 3.69 (s, 3H), 3.42 (m, 1H). ^{13}C NMR (100 MHz, CDCl$_3$): δ 176.9, 142.4, 137.7, 131.2, 129.6, 127.7, 125.4, 124.9, 123.1, 122.1, 120.8, 119.7, 117.8, 113.8, 109.5, 109.4, 75.5, 42.5, 32.8. IR (KBr): υ 3349, 3046, 1699, 1606, 1462, 1412, 1365, 1329, 1208, 1178, 1137, 1113, 1072, 984, 934, 888, 743, 669 cm^{-1}. HRMS: Calcd for $C_{20}H_{18}O_2N_2$ $[M + Na]^+$ 341.1260; found 341.1261.

3.1.6. N-allyl-3-hydroxy-3-(2-methyl-indol-3-yl)indolin-2-one (7c)

White solid; mp: 164°C - 166°C (lit[26]: 164°C - 166°C). ^{1}H NMR (400 MHz, DMSO-d_6): δ 10.91 (s, 1H), 7.32 (t, J = 7.6 Hz, 1H), 7.26 (d, J = 7.2 Hz, 1H), 7.19 (d, J = 8 Hz, 1H), 7.01 (t, J = 7.6 Hz, 2H), 6.92 - 6.85 (m, 2H), 6.71 (t, J = 7.2 Hz, 1H), 6.40 (s, 1H), 5.91 - 5.81 (m, 1H), 5.27 - 5.16 (m, 2H), 4.39 - 4.26 (m, 2H), 2.38 (s, 3H). ^{13}C NMR (100 MHz, DMSO-d$_6$): δ 176.6, 142.2, 134.8, 133.7, 133.3, 131.9, 129.0, 126.6, 124.7, 122.4, 119.8, 119.2, 118.2, 117.1, 110.3, 109.1, 75.5, 41.5, 13.3. IR (KBr): υ 3401, 1703, 1607, 1523, 1488, 1461, 1431, 1370, 1331, 1304, 1222, 1181, 982, 931, 895, 759, 673 cm^{-1}. HRMS: Calcd for $C_{20}H_{18}O_2N_2$ $[M + Na]^+$ 341.1261; found 341.1266.

3.1.7. N-allyl-3-hydroxy-3-(5-methoxy-indol-3-yl)indolin-2-one (8c)

White solid; mp: 102°C - 104°C. ^{1}H NMR (400 MHz, DMSO-d_6): δ 10.88 (s, 1H), 7.34 (brs, 2H), 7.24 (d, J = 8.8 Hz, 1H), 7.08 - 7.06 (m, 3H), 6.78 (s, 1H), 6.70 (d, J = 8.4 Hz, 1H), 6.48 (s, 1H), 5.89 - 5.82 (m, 1H), 5.25 (s, 1H), 5.18 (t, J = 10.8 Hz, 1H), 4.32 (dd, J = 14 Hz, J = 35.6 Hz, 2 H), 3.60 (s, 3 H). ^{13}C NMR (100 MHz, DMSO-d$_6$): δ 176.4, 152.8, 142.2, 132.6, 132.0, 131.9, 129.1, 125.3, 124.6, 124.3, 122.4, 117.0, 114.7, 112.1, 111.2, 109.0, 102.2, 74.6, 55.1, 41.5. IR (KBr): υ 3324, 1702, 1611, 1485, 1463, 1435, 1361, 1210, 1177, 1109, 1064, 990, 9224, 899, 840, 805, 752, 704, 674, 631 cm^{-1}. HRMS: Calcd for $C_{20}H_{18}O_3N_2$ $[M + Na]^+$ 357.1215; found 357.1202.

3.1.8. N-benzyl-3-hydroxy-3-(1*H*-indol-3-yl)indolin-2-one (9c)

White solid; mp: 120°C - 124°C (lit[26]: 120°C - 124°C). ^{1}H NMR (400 MHz, DMSO-d_6): δ 11.03 (s, 1H), 7.37 - 7.24 (m, 9H), 7.10 (s, 1H), 7.05 - 6.96 (m, 3H), 6.82 (t, J = 7.6 Hz, 1H), 6.57 (s, 1H), 4.92 (s, 2H). ^{13}C NMR (100 MHz, DMSO-d$_6$): δ 176.8, 142.1, 136.8, 136.4, 132.7, 129.0, 128.5, 127.4, 124.8, 124.5, 124.5, 123.6, 122.5, 121.1, 120.4, 118.4, 115.1, 111.5, 109.1, 74.7, 42.7. IR (KBr): υ 3296, 3029, 1697, 1607, 1488, 1458, 1428, 1342, 1238, 1212, 1166, 1067, 991, 901, 738, 694 cm^{-1}. HRMS: Calcd for $C_{23}H_{18}O_2N_2$ $[M + Na]^+$ 377.1266; found 377.1251.

3.1.9. N-benzyl-3-hydroxy-3-(1-methyl-indol-3-yl)indolin-2-one (10c)

White solid; mp: 126°C - 128°C. ^{1}H NMR (400 MHz, DMSO-d_6): δ 7.38 - 7.25 (m, 9H), 7.12 - 7.10 (m, 2H),

7.02 (t, $J = 7.6$ Hz, 1H), 6.97 (d, $J = 8$ Hz, 1H), 6.87 (t, $J = 7.6$ Hz, 1H), 6.54 (s, 1H), 4.91 (s, 2H), 3.73 (s, 3H). ^{13}C NMR (100 MHz, DMSO-d$_6$): δ 176.6, 142.1, 137.2, 136.3, 132.6, 129.0, 128.4, 127.9, 127.3, 125.1, 124.5, 122.4, 121.1, 120.6, 118.5, 114.3, 109.6, 109.1, 74.5, 42.7, 32.3. IR (KBr): υ 3302, 1693, 1610, 1549, 1486, 1464, 1373, 1342, 1211, 1164, 1112, 1077, 990, 928, 893, 855, 747, 702, 669, 628 cm^{-1}. HRMS: Calcd for $C_{24}H_{20}O_2N_2$ $[M + Na]^+$ 391.1422; found 391.1420.

3.1.10. N-benzyl-3-hydroxy-3-(2-methyl-indol-3-yl)indolin-2-one (11c)

White solid; mp: 96°C - 98°C. ^{1}H NMR (400 MHz, DMSO-d_6): δ 10.91 (s, 1H), 7.36 - 7.18 (m, 7H), 7.19 (d, $J =$ 8 Hz, 1H), 6.99 (t, $J = 7.2$ Hz, 2H), 6.92 - 6.88 (m, 1H), 6.80 (d, $J = 8$ Hz, 1H), 6.68 - 6.64 (m, 1H), 6.47 (s, 1H), 4.92 (ABq, $J = 16$ Hz, $J = 20$ Hz, 2H), 2.37 (s, 3H). ^{13}C NMR (100 MHz, DMSO-d$_6$): δ 177.0, 142.1, 136.3, 134.8, 133.7, 133.4, 129.0, 128.5, 127.5, 126.5, 124.8, 122.5, 119.8, 119.2, 118.1, 110.3, 109.1, 109.2, 109.1, 75.6, 42.7, 13.3. IR (KBr): υ 3314, 3288, 3092, 1700, 1610, 1551, 1530, 1486, 1461, 1430, 1350, 1302, 1237, 1173, 1074, 993, 917, 745, 698, 634 cm^{-1}. HRMS: Calcd for $C_{24}H_{20}O_2N_2$ $[M + Na]^+$ 391.1417; found 391.1418.

3.1.11. N-benzyl-3-hydroxy-3-(5-methoxy-indol-3-yl)indolin-2-one (12c)

White solid; mp: 204°C - 206°C. ^{1}H NMR (400 MHz, DMSO-d_6): δ 10.89 (s, 1H), 7.30 - 7.23 (m, 8H), 7.05 - 7.02 (m, 2H), 6.97 (d, $J = 7.6$ Hz, 1H), 6.70 - 6.67 (m, 2H), 6.55 (s, 1H), 4.91 (dd, $J = 15.6$ Hz, $J = 22$ Hz, 2H), 3.49 (s, 3 H). ^{13}C NMR (100 MHz, DMSO-d$_6$): δ 176.8, 152.7, 142.2, 136.4, 132.7, 131.9, 129.1, 128.5, 127.3, 125.2, 124.7, 124.3, 122.5, 114.7, 112.1, 111.2, 109.1, 102.2, 74.7, 54.9, 42.6. IR (KBr): υ 3354, 3283, 3047, 1702, 1611, 1582, 1527, 1486, 1461, 1350, 1306, 1211, 1174, 1065, 995, 931, 902, 854, 751, 698, 673 cm^{-1}. HRMS: Calcd for $C_{24}H_{20}O_3N_2$ $[M + Na]^+$ 407.1372; found 407.1357.

3.1.12. N-propargyl-3-hydroxy-3-(1-methyl-indol-3-yl)indolin-2-one (13c)

White solid; mp: 158°C - 162°C. ^{1}H NMR (400 MHz, DMSO-d_6): δ 7.41 - 7.33 (m, 4H), 7.19 (d, $J = 8$ Hz, 1H), 7.13 - 7.08 (m, 3H), 6.91 (t, $J = 7.2$ Hz, 1H), 6.60 (s, 1H), 4.56 (ABq, $J = 16.4$ Hz, $J = 33.2$ Hz, 2 H), 3.72 (s, 3 H), 3.31 (s, 1H). ^{13}C NMR (100 MHz, DMSO-d$_6$): δ 175.7, 141.2, 137.2, 132.5, 129.2, 127.9, 125.1, 124.6, 122.8, 121.3, 120.6, 118.7, 114.2, 109.7, 109.3, 78.0, 74.5, 32.4, 28.8. IR (KBr): υ 3322, 3254, 1711, 1609, 1535, 1463, 1426, 1369, 1335, 1244, 1213, 1170, 1110, 1071, 997, 934,892, 744, 671, 628. HRMS: Calcd for $C_{20}H_{16}O_2N_2$ $[M + Na]^+$ 339.1109; found 339.1097.

3.1.13. N-propargyl-3-hydroxy-3-(5-methoxy-indol-3-yl)indolin-2-one (14c)

White solid; mp: 181°C - 184°C. ^{1}H NMR (400 MHz, DMSO-d_6): δ 10.9 (s, 1H), 7.41 (t, $J = 7.6$ Hz, 1H), 7.33 (d, $J = 7.6$ Hz, 1H), 7.25 - 7.18 (m, 2H), 7.12 - 7.07 (m, 2H), 6.75 (s, 1H), 6.70 (d, $J = 8.4$ Hz, 1H), 6.55 (s, 1H), 4.57 (ABq, $J = 18$ Hz, $J = 45.2$ Hz, 2H). ^{13}C NMR (100 MHz, DMSO-d$_6$): δ175.8, 152.9, 141.3, 132.5, 131.9, 129.1, 125.1, 124.6, 124.3, 122.8, 114.5, 112.1, 111.3, 109.2, 102.0, 78.1, 74.7, 74.5, 55.2 28.8. IR (KBr): υ 3411, 3325, 3276, 3089, 3052, 1705, 1611, 1583, 1530, 1486, 1463, 1438, 1360, 1291, 1242, 1209, 1177, 1111, 1059, 991, 924, 894, 835, 752, 702, 667 cm^{-1}. HRMS: Calc. for $C_{20}H_{16}O_3N_2$ $[M + Na]^+$ Calcd 355.1059; found 355.1047.

3.1.14. N-methyl-3-hydroxy-3-(5-methoxy-indol-3-yl)indolin-2-one (15c)

White solid; mp: 102°C - 104°C. ^{1}H NMR (400 MHz, DMSO-d_6): δ 10.85 (s, 1H), 7.39 - 7.31 (m, 2H), 7.23 (d, $J = 8.8$ Hz, 1H), 7.09 - 7.06 (m, 2H), 7.01 (s, 1H), 6.81 (s, 1H), 6.70 (d, $J = 8.8$Hz, 1H), 6.40 (s, 1H), 3.62 (s, 3H), 3.16 (s, 3H). ^{13}C NMR (100 MHz, DMSO-d$_6$): δ176.6, 152.7, 143.1, 132.7, 131.9, 129.2, 125.3, 124.4, 124.2, 122.4, 114.7, 112.1, 111.0, 108.4, 102.3, 74.7, 55.1, 25.9. IR (KBr): υ 3292, 3060, 1700, 1611, 1466, 1346, 1301, 1212, 1172, 1066, 997, 934, 903, 848, 752, 695, 668, 635 cm^{-1}. HRMS: Calcd for $C_{18}H_{16}O_2N_2$ $[M + Na]^+$ 331.1053; found 315.1064.

3.2. General Procedure for the Synthesis of Unsymmetrical 3,3Di(indolyl)indolin-2-ones (1e to 5e)

A mixture of 3-indolyl-3-hydroxy oxindoles (0.3400 mmol, 100 mg) and indole (0.4100 mmol, 48 mg) was added to a microwave-oven reaction vial, and then irradiated for 10 minutes at 150°C. Then the residue as washed with 2 mL of EtOAc and mixed with silicagel and then evaporated the solvent under reduced pressure, and the mixture was purified by column chromatography using EtOAc in hexanes.

3.2.1. 3-(5-Methoxy-1*H*-indol-3-yl)-3-(1*H*-indol-3-yl)indolin-2-one (1e)

White solid; mp: 280°C - 282°C. ^{1}H NMR (400 MHz, DMSO-d_6): δ 10.95 (s, 1H), 10.79 (s, 1H), 10.59 (s, 1H), 7.35 (d, J = 8 Hz, 1H), 7.25 - 7.22 (m, 4H), 7.03 - 6.98 (m, 2H), 6.93 (t, J = 7.2 Hz, 1H), 6.89 - 6.81 (m, 3H), 6.70 - 6.67 (m, 2H), 3.51 (s, 3H). ^{13}C NMR (100 MHz, DMSO-d$_6$): δ 178.7, 152.4, 141.3, 136.9, 134.5, 132.1, 127.8, 126.0, 125.7, 125.0, 124.9, 124.3, 124.2, 121.4, 120.9, 120.7, 118.2, 112.0, 111.5, 110.4, 109.5, 103.3, 55.1, 52.5. IR (KBr): υ 3371, 1737, 1676, 1620, 1577, 1472, 1418, 1372, 1339, 1291, 1208, 1174, 1129, 1101, 1053, 1015, 958, 930, 889, 837, 796, 746, 679, 637 cm^{-1}. HRMS: Calcd for $C_{25}H_{19}O_2N_3$ $[M + Na]^+$ 416.1369; found 416.1383.

3.2.2. 3-1-Methyl(5-methoxy-1*H*-indol-3-yl)-3-(1*H*-indol-3-yl)indolin-2-one (2e)

White solid; mp: 280°C - 282°C. ^{1}H NMR (400 MHz, DMSO-d_6): δ 10.97 (s, 1H), 10.82 (s, 1H), 7.37 - 7.32 (m, 2H), 7.28 - 7.23 (m, 2H), 7.19 - 7.16 (m, 2H), 7.04 - 7.00 (m, 2H), 6.89 (d, J = 2.4 Hz, 1 H), 6.82 - 6.77 (m, 2H), 6.69 (dd, J = 2.4 Hz, J = 8.8 Hz, 1H), 6.58 (d, J = 2.4 Hz, 1H), 3.50 (s, 3H), 3.26 (s, 3H). ^{13}C NMR (100 MHz, DMSO-d$_6$): δ 177.0, 152.5, 142.8, 137.0, 133.7, 132.2, 128.1, 126.0, 125.7, 125.1, 124.7, 124.5, 122.3, 121.0, 120.7, 118.4, 113.9, 113.6, 112.2, 111.7, 110.6, 108.6, 103.0, 55.1, 52.2, 26.3. IR (KBr): υ 3396, 3345, 3314, 2924, 2854, 2363, 2337, 1734, 1692, 1608, 1462, 1423, 1350, 1292, 1212, 1172, 1123, 1086, 1038, 1016, 913, 891, 844, 795, 743, 692, 650 cm^{-1}. HRMS: Calcd for $C_{26}H_{21}O_2N_3$ $[M + Na]^+$ 430.1531; found 430.1538.

3.2.3. 3-1-Allyl(5-methoxy-1*H*-indol-3-yl)-3-(1*H*-indol-3-yl)indolin-2-one (3e)

White solid; mp: 266 - 268. ^{1}H NMR (400 MHz, DMSO-d_6): δ 10.99 (s, 1H), 10.83 (s, 1H), 7.36 (d, J = 8.4 Hz, 1 H), 7.33 - 7.29 (m, 2H), 7.24 (d, J = 8.4 Hz, 1H), 7.20 (d, J = 8 Hz, 1H), 7.11 (d, J = 7.6 Hz, 1H), 7.02 (t, J = 7.6 Hz, J = 14.8 Hz, 2H), 6.89 (d, J = 2.4 Hz, 1H), 6.82 (d, J = 2.8 Hz, 1H), 6.79 (d, J = 7.2 Hz, 1H), 6.68 (dd, J = 2.4 Hz, J = 8.4 Hz, 1H), 6.55 (d, J = 2.4 Hz, 1H), 5.95 - 5.86 (m, 1H), 5.21 (d, J = 1.6 Hz, 1H), 5.18 - 5.15 (m, 1H), 4.43 (d, J = 4.8 Hz, 2H), 3.48 (s, 3H). ^{13}C NMR (100 MHz, DMSO-d$_6$): δ 176.8, 152.6, 141.8, 137.0, 133.8, 132.2, 128.0, 126.0, 125.7, 125.2, 124.8, 124.4, 122.3, 121.1, 120.8, 118.4, 117.2, 114.0, 113.6, 112.3, 111.8, 110.8, 109.3, 102.9, 55.1, 52.2, 41.8. IR (KBr): υ 3754, 3708, 3345, 3010, 1736, 1669, 1606, 1540, 1478, 1456, 1358, 1293, 1208, 1171, 1093, 1018, 926, 841, 798, 744, 702, 632 cm^{-1}. HRMS: Calcd for $C_{28}H_{23}O_2N_3$ $[M + Na]^+$ 456.1688; found 456.1701.

3.2.4. 3-1-Benzyl(5-methoxy-1*H*-indol-3-yl)-3-(1*H*-Indol-3-yl)indolin-2-one (4e)

White solid; mp: 260°C - 262°C. ^{1}H NMR (400 MHz, DMSO-d_6): δ 11.00 (s, 1H), 10.9 (s, 1H), 7.38 - 7.37 (m, 3H), 7.32 - 7.25 (m, 6H), 7.15 - 7.09 (m, 2H), 7.04 - 6.98 (m, 2H), 6.93 (d, J = 2 Hz, 1H), 6.85 (d, J = 2.4 Hz, 1H), 6.75 (t, J = 7.6 Hz, J = 15.2 Hz, 1H), 6.69 (dd, J = 2 Hz, J = 8.8 Hz, 1H), 6.56 (d, J = 2 Hz, 1H), 5.02 (s, 2H), 3.42 (s, 3H).^{13}C NMR (100 MHz, DMSO-d$_6$): δ 177.2, 152.5, 141.8, 137.0, 133.8, 132.2, 128.6, 127.9, 127.5, 126.0, 125.6, 125.2, 124.9, 124.5, 122.4, 121.1, 120.7, 118.3, 113.9, 113.4, 112.2, 111.7, 110.8, 109.3, 103.0, 55.0, 52.3, 43.0. IR (KBr): υ 3347, 3025, 2850, 1699, 1676, 1607, 1535, 1483, 1458, 1358, 1291, 1209, 1172, 1133, 1099, 1016, 930, 847, 799, 745, 698, 63 cm^{-1}. HRMS: Calcd for $C_{32}H_{25}O_2N_3$ $[M + Na]^+$ 506.1839 found 506.1848.

3.2.5. 3-1-(Prop-2-ynyl)(5-methoxy-1*H*-indol-3-yl)-3-(1*H*-indol-3-yl)indolin-2-one (5e)

White solid; mp: 266°C - 268°C. ^{1}H NMR (400 MHz, DMSO-d_6): δ 10.99 (s, 1H), 10.85 (s, 1H), 7.38 - 7.36 (m, 2H), 7.27 - 7.25 (m, 4H), 7.08 - 7.01 (m, 2H), 6.87 (m, 1H), 6.81 - 6.79 (m, 2H), 6.71 - 6.69 (m, 1H), 6.63 (s, 1H), 4.65 (s, 2H), 3.51 (s, 3H), 3.31 (s, 1H). ^{13}C NMR (100 MHz, DMSO-d$_6$): δ 176.3, 152.6, 140.9, 137.0, 133.6, 132.2, 128.0, 125.9, 125.6, 125.3, 124.9, 124.6, 122.7, 121.2, 120.9, 118.4, 113.8, 113.2, 112.3, 111.7, 111.0, 109.4, 103.0, 78.2, 74.5, 55.2, 52.3, 29.0. IR (KBr): υ 3388, 3326, 3267, 1689, 1605, 1580, 1482, 1460, 1427, 1378, 1356, 1338, 1295, 1250, 1209, 1173, 1128, 1100, 1037, 1013, 932, 911, 888, 857, 800, 748, 703, 660 cm^{-1}. HRMS: Calcd for $C_{28}H_{21}O_2N_3$ $[M + Na]^+$ 454.1531; found 454.1538.

4. Conclusion

A simple and green alternative protocol for the synthesis 3-indolyl-3-hydroxy oxindoles in moderate to excellent yields is reported here. The unsymmetrical 3,3-di(indolyl)indolin-2-ones are also obtained in moderate to good yields under microwave irradiation. The highlights of the method are that no solvent, no catalysts are needed, and the reaction times are very short, *i.e.* five minutes for 3-indolyl-3-hydroxy oxindoles and ten minutes for unsymmetrical 3,3-di(indolyl)indolin-2-ones. Hence, this methodology can be conveniently used to synthesize the hybrid molecules of isatin and indoles in a short reaction time.

Acknowledgements

We thank the Department of Biotechnology and IIT-Madras for infrastructure.

References

[1] Paira, P., Hazra, A., Kumar, S., Paira, R., Sahu, K.B., Naskar, S., Saha, P., Mondal, S., Maity, A., Banerjee, S. and Mondal, N.B. (2009) Efficient Synthesis of 3,3-Diheteroaromatic Oxindole Analogues and Their *in Vitro* Evaluation for Spermicidal Potential. *Bioorganic & Medicinal Chemistry Letters*, **19**, 4786-4789. http://dx.doi.org/10.1016/j.bmcl.2009.06.049

[2] Subba Reddy, B.V., Rajeswari, N., Sarangapani, M., Prashanthi, Y., Ganji, R.J. and Addlagatta, A. (2012) Iodine-Catalyzed Condensation of Isatin with Indoles: A Facile Synthesis of Di(indolyl)indolin-2-ones and Evaluation of Their Cytotoxicity. *Bioorganic & Medicinal Chemistry Letters*, **22**, 2460-2463. http://dx.doi.org/10.1016/j.bmcl.2012.02.011

[3] Pajouhesh, H., Parson, R. and Popp, F.D. (1983) Potential Anticonvulsants VI: Condensation of Isatins with Cyclohexanone and Other Cyclic Ketones. *Journal of Pharmaceutical Sciences*, **72**, 318-321. http://dx.doi.org/10.1002/jps.2600720330

[4] DeLorbe, J.E., Jabri, S.Y., Mennen, S.M., Overman, L.E. and Zhang, F.L. (2011) Enantioselective Total Synthesis of (+)-Gliocladine C: Convergent Construction of Cyclotryptamine-Fused Polyoxopiperazines and a General Approach for Preparing Epidithiodioxopiperazines from Trioxopiperazine Precursors. *Journal of the American Chemical Society*, **133**, 6549-6552. http://dx.doi.org/10.1021/ja201789v

[5] Guo, C., Song, J., Huang, J.-Z., Chen, P.-H., Luo, S.-W. and Gong, L.-Z. (2012) Core-Structure-Oriented Asymmetric Organocatalytic Substitution of 3-Hydroxyoxindoles: Application in the Enantioselective Total Synthesis of (+)-Folicanthine. *Angewandte Chemie-International Edition*, **51**, 1046-1050. http://dx.doi.org/10.1002/anie.201107079

[6] Berens, U., Brown, J.M., Long, J. and Selke, R.D. (1996) Synthesis and Resolution of 2,2'-Bis-diphenylphosphino [3,3']biindolyl, a New Atropisomeric Ligand for Transition Metal Catalysis. *Tetrahedron*: *Asymmetry*, **7**, 285-292. http://dx.doi.org/10.1016/0957-4166(95)00447-5

[7] Kumar, V.P., Reddy, V.P., Sridhar, R., Srinivas, B., Narender, M. and Rao, K.R. (2008) Supramolecular Synthesis of 3-Indolyl-3-hydroxy Oxindoles under Neutral Conditions in Water. *Journal of Organic Chemistry*, **73**, 1646-1648. http://dx.doi.org/10.1021/jo702496s

[8] Jafarpour, M., Rezaeifard, A., Gazkar, S. and Danehchin, M. (2011) Catalytic Activity of a Zirconium (IV) Schiff Base Complex in Facile and Highly Efficient Synthesis of Indole Derivatives. *Transition Metal Chemistry*, **36**, 685-690. http://dx.doi.org/10.1007/s11243-011-9519-6

[9] Sarrafi, Y., Alimohammadi, K., Sadatshahabi, M. and Norozipoor, N. (2012) An Improved Catalytic Method for the Synthesis of 3,3-Di(indolyl)oxindoles Using Amberlyst 15 as a Heterogeneous and Reusable Catalyst in Water. *Monatshefte für Chemie*, **143**, 1519-1522. http://dx.doi.org/10.1007/s00706-012-0723-7

[10] Jafarpour, M., Rezaeifard, A. and Gorzin, G. (2011) Enhanced Catalytic Activity of Zr(IV) Complex with Simple Tetradentate Schiff Base Ligand in the Clean Synthesis of Indole Derivatives. *Inorganic Chemistry Communications*, **14**, 1732-1736. http://dx.doi.org/10.1016/j.inoche.2011.07.017

[11] Kamal, A., Srikanth, Y.V.V., Khan, M.N.A., Shaik, T.B. and Ashraf, M. (2010) Synthesis of 3,3-Diindolyl Oxyindoles Efficiently Catalysed by $FeCl_3$ and Their *in Vitro* Evaluation for Anticancer Activity. *Bioorganic & Medicinal Chemistry Letters*, **20**, 5229-5231. http://dx.doi.org/10.1016/j.bmcl.2010.06.152

[12] Azizian, J., Mohammadi, A.A., Karimi, N., Mohammadizadeh, M.R. and Karimi, A.R. (2006) Silica Sulfuric Acid a Novel and Heterogeneous Catalyst for the Synthesis of Some New Oxindole Derivatives. *Catalysis Communications*, **7**, 752-755. http://dx.doi.org/10.1016/j.catcom.2006.01.026

[13] Alinezhad, H., Haghighi, A.H. and Salehian, F. (2010) A Green Method for the Synthesis of Bis-Indolylmethanes and 3,3'-Indolyloxindole Derivatives Using Cellulose Sulfuric Acid under Solvent-Free Conditions. *Chinese Chemical Letters*, **21**, 183-186. http://dx.doi.org/10.1016/j.cclet.2009.09.001

[14] Saffar-Teluri, A. (2014) Boron Trifluoride Supported on Nano-SiO_2: An Efficient and Reusable Heterogeneous Catalyst for the Synthesis of Bis(indolyl)methanes and Oxindole Derivatives. *Research on Chemical Intermediates*, **40**, 1061-1067. http://dx.doi.org/10.1007/s11164-013-1021-7

[15] Sarrafi, Y., Alimohammadi, K., Sadatshahabi, M. and Norozipoor, N. (2012) An Improved Catalytic Method for the Synthesis of 3,3-Di(indolyl)oxindoles Using Amberlyst 15 as a Heterogeneous and Reusable Catalyst in Water. *Monatshefte für Chemie—Chemical Monthly*, **143**, 1519-1522. http://dx.doi.org/10.1007/s00706-012-0723-7

[16] Karimil, N., Oskooil, H., Heravi, M., Saeedi, M., Zakeri, M. and Tavakoli, N. (2011) On Water: Bronsted Acidic Ionic Liquid [$(CH_2)_4SO_3$HMIM][HSO_4] Catalysed Synthesis of Oxindoles Derivatives. *Chinese Journal of Chemistry*, **29**, 321-323. http://dx.doi.org/10.1002/cjoc.201190085

[17] Azizian, J., Mohammadi, A.A., Karimi, A.R. and Mohammadizadeh, M.R. (2004) $KAl(SO_4)_2 \cdot 12H_2O$ as a Recyclable Lewis Acid Catalyst for Synthesis of Some New Oxindoles in Aqueous Media. *Journal of Chemical Research*, **2004**, 424-426. http://dx.doi.org/10.3184/0308234041423600

[18] Yadav, J.S., SubbaReddy, B.V., Uma, G.K., Meraj, S. and Prasad, A.R. (2006) Bismuth (III) Triflate Catalyzed Condensation of Isatin with Indoles and Pyrroles: A Facile Synthesis of 3,3-Diindolyl- and 3,3-Dipyrrolyl Oxindoles. *Synthesis*, **2006**, 4121-4123. http://dx.doi.org/10.1055/s-2006-950373

[19] Feng, G.L., Geng, L.J. and Zhang, H.L. (2009) Facile Synthesis of 3,3-Di(indolyl)indolin-2-one Derivatives Catalyzed by $ZrO_2/S_2O_8^{2-}$ Solid Superacid under Grinding Condition. *Chemical Journal on Internet*, **11**, Article ID: 111001pe.

[20] Chakrabarty, M., Sarkar, S. and Harigaya, Y. (2005) A Facile Clay-Mediated Synthesis of 3,3-Diindolyl-2-indolinones from Isatins. *Journal of Chemical Research*, **8**, 540-542. http://dx.doi.org/10.3184/030823405774663264

[21] Deb, M.L. and Bhuyan, P.J. (2009) Water-Promoted Synthesis of 3,3'-Di(indolyl)oxindoles. *Synthetic Communications*, **39**, 2240-2243. http://dx.doi.org/10.1080/00397910802654690

[22] Praveen, C., Ayyanar, A. and Perumal, P.T. (2011) Practical Synthesis, Anticonvulsant, and Antimicrobial Activity of N-Allyl and N-Propargyl Di(indolyl)indolin-2-ones. *Bioorganic & Medicinal Chemistry Letters*, **21**, 4072-4077. http://dx.doi.org/10.1016/j.bmcl.2011.04.117

[23] Wang, S.Y. and Ji, S.J. (2006) Facile Synthesis of 3,3-Di(heteroaryl)indolin-2-one Derivatives Catalyzed by Ceric Ammonium Nitrate (CAN) under Ultrasound Irradiation. *Tetrahedron*, **62**, 1527-1535. http://dx.doi.org/10.1016/j.tet.2005.11.011

[24] Moghadam, K.R., Kiasaraie, M.S. and Amlashi, H.T. (2010) Synthesis of Symmetrical and Unsymmetrical 3,3-Di(indolyl)-indolin-2-ones under Controlled Catalysis of Ionic Liquids. *Tetrahedron*, **66**, 2316-2321. http://dx.doi.org/10.1016/j.tet.2010.02.017

[25] Nikpassand, M., Mamaghani, M., Tabatabaeian, K. and Samimi, H.A. (2010) An Efficient and Clean Synthesis of Symmetrical and Unsymmetrical 3,3-Di(indolyl)indolin-2-ones Using KSF. *Synthetic Communications*, **40**, 3552-3560. http://dx.doi.org/10.1080/00397910903457399

[26] Shanthi, G., Lakshmi, N.V. and Perumal, P.T. (2009) A Simple and Eco-Friendly Synthesis of 3-Indolyl-3-hydroxy Oxindoles and 11-Indolyl-11H-indeno[1,2-b]quinoxalin-11-ols in Aqueous Media. *ARKIVOC*, **2009**, 121-130. http://dx.doi.org/10.3998/ark.5550190.0010.a12

[27] Meshram, H.M., Kumar, D.A., Goud, P.R. and Reddy, B.C. (2010) ChemInform Abstract: Triton B Assisted, Efficient, and Convenient Synthesis of 3-Indolyl-3-hydroxy Oxindoles in Aqueous Medium. *Synthetic Communications*, **40**, 39-45. http://dx.doi.org/10.1002/chin.201025093

[28] Hosseini-Sarvari, M. and Tavakolian, M. (2012) Preparation, Characterization, and Catalysis Application of Nano-Rods Zinc Oxide in the Synthesis of 3-Indolyl-3-hydroxy Oxindoles in Water. *Applied Catalysis A: General*, **441-442**, 65-71. http://dx.doi.org/10.1016/j.apcata.2012.07.009

[29] Khorshidi, A. and Tabatabaeian, K.J. (2011) An Ultrasound-Promoted Green Approach for the Synthesis of 3-(Indol-3-yl)-3-hydroxyindolin-2-ones Catalyzed by Fe(III). *Journal of the Serbian Chemical Society*, **76**, 1347-1353. http://dx.doi.org/10.2298/JSC110420120K

[30] Makarem, S., Fakhari, A.R. and Mohammadi, A.A. (2012) Electro-Organic Synthesis of Nanosized Particles of 3-Hydroxy-3-(1H-indol-3-yl)indolin-2-one Derivatives. *Monatshefte für Chemie—Chemical Monthly*, **143**, 1157-1160. http://dx.doi.org/10.1007/s00706-011-0693-1

[31] Srihari, G. and Murthy, M.M. (2011) Kaolin/KOH Is an Efficient Heterogeneous Catalyst for the Synthesis of 3-Hydroxy-3-indolyl Oxindoles. *Synthetic Communications*, **41**, 2684-2692. http://dx.doi.org/10.1080/00397911.2010.515342

[32] Deng, J., Zhang, S., Ding, P., Jiang, H., Wang, W. and Li, J. (2010) Facile Creation of 3-Indolyl-3-hydroxy-2-oxindoles by an Organocatalytic Enantioselective Friedel-Crafts Reaction of Indoles with Isatins. *Advanced Synthesis & Catalysis*, **352**, 833-838. http://dx.doi.org/10.1002/adsc.200900851

[33] Hanhan, N.V., Sahin, A.H., Chang, T.W., Fettinger, J.C. and Franz, A.K. (2010) Catalytic Asymmetric Synthesis of Substituted 3-Hydroxy-2-oxindoles. *Angewandte Chemie International Edition*, **49**, 744-747. http://dx.doi.org/10.1002/anie.200904393

[34] Prathima, P.S., Rajesh, P., Rao, J.V., Kailash, U.S., Sridhar, B. and Rao, M.M. (2014) "On Water" Expedient Synthesis of 3-Indolyl-3-hydroxy Oxindole Derivatives and Their Anticancer Activity *in Vitro*. *European Journal of Medicinal Chemistry*, **84**, 155-159. http://dx.doi.org/10.1016/j.ejmech.2014.07.004

[35] Lakhdar, S., Westermaier, M., Terrier, F., Goumont, R., Boubaker, T., Ofial, A.R. and Mayr, H. (2006) Nucleophilic Reactivities of Indoles. *The Journal of Organic Chemistry*, **71**, 9088-9095. http://dx.doi.org/10.1021/jo0614339

18

Microwave/Thermal Analyses for Human Bone Characterization

Vinay Kumar Suryadevara[1], Suyog Patil[1], James Rizkalla[2], Ahdy Helmy[2], Paul Salama[1], Maher Rizkalla[1,3]

[1]Department of Electrical and Computer Engineering, Indiana University-Purdue University Indianapolis, Indianapolis, USA
[2]Indiana University School of Medicine, Indianapolis, USA
[3]Integrated Nanosystems Development Institute (INDI), Indiana University-Purdue University Indianapolis, Indianapolis, USA
Email: mrizkall@iupui.edu

Abstract

A novel imaging approach utilizing microwave scattering was proposed in order to analyze various properties of bone. Microwave frequencies of 900 MHz, 1 GHz, and 2.4 GHz were used during this study. This investigation's objectives were to emphasize characteristics of abnormalities in human bones and to detect fine fractures through contrasts in bone density. The finite element method (FEM) presented here is generated from COMSOL software at different frequencies. The study identified the optimum transmission directed at the interface layers from an external microwave source. It was found that approximately 900 MHz microwave power was ideal for this application. This can be attributed to the penetration depth where the power dissipation is analyzed based on bone condition. The microwave energy was generated from an exterior antenna that was interfaced, via catheter, to skeletal bone. The power transmitted to bone was converted into thermal energy, and has led to a visible temperature distribution pattern, which reflects the bone density level, and accordingly, the type of bone under investigation. The electrical and thermal properties, including the dielectric permittivity, thermal conductivity, and heat flux absorption through the bone substance, have great implications on the FEM distribution. The boundary conditions using tangential matching of field components at the tissue-bone interface were incorporated into the finite element method. The average power from the electromagnetic fields (estimated from the Poynting's vector, $P = E*H$), was assumed to be fully absorbed as heat due to the conductivity of the bone material. Furthermore, microwave energy was applied as a delta function and the thermal distributions have been analyzed in order to distinguish between normal healthy bone and bones with structural or metabolic abnormalities. The latter was emulated by different bone density to contrast normal bone anatomy. The FEM simulation suggests that thermography microwave imaging could be a good tool for bone characterization in order to detect skeletal abnormalities. This approach could be advantageous over other existing methods such as X-ray imaging.

Keywords

Microwave, Human Bones, Simulation, FEM, COMSOL, Thermal, Diagnosis

1. Introduction

The health needs of patients, across the world, are ever-evolving and changing with each patient's lifestyle and habits. The healthcare industry is constantly faced with new medical issues and must always be advancing in order to maximize patient care. With obesity becoming more and more prevalent across the world, serious skeletal abnormalities are presenting themselves earlier and earlier in society. Increased stress on bone, joints, and cartilage are paving the way to increasing forms of bone pathology. Taking a preventive approach to orthopedic care is vital to optimizing the prevention of ailments seen in these patients. Detecting abnormalities at earlier stages is a current challenge in orthopedics given the limited imaging modalities that are conducive for visualizing the bone's properties. Several examples, such as subtle or micro-fractures, early neoplastic growth in bones, and micro damage to joints may not be adequately diagnosed early due to difficult visualization with the current imaging modalities. Standard of care for most orthopedic pathology consists of radiographic imaging to diagnose bone fractures, but is limited in visualizing dislocations or injuries to ligaments or muscles. Furthermore, radiographic imaging is also dependent on the size of the fracture seen. Challenging the current standard of care with new technologies may prove to be beneficial and more effective in the long-term. Therefore, research is a major key to success in this field and remaining at the forefront of research is essential for patients suffering from bone and joint injury.

In this study, we concentrated our efforts on the FEM simulation of human bones. The deterioration of bone structure, for instance, may lead to bone fragility and osteoporosis-related fractures that may occur to nearly 50% of women and 25% of men. Some of these diseases lead to long-term care and hospitalization, which is both difficult and costly for the patient. More serious sequelae, such as spinal cord damage, may result from osteoporotic spinal fractures in the elderly ultimately leading to severe long-term disability or even death due to additional complications [1] [2]. Orthopedic computer simulation may serve as a preliminary means of testing the performance of such methods within the human body. Furthermore, this investigation may enhance the study of correlated parameters such as the size of the pelvis, weight, age, gender, etc on the future occurrence of orthopedic diseases. Currently, first line imaging of bone is X-ray or CT scans. However, both of these imaging modalities expose patients to small amounts of radiation. Radiation exposure in children and pregnant women opens the door to lots of other potential pathologies in the future. Therefore, other imaging modalities are oftentimes preferred [5].

2. Simulation Model

Electromagnetic wave transmission and interaction with biological materials plays an important role for biomedical applications. The role of electromagnetics within the human body has been emphasized and researched by investigators in recent years [3] [4]. This association can be further analyzed by absorption and scattering properties through multiple layer structures that mimic human tissue. For the purpose of this investigation, human tissue was simplified into a bilayer interface of skin (epidermal and sub-dermal layers) and bone. Additionally, the EM model incorporates power consumption from the lossy medium of human bone as waves pass through it, resulting in power losses and altering temperatures. Any noise that resulted from electromagnetic interference was neglected in the study.

The dielectric property of bone was a controlling factor in determining the temperature distribution after wave energy was applied. These properties have been studied as function of the frequency [6] in past research. Dielectric properties are subject to dynamic behavior and change with age, genetics, gender, etc. which may impact the condition of bone. As a result, the generated heat is attributed to the conductivity effect of the dielectric constant at a given frequency, as described in the heat equation created for this bilayered structure. The equations presented below, were generated using COMSOL software for FEM simulation.

The antenna's structure, as defined by COMSOL, was enclosed in a catheter sleeve and placed adjacent to skin and bone. **Figure 1** describes the structure of the multilayered composition used in this simulation. Given

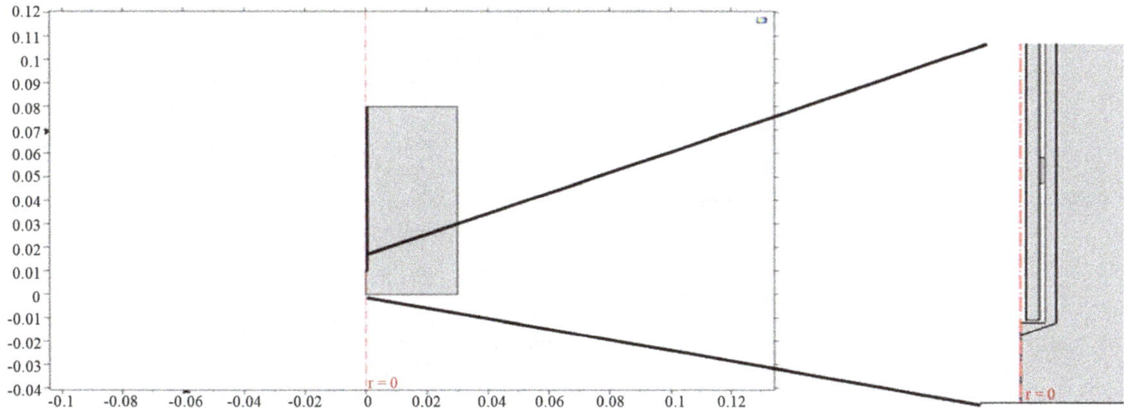

Figure 1. Antenna inserted in human body part. From the left, the first layer is antenna, air, catheter, and bone material. The catheter serves as a protective covering for the EM antenna.

the limitations of simulation software, the catheter sleeve is representative of human skin. As EM waves are transmitted to the catheter, bio-heat is generated accordingly along the catheter sleeve. The bio-heat is then transferred onto the simulated bone material and thermographic changes are monitored. As stated previously, the operational frequencies used for this investigation were 900 MHz, 1 GHz, and 2.4 GHz. The following are the steps to design model in COMSOL:

- The model is constructed with the antenna enclosed in a catheter made of Polytetrafluoroethylene (PTFE). This may serve as membrane filter that optimizes the heat transfer from the antenna into the bone material.
- The RF and Bio-heat transfer COMSOL modules were used for analysis.
- The Electromagnetic antenna waves were modeled in the COMSOL RF module, while the heat transfer was modeled via the bio-heat transfer. Within the RF module, the frequency domain was used.

Bones are typically categorized into 5 types based on their physical and functional properties. There are long bones, short bones, flat bones, irregular bones, and sesamoid bones. Long bones possess a hard outer layer of dense bone with a spongy inner (cancellous) layer, in which bone marrow resides. Short bones, generally known as carpals, have a thin outer layer of compact bone along with a thin cancellous inner layer and comparatively higher bone marrow. Flat bones consist of higher cortical bone for protection with the core consisting of spongy bone. This type of bone possesses varying amounts of bone marrow. Irregular bones consist of a thin outer compact bone and a larger portion of cancellous bone. These are generally present in vertebrae.

Each kind of bone performs a different function, such as mechanical support, protection, movement, or mineral storage [7] [9]. Additionally, each type of bone is composed in an alternative way that will affect the dielectric and thermal properties. As mentioned, some types of bone consist of a thick cortical layer (flat bones) and function mainly as mechanical support and protection. In contrast, other kinds of bones, such as long bones, have a larger central spongy area providing more of a reservoir for bone marrow. As expected, the differing compositions of these bones result in different dielectric and thermal properties that can be exploited by electric fields through microwave energy. These properties were incorporated into the simulation model.

3. The Microwave/Heat Transfer Model

The scalar equation for the electric field, E is given from the wave equation as:

$$-\nabla\cdot\left(\frac{1}{\mu}\nabla E\right)+\left(j\omega\sigma-\omega^{2}\varepsilon\right)E=0 \tag{1}$$

where ω is the angular frequency, μ is the magnetic permeability, σ is the conductivity, and ε is the dielectric constant of the material. The boundary conditions that are associated with the above wave equation are given by matching the tangential component of the magnetic field H at the interface. Considering the relation between E

and H through the wave impedance, and Maxwell equation:

$$\nabla \times E = -\frac{\partial B}{\partial t} \tag{2}$$

$$E = \eta H \tag{3}$$

where η is the wave impedance given as

$$\eta = \sqrt{\left(\frac{\mu}{\varepsilon}\right)} \tag{4}$$

The conductive media from the EM fields can be incorporated via the equation of continuity given below:

$$\nabla \cdot J = Q \tag{5}$$

where Q is the current source.

With the electric potential given by:

$$-\sigma\nabla^2 V = Q \tag{6}$$

The outer boundary was given by $\frac{\partial V}{\partial n} = 0$.

The Ohmic losses in the wave equation will result in thermal propagation, following the heat transfer equation:

$$\rho C\frac{\partial T}{\partial t} - \nabla \cdot (k\nabla T) = Q + h \cdot \Delta T \tag{7}$$

where ρ is the mass density, C is the heat capacity, k is the heat conduction coefficient, Q is the source of heat, h is the convection heat transfer coefficient, k is the thermal conductivity, and ΔT is the difference if temperature from external to internal. The $h \cdot \Delta T$ term models the transversal heat transfer from the surroundings. The heat flux is defined as $k \cdot \Delta T$. The above equations were combined utilizing COMSOL heat equation and rf modules for human bone that is characterized with bone density to emulate the difference between the healthy and abnormal bone characteristics.

4. COMSOL Model

The mesh distribution for the finite element simulation was given between 3 mm as shown in **Figure 2**, with 4 point locations as shown in **Figure 3**. The distribution of the four locations w.r.t. the antenna positions show the differential temperature throughout the bone materials. **Table 1** presents the material properties used in the simulation.

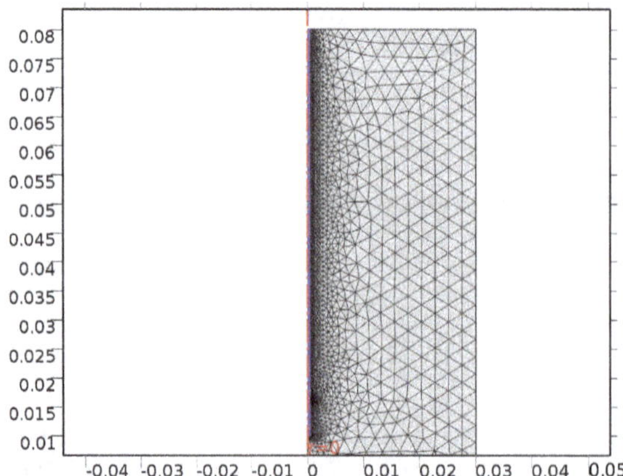

Figure 2. The COMSOL Mesh model within 3 mm.

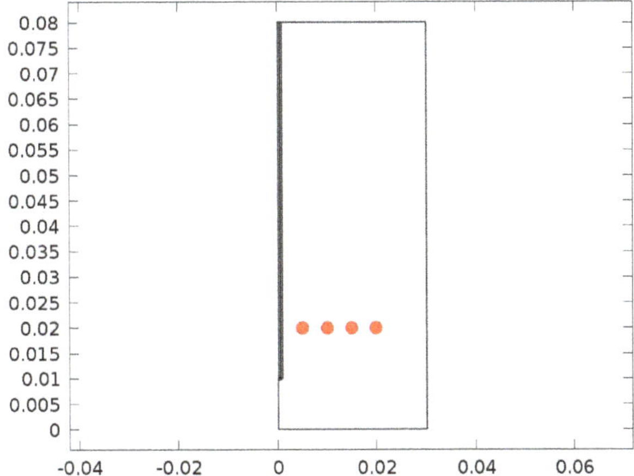

Figure 3. The COMSOL model for the four point locations to provide the temperature distributions throughout the bone materials.

Table 1. Electrical properties of bone [8].

Bone Type	Frequency	Permittivity	Elec. Cond. (S/m)	Density (kg/m³)	Heat Capacity (J/kg/°C)	Therm. Cond. W/m/°C	Heat Transfer Rate (ml/min/kg)	Heat Generation Rate (W/kg)
	900 MHz	2.08E+1	3.40E−1	1178	2274	0.31	30	0.46
Cancellous	1 GHz	2.06E+01	3.64E−01	1178	2274	0.31	30	0.46
	2.4 GHz	1.86E+01	7.88E−01	1178	2274	0.31	30	0.46
	900 MHz	1.25E+1	1.43E−1	1908	1313	0.32	10	0.15
Cortical	1 GHz	1.24E+01	1.56E−01	1908	1313	0.32	10	0.15
	2.4 GHz	1.14E+01	3.85E−01	1908	1313	0.32	10	0.15
	900 MHz	1.13E+1	2.28E−1	1029	2666	0.28	135	2.09
Bone Marrow red	1 GHz	1.12E+01	2.39E−01	1029	2666	0.28	135	2.09
	2.4 GHz	1.03E+01	4.50E−01	1029	2666	0.28	135	2.09
	900 MHz	5.50E+0	4.07E−2	980	2065	0.19	30	0.46
Bone Marrow Yellow	1 GHz	5.49E+00	4.33E−02	980	2065	0.19	30	0.46
	2.4 GHz	5.30E+00	9.33E−02	980	2065	0.19	30	0.46

5. Simulation Results

The following simulations were conducted at three different frequencies: 900 MHs, 1 GHz, and 2.4 GHz. The temperature distributions obtained for the four different locations in the bones were obtained. The input microwave power was 10:00 Watts, and in one minute simulation time, dissipation was 9.354 Watts. **Figure 4** shows the differential temperature distributions for four different frequencies within one minute change. **Figure 5** gives the surface power density dissipation at 900 MHz for a one minute simulation. **Figure 6** gives the normalized distribution of a damaged tissue. This is given for comparison between homogenous bone materials. **Figure 7** demonstrates the temperature distribution for a second type of bone material at 900 MHz. **Figure 8** and **Figure 9** display the temperature distribution for a damaged tissue at 900 MHz and 1 GHz, respectively. The characteristic difference of the kinds of bones are represented in **Figures 10-13**.

6. Conclusion

The simulation results show differential power dissipation over the bone materials with different temperatures within 2 - 4 degree change for various frequencies. This simulation also shows the distinction between normal

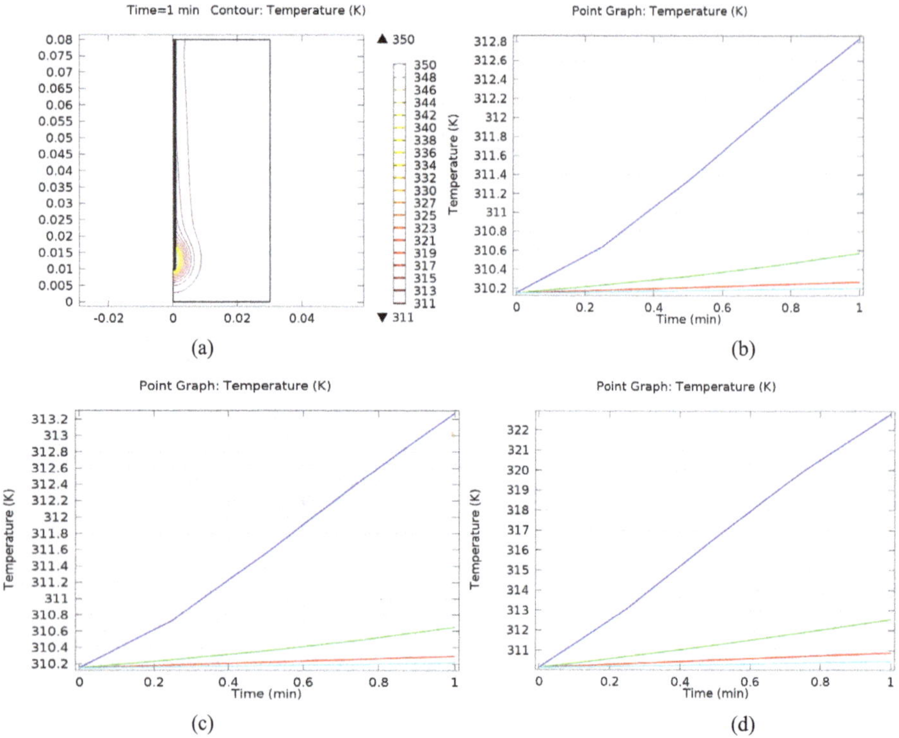

Figure 4. The temperature Distributions for various frequencies given for one minute simulation; (a) Data cut for 2D, giving the temperature distribution for the blue curve at 900 MHz, (b) Gives the differential distribution for the four location within the bone materials at 900 MHz, (c) The temperature distribution at the four locations throughout the bone materials at 1 GHz, and (d) The temperature distribution within the bone materials at 2.4 GHz.

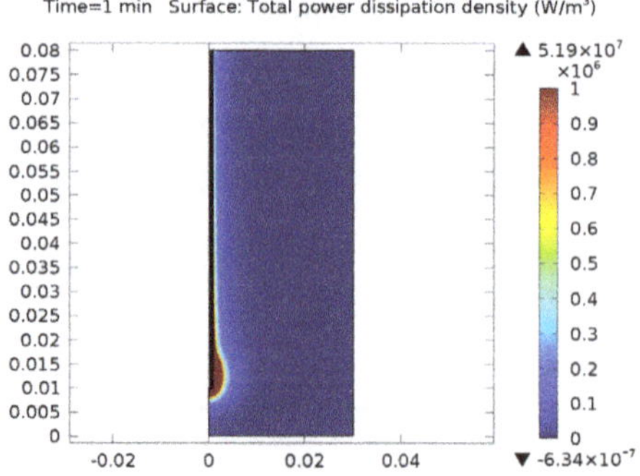

Figure 5. One Minute total power dissipation density (W/m^3).

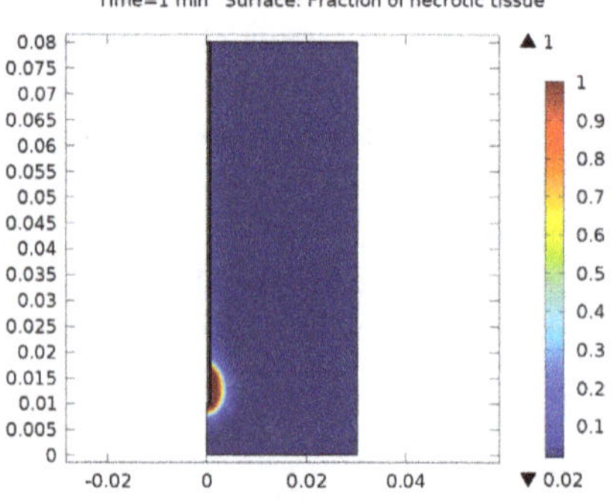

Figure 6. Power distribution for damaged tissue at 900 MHZ.

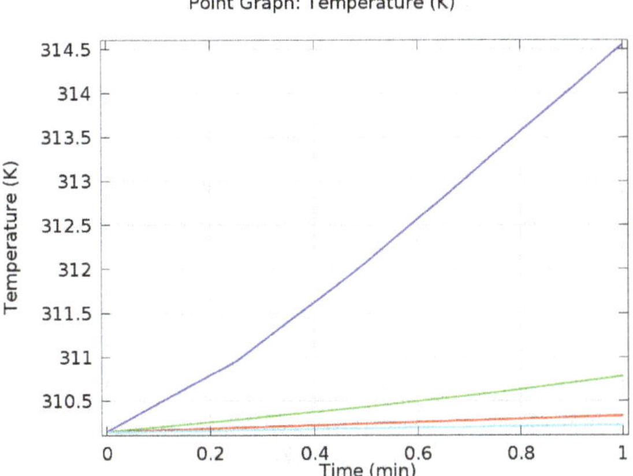

Figure 7. Four point distribution for the bone read throughout the bone materials at 900 MHz, demonstrating temperature distribution throughout the depth of the bone thickness. The curve also shows the thermal energy transmitted for one minute maximum.

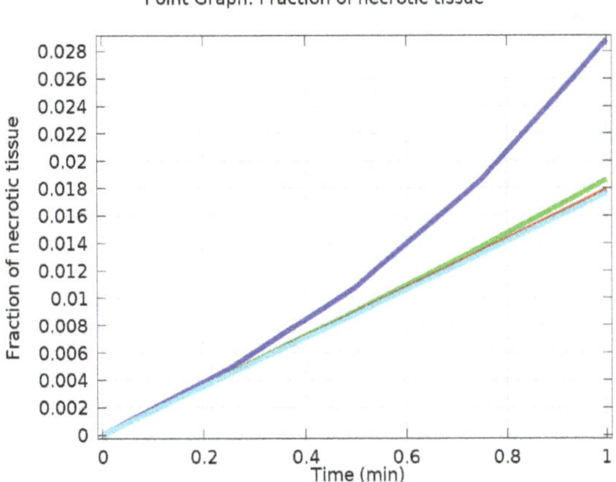

Figure 8. Temperature distribution for damaged tissue at 900 MHz. The data shows different distribution for a given bone density, representing a damaged tissue.

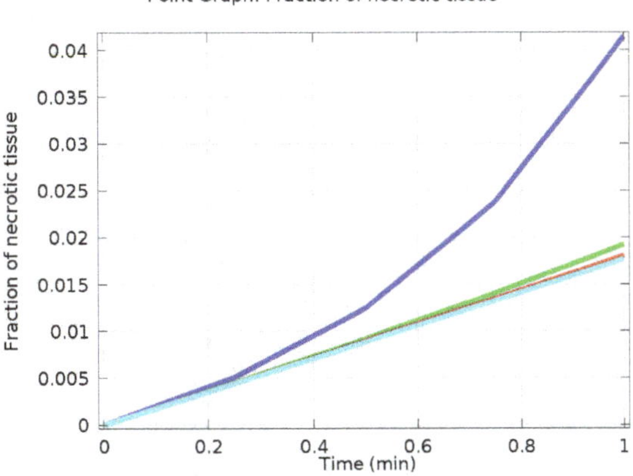

Figure 9. Temperature distribution for damaged tissue at 1 GHz. The curve shows the differential distribution as compared to the 900 MHz data.

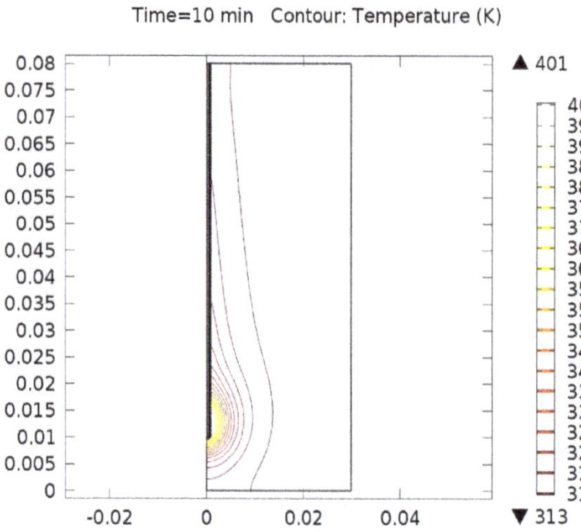

(a)

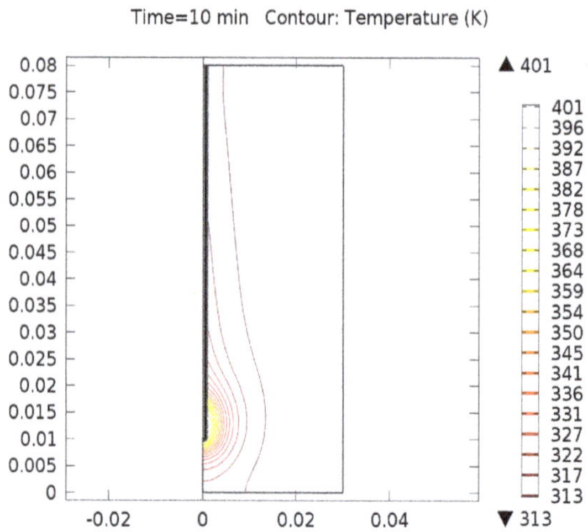

(b)

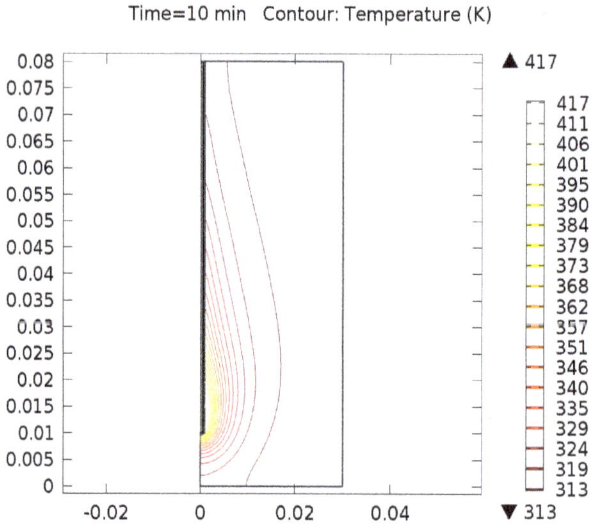

(c)

Figure 10. Isothermal contours of red bone marrow at (a) 900 MHz; (b) 1 GHz; (c) 2.4 GHz.

(a)

(b)

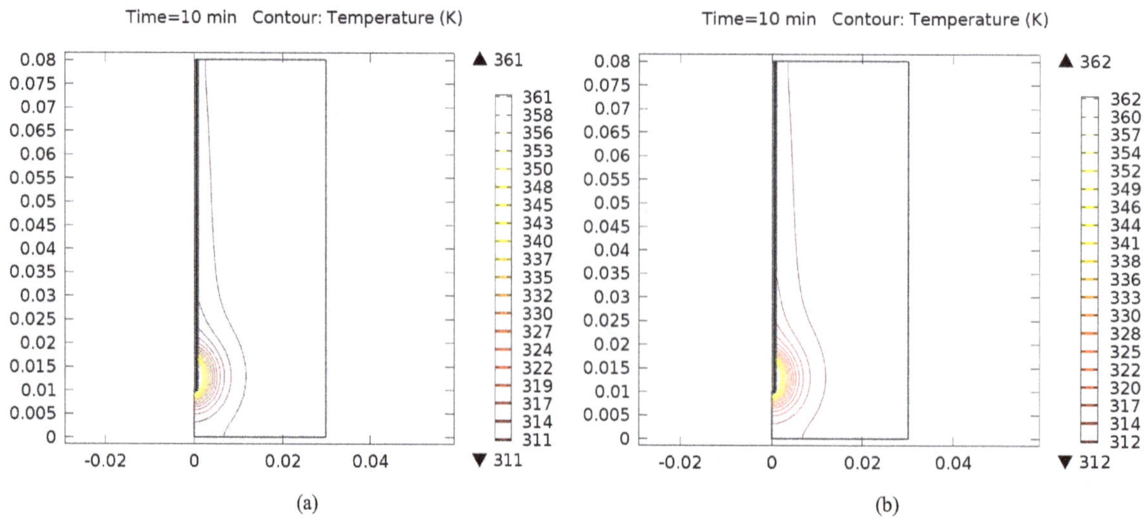

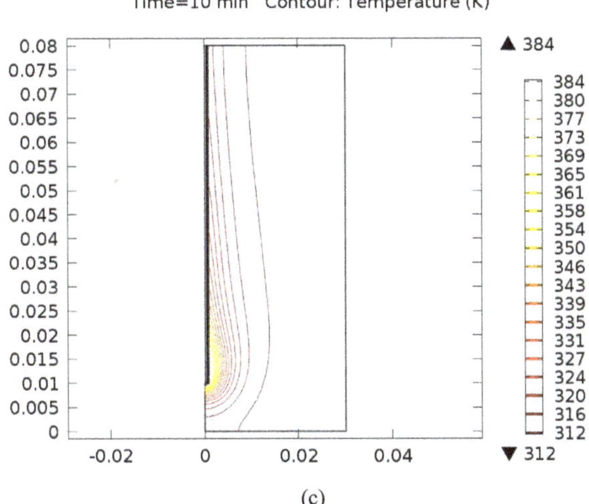

(c)

Figure 11. Isothermal contours of yellow bone marrow at (a) 900 MHz; (b) 1 GHz; (c) 2.4 GHz.

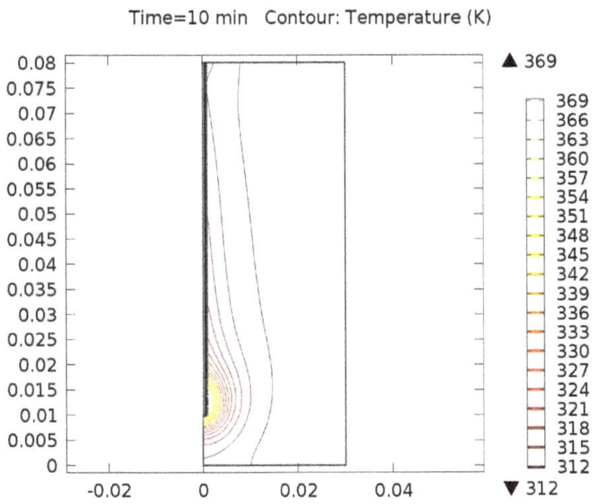

Time=10 min Contour: Temperature (K)

(a)

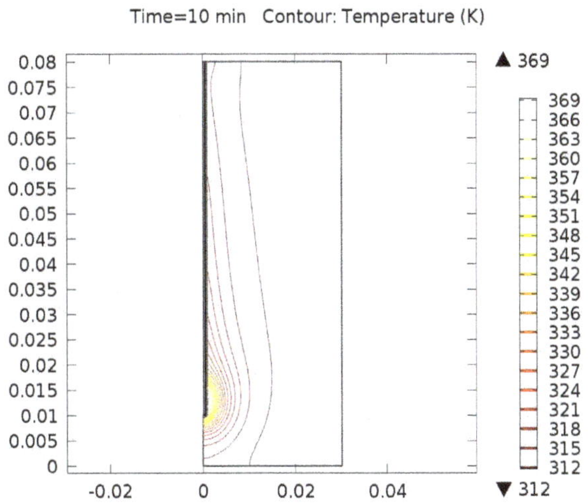

Time=10 min Contour: Temperature (K)

(b)

Time=10 min Contour: Temperature (K)

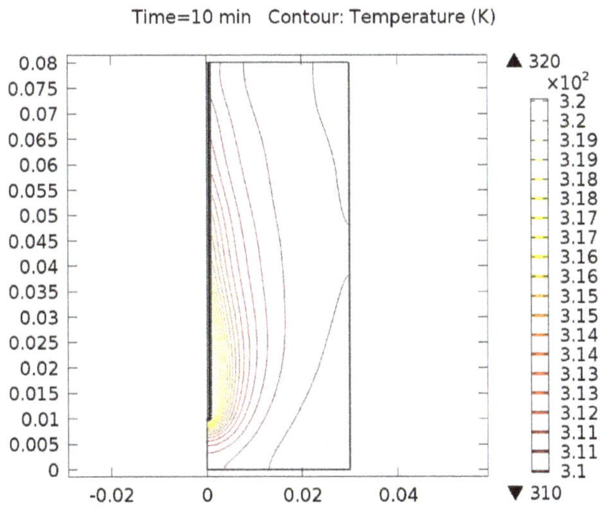

(c)

Figure 12. Isothermal contours of cancellous bone at (a) 900 MHz; (b) 1 GHz; (c) 2.4 GHz.

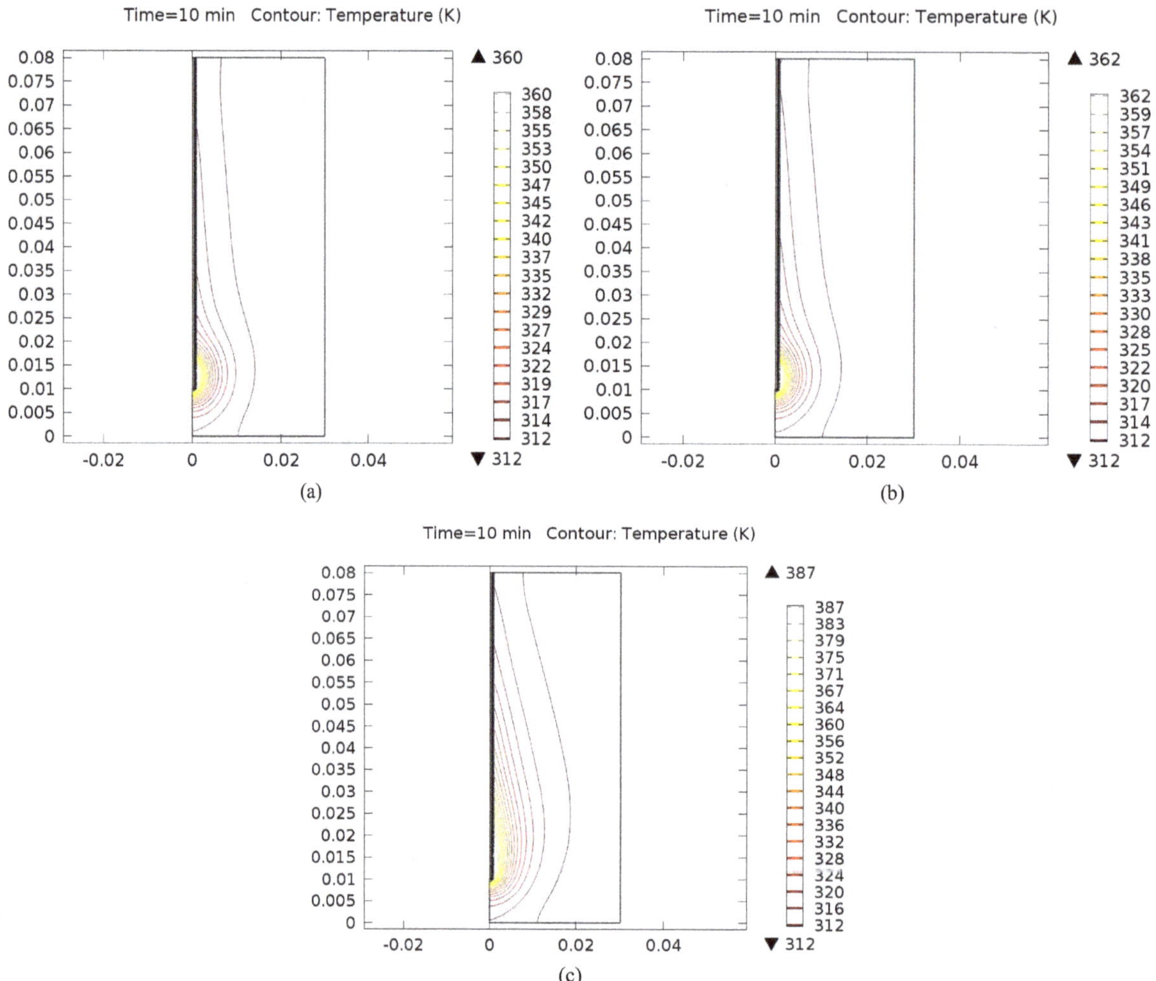

Figure 13. Isothermal contours of cortical bone at (a) 900 MHz; (b) 1 GHz; (c) 2.4 GHz.

bone materials and damaged tissues, indicating that this is an effective method for diagnosing bone and tissue pathology. Future consideration will be given to a practical model for a medical setting.

Acknowledgements

The team of authors and Co-Authors acknowledge the INDI facility at IUPUI.

References

[1] Krug, R., Burghardt, A.J., Majumdar, S. and Link, T.M. (2010) High-Resolution Imaging Techniques for the Assessment of Osteoporosis. *Radiologic Clinics of North America*, **48**, 601-621. http://dx.doi.org/10.1016/j.rcl.2010.02.015

[2] Johnell, O. and Kanis, J. (2005) Epidemiology of Osteoporotic Fractures. *Osteoporosis International*, **16**, S3-S7. http://dx.doi.org/10.1007/s00198-004-1702-6

[3] Wong, M.F. and Wiart, J. (2005) Modelling of Electromagnetic Wave Interactions with the Human Body. *Comptes Rendus Physique*, **6**, 585-594. http://dx.doi.org/10.1016/j.crhy.2005.07.003

[4] Rouf, H.K., Costen, F. and Fujii, M. (2011) Modeling EM Wave Interactions with Human Body in Frequency Dependent Crank Nicolson Method. *Journal of Electromagnetic Waves and Applications*, **25**, 2429-2441. http://dx.doi.org/10.1163/156939311798806185

[5] (ACR), Radiological (2016) Brain Imaging, Functional (Fmri). *Radiologyinfo.org*. N.p.

[6] Meaney, P.M., Zhou, T., Goodwin, D., Golnabi, A., Attardo, E.A. and Paulsen, K.D. (2012) Bone Dielectric Property Variation as a Function of Mineralization at Microwave Frequencies. *International Journal of Biomedical Imaging*,

2012, Article ID: 649612.

[7] Types of Bones in the Human Body. (n.d.). http://www.teachpe.com/anatomy/types_of_bones.php

[8] Itis.ethz.ch. (2016) Dielectric Tissue Properties|Average EM Tissue Parameter|Average Thermal Tissue Parameter| Frequency-Dependent Tissue Parameters|Tissue-Specific Parameters|Interactive Tissue Parameter Table IT' Is Foundation. N.p.

[9] Safadi, F.F., *et al.* (2009) Bone Structure, Development and Bone Biology. *Bone Pathology*, Humana Press, 1-50.

19

The Effect of High Power at Microwave Frequencies on the Linearity of Non-Polar Dielectrics in Space RF component

Tawfik Elsayed Khattab

Department of Electrical Engineering, Engineering College, King Khalid University, Abha, Saudi Arabia
Email: tkhtab@kku.edu.sa

Abstract

In satellite communication systems, there are high power multichannel transmitters and wide-band receivers that have shared RF antenna transmission lines because of: 1) large power level difference between the transmitted and received signal; 2) limited frequency channels. The harmonics and Passive Intermodulation (PIM) Interference will be generated due to passive non-linearities in the high power transmission path. This can be a serious problem. This paper describes how to determine the signal levels and dominant mechanisms that are associated with non-linear dielectric behavior in this context. A novel measurement system for testing dielectric samples is described and measurement results are provided for commonly used microwave dielectrics.

Keywords

Passive Intermodulation, PIM, Nonlinear, Polar Dielectric, Harmonics, Cavity Resonator

1. Introduction

Passive Intermodulation, PIM, is a form of intermodulation distortion that can occur even when no active components are present. PIM can occur in a variety of areas from coaxial connectors to cables, even rusty bolts or any joint where dissimilar metals occur [1]. Although harmonics generation is a ubiquities phenomenon, very little is known of the underlying causative mechanisms, relating to either conductors or dielectrics. There are a large number of candidate mechanisms and it is likely that several may contribute at any given site of occurrence. There currently exists no data on the levels of the harmonics to be expected from commonly used dielectric materials nor have any studies identified the dominant mechanisms. As greater demand is placed under system performance, it is becoming increasingly difficult to provide adequate harmonic generation immunity in the absence of data. The purpose of this work is to provide an improved understanding of causes and effects of the non-linearity of dielectrics leading to improvements in RF system design. The specific objectives are as follows:

- To design and develop a measurement system whereby the non-linear behavior of dielectric materials under test can be measured.
- To characterize the non-linear behavior of dielectrics in common use in microwave RF engineering, such as PTFE and Polystyrene.

2. The Measurement System

The measurement system is a new design of harmonic detector, implemented in coaxial components. The system is designed for 1.56 GHz to detect the 3rd harmonic at 4.68 GHz, due to any significant non-linear behavior of dielectric test samples. Dielectric samples are tested in a coaxial cavity, where they are exposed to a high uniform electric field. Because the level of system residual 3rd harmonic must be kept to minimum, all components are in-house designs in which metal-to-metal junctions are absent from critical sections in the current path. Important features are the use of contactless cable entry for critical components and fully demountable contactless coaxial connectors on test cavity.

2.1. System Configuration

A block diagram of the 3rd harmonic measurement setup is given in **Figure 1**. A source signal of 1.56 GHz, originated from a phase locked oscillator, is amplified by a separate power amplifier to give an output power up to 49 dBm (80 Watts).

The source signal passes through a diplexer using two hybrids and two low pass filters and hence to the input port of the test cavity. The low pass filters pass only the signal at the source frequency and attenuate any harmonics which may be generated in the oscillator or in the power amplifier. The diplexer provides a test signal to the cavity and couples the 3rd harmonic which may be generated in the test chamber due to any non-linear behavior of the sample under test.

This consists of a high pass filter which attenuates the source frequency and passes the 3rd harmonic to the input port of the analyzer. Source signal power is absorbed in a quiet load. This consists of two semi rigid coaxial cables, UT250 with length 45 m, followed by UT-141 with length 20 m, terminated by a lumped load. The total one way attenuation for the quiet load is 30 dB.

The test chamber is a tunable quarter-wave coaxial cavity resonator which satisfies the following requirements:

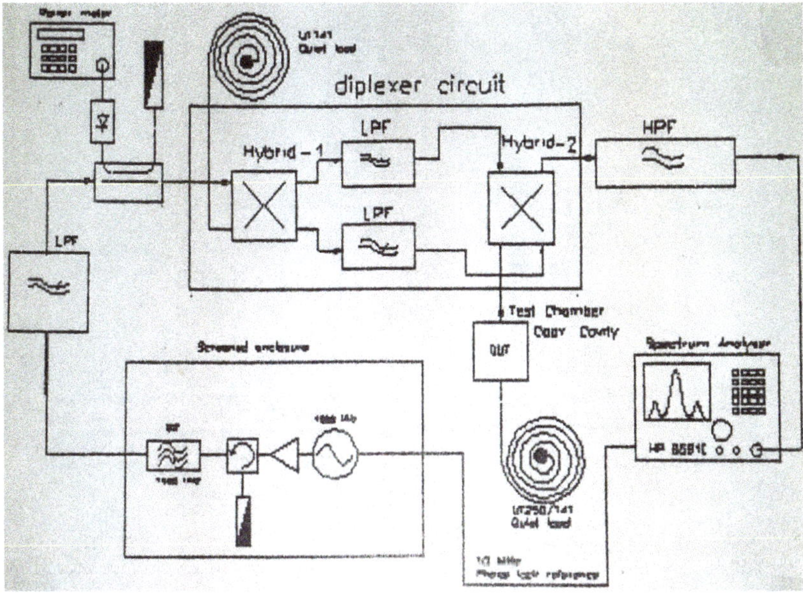

Figure 1. The Measurement system block diagram.

- Flexibility to accommodate and exchange test samples.
- The test samples can be tested under controlled conditions.
- Good matching between input and output ports. This is important since the system residual can be degraded by current standing waves.
- Good linearity is essential. If the test chamber construction includes non-linear materials or vulnerable junctions, consequently the system residual intermodulation will be increased, and this will mask the intermodulation signal produced by the test samples, which affect the dynamic range of the measurement system.
- A high uniform field at the position of the test sample.

The test cavity is machined in brass and is shown in **Figure 2**. It incorporates contactless connection at cable entry, one piece body/resonator constriction and a contactless plunger to control the gap capacitance. Past experience in the design of low passive intermodulation coaxial components has shown that these features are important in achieving a low residual. The test field is determined from measurement of the cavity Q factor and the source signal power, given the cavity geometry. The cavity loaded Q is in the region of 130.

2.2. The System Performance

The electrical performance of the system has been tested using network analyzer. **Figure 3** shows the transmission of the source signal and absolute attenuation for the harmonics at the output of the test chamber. **Figure 4** shows the absolute attenuation for the source signal and transmission of the 3rd harmonic at the output of the high pass filter.

The system noise flour is −125 dBm and the 3rd order residual is −115 dBm at a source signal power of 32 W and increases at 3 dB rate. It should be noted that these figures should be related to the field in the test cavity, rather than the source power, as in a conventional two tone system.

3. Dielectric Non-Linear Mechanisms

The candidate non-linear mechanisms which may reasonably be suspected as significant causes of dielectric harmonics generation are considered to be non-linear permittivity, electrostriction and temperature coefficient of

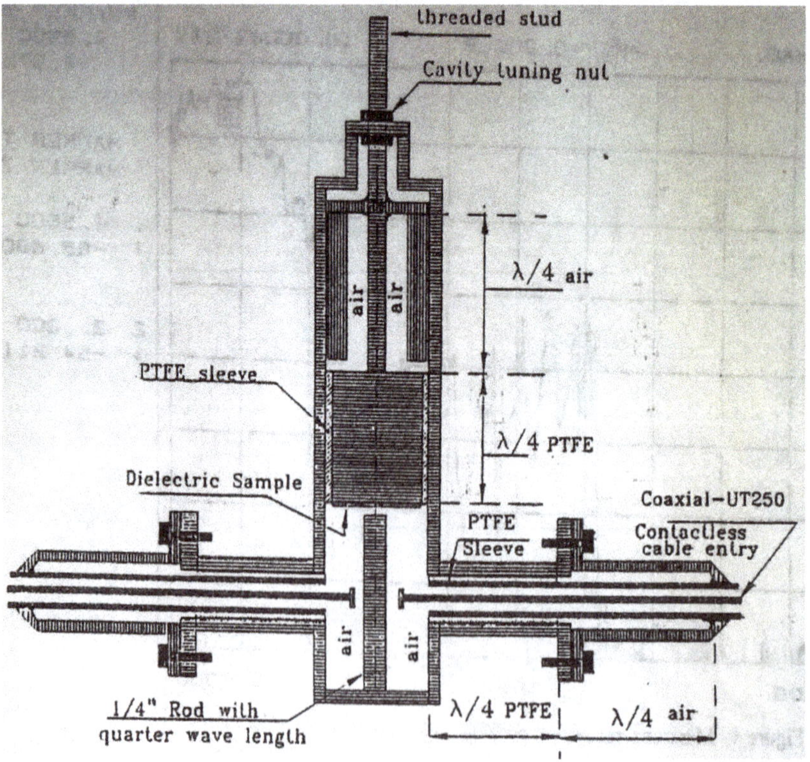

Figure 2. Test cavity.

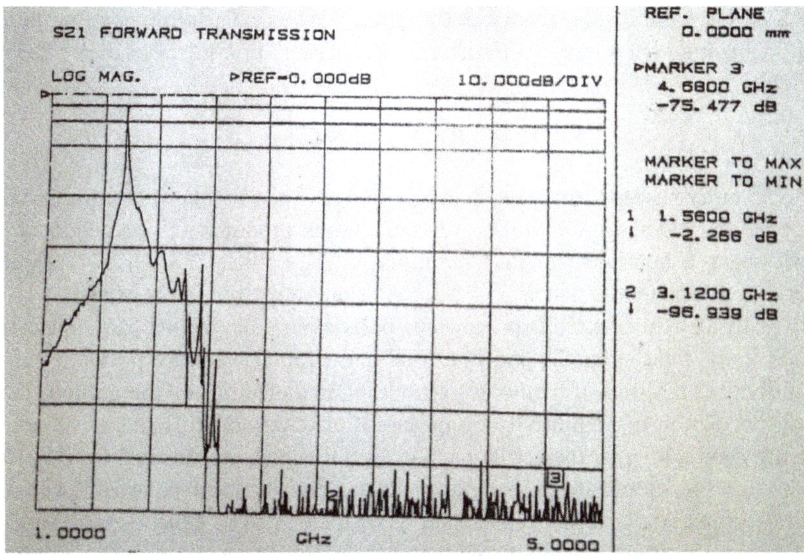

Figure 3. Measurement system performance in transmission mode.

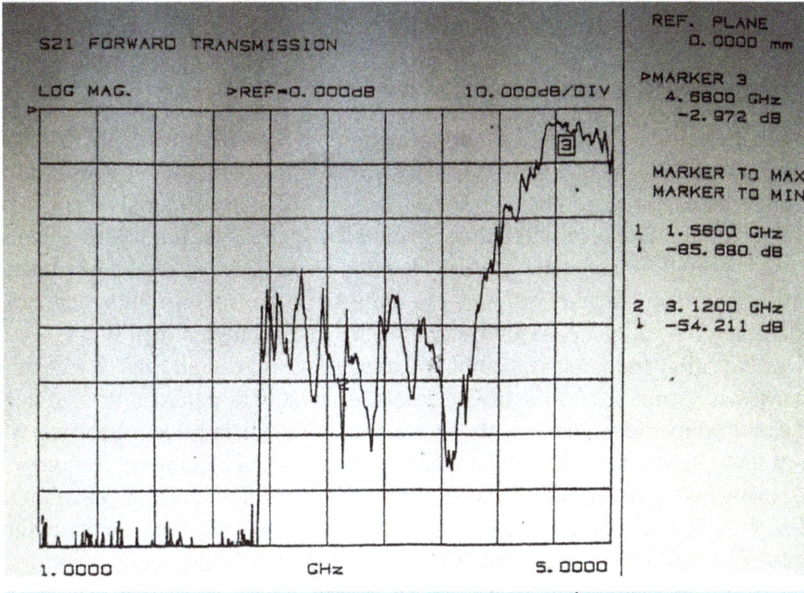

Figure 4. Measurement system performance for detecting 3rd Harmonic.

permittivity. Many potential causes of harmonics generation arise from intersection of two or more linear mechanisms which may give rise to multi-order non-linear behavior. This results in a bewildering number of possible combinations. An outline theoretical analysis is provided for candidate mechanisms and their relative importance is estimated based on theoretical considerations.

3.1. Non-Linear Permittivity

The linear relationship between Polarization, *P*, and electric field strength, *E*, in dielectrics is expressed by:

$$P = \varepsilon_0 \gamma E \tag{1}$$

where γ is the susceptibility and ε_0 is the permittivity of free space [2]-[4].

The value for the field at which deviation from linearity becomes apparent is given as around 1000 v/mm [5]. This implies high, but not unreasonable, current densities in the dielectric at microwave frequencies. The non-linear polarization may be modeled as power series of susceptibility terms. Only odd powers will be present for dielectrics which do not exhibit spontaneous polarization, since a polarity reversal in the electric field reverses the direction of polarization but does not alter its intensity. The 1st non-linear term to become significant is cubic in the electric field:

$$P = \varepsilon\left(\gamma E + \aleph E^3\right) \tag{2}$$

where $\aleph$ is the 3rd order susceptibility coefficient. This gives rise to a current term that is also proportional to the cube of the electric field and the 3rd order intermodulation products will be generated. Alternatively, if this behavior is expressed as a non-linear relative permittivity ε_r, so that $P = \varepsilon_0\left(\varepsilon_r - 1\right)E$, then the 1st non-linear term may be described as quadratic (since $\varepsilon_r = \gamma + \aleph E^2$) and only even terms exist.

For dielectric materials which exhibit spontaneous polarization, deviations from linear behavior may become apparent at much lower field strengths and both odd and even powers are possible. Stauss G.H. [5] dismiss non-linear permittivity as a source of harmonics generation in dielectrics on the grounds that it is less significant than the indirect modulation of permittivity by means of electrostriction it should be noted; however, that the examples given are good non-polar dielectrics such as Teflon which are selected for their desirable properties as insulators at radio frequencies and are of high purity. Thus only electronic polarization contributes to the relative permittivity and these materials are not representative of dielectrics in general. Ionization loss occurs in solids which contain trapped gas. It only occurs when the electric field strength exceeds a critical value, above which the loss tangent rises rapidly with any increase in the field. Thus the imaginary component of the complex permittivity becomes non-linear and intermodulation components could appear in the conduction current. The increased power dissipation can also result in significant greater heating of the dielectric.

3.2. Temperature Coefficient of Permittivity

The permittivity of dielectrics which only exhibits electronic polarization typically have small negative temperature coefficient of around 0.01% C^{-1}. This is primarily due to the reduction of density as the material expands. The same effect also tends to reduce atomic polarization; but for many materials this is more than compensated by the reduced inter-atomic forces in a less dense medium. This results in a positive temperature coefficient of typically 0.01% C^{-1}. There are exceptions to this: Titanium dioxide, for example, has negative temperature coefficient of permittivity at room temperature. As these forms of polarization result in very small dielectric losses at microwave frequencies, they do not contribute significantly to loss tangent [6]-[8].

Orientational polarization and ionic relaxation polarization also generally make a positive contribution to the temperature coefficient permittivity. In some materials, orientational polarization that can exhibit much larger positive and negative temperature coefficients over a certain ranges is often associated with the melting point, but may begin at lower temperature while the material is still a solid. Ionic relaxation polarization exhibits an approximately exponential increase in the loss tangent with temperature. For orientation polarization, lose tangent typically rises to a maximum (at a temperature which depends on the frequency) and thin falls, before rising again as the increase in conductivity with temperature becomes significant.

Interfacial polarization is likely to be strongly influenced by temperature since it relies on the presence of free charges, but the nature of this effect will depend on the particular materials involved. In dielectrics which do not exhibit interfacial polarization, (typically exponential) increase in carrier density and conductivity with temperature is responsible for positive contribution to the overall temperature coefficient of the loss tangent. This is more significant for those materials which are poor insulators and at high temperatures. It has less effect on the loss tangent at higher frequencies, where a greater proportion of the current is conveyed by polarization mechanisms.

3.3. Electrostriction

Electrostriction is the volume change due to the variation of energy density in a dielectric under RF excitation. The change in volume causes a quadratic dependence of permittivity upon the electric field. Resulting permittivity variations at the fundamental and harmonics of the input frequency modulate the primary fields leading to the generation of harmonics.

If the high field polarization is non-linear in the field strength, the dependence of the dielectric displacement, *D*, on the field strength will also be non-linear [9].

$$D = \varepsilon_0 E + P = \varepsilon_0 E + \varepsilon_0 \left(\gamma E + \aleph E^3\right) \quad (3)$$

When measurements of the non-linear effects are made by superposing a low intensity alternating field on static field of high intensity, the measurement results in the field dependent incremental dielectric permittivity ε_E which is given by:

$$\varepsilon_E = \frac{\partial D}{\partial E} = \varepsilon + 3\varepsilon_0 \aleph E^2 \quad (4)$$

where $\varepsilon = \varepsilon_0 \varepsilon_r$ and $\varepsilon_r = (1 + \aleph)$, hence

$$\frac{\Delta\varepsilon}{E^2} = \frac{\varepsilon E - \varepsilon}{E^2} = 3\varepsilon_0 \aleph \quad (5)$$

The non-linear effect is then characterized by the quantity $\frac{\Delta\varepsilon}{E^2}$ resonator rod and the top wall of the cavity as shown in **Figure 1**. Samples consist of 1 mm thick discs, lightly held in place, *i.e.* under constant pressure (p). The variation of the dielectric permittivity is given by:

$$\varepsilon_E = \left(\frac{\mathrm{d}D}{\mathrm{d}E}\right)_{T,d} + \left(\frac{\mathrm{d}D}{\mathrm{d}d}\right)_{T,E} \left(\frac{\mathrm{d}d}{\mathrm{d}E}\right)_{T,P} \quad (6)$$

where, *T*, *d* denote to the temperature and density of the sample under test. The first term of Equation (6) is the contribution to ε_E due to the electrostriction; denoting this term by $\Delta\varepsilon_e$ [5].

$$\Delta\varepsilon_e = \left(\frac{\mathrm{d}D}{\mathrm{d}E}\right)_{T,d} = E^2 \frac{1}{K} d^2 \left(\frac{\partial\varepsilon}{\partial p}\right)_T \quad (7)$$

Thus, the magnitude of the contribution to dielectric permittivity due to electrostriction can be calculated from the bulk modulus (*k*) and the dependence of the permittivity on either the density or the pressure.

Intermodulation powers developed by this mechanism vary in inverse proportion to the square of the bulk modulus of the dielectric. The modulus, *k*, is related to the elasticity modulus and Poission's ratio by the following equation [10].

Where E_s = elasticity modules, and v = Poission's ratio. There is an associated thermal effect so that the volume change can also result in temperature variation. Electrostriction is expected to be the principal source of non-linearity in good non-polar dielectrics such as PTFE and cross-linked polystyrene because the bulk modules of these materials are very small.

4. Results Analysis and Discussion

Any variation in dielectric properties with the applied field will serve to modulate in coming signal currents. The 3rd harmonic intermodulation level generated due to the non-linear mechanisms of the dielectric sample under test, and excited by a uniform field that can be derived based on power series as follows [11].

$$D = \varepsilon_0 \varepsilon_r \left(E + \alpha E^3 + \beta E^5 + \cdots\right)$$

where α, β are coefficients of non-linear dielectric behavior.

$$E = E_0 \sin \omega t$$

$$\begin{aligned} i &= \varepsilon_0 \varepsilon_r \frac{\partial}{\partial t}\left(E_0 \sin \omega t + \alpha \left(E_0 \sin \omega t\right)^3 + \beta \left(E_0 \sin \omega t\right)^5 + \cdots\right) \\ &= \varepsilon_0 \varepsilon_r \omega \left[\left(E_0 + \frac{3}{4}\alpha E_0^3 + \frac{5}{8}\beta E_0^5 + \cdots\right)\cos \omega t \right. \\ &\quad \left. + \left(-\frac{3}{4}\alpha E_0^3 - \frac{5}{16}\beta E_0^5 + \cdots\right)\cos 3\omega t + \left(\frac{5}{16}\beta E_0^5 + \cdots\right)\cos 5\omega t + \cdots\right] \end{aligned}$$

Hence the 3rd harmonic transfer function is given by:

$$IM_{H3} = 20\log\left[\frac{-\frac{3}{4}\alpha E_0 - \frac{5}{16}\beta E_0 + \cdots}{E_0 + \frac{3}{4}\alpha E_0 + \frac{5}{8}\beta E_0 + \cdots}\right] = 20\log\left(\frac{3}{4}\alpha E_0^2\right)$$

The value of α determines the level of non-linearity of each dielectric material and will vary from one material to another according to the operative mechanism. The main contribution of this paper is the experimental results shown in **Figure 5** and **Table 1**, which show non-linearity become significant in Nylon-66 at 700 V/mm and polythene at 2000 V/mm respectively. Hence, non-linear permittivity is the likely mechanism. The polystyrene shows a much better performance and the most linear material is PTFE. Hence, electrostriction is expected to be the main source of non-linearity. Good non-polar materials (such as PTFE, AL(AL_2O_2)) need much higher powers to excite significant non-linearity.

5. Conclusion

These results demonstrate that care is needed in the use of dielectrics in any high field environment where 3rd harmonic must kept to minimum. No measurements have been made on futile loaded materials, but these can be expected to demonstrate significant non-linearity due to non-linear permittivity. It must also be appreciated that any trace of moisture will cause similar behavior. This can be due to moisture absorption, which occurs to a varying extent, in most engineering dielectrics. Sample impurity is a farther factor and can cause significant degradation particularly where good performance is important.

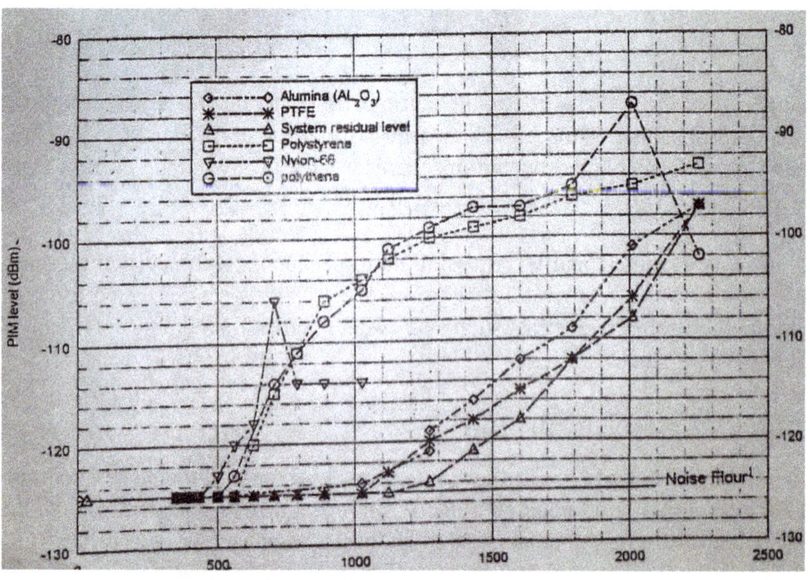

Maximum applied field on the sample under test (V/mm)

Figure 5. Measurement results.

Table 1. 3rd harmonic level for the measured dielectric samples.

Dielectric Material	Field Strength (V/mm)	3rd Harmonic Level (dBm)
Nylon-66	700	−106
Polythene	2000	−88
Polystyrene	2250	−94
Al.(AL_2O_2)	2250	Not Detected
PTFE	2250	Not Detected

Future Work

The rest of the common used Dielectric materials (such as, PTFE, AL_2O_2, etc.) need to be tested by using different technique and/or different design for the measurement system to be able to detect the nonlinear behavior.

Acknowledgments

This project is funded by King Khalid University, Deanship of Scientific Research, project number (kku 151/2). I express my warm thanks to all academic staff of Deanship of Scientific Research for their support and guidance throughout project time.

References

[1] Khattab, T. and Rawlins, A.D. (1996) Principle of Low PIM Hardware Design. 13*th National Radio Science Conference*, *NRSC*'96, Cairo, 19-21 March 1996, 355-362. http://dx.doi.org/10.1109/NRSC.1996.551127

[2] Foord, A.P. and Rawlins, A.D. (1992) A Study of Passive Intermodulation Interference in Spce RF Hardware. ESA Report, University of Kent, Canterbury.

[3] Burfoot, J.C. and Taylor (1979) Polar Dielectrics and Their Applications. Macmillan Press, London.

[4] Landau, L.D. and Lifshitz, E.M. (1984) Electrodynamics of Continuous Media. Pregamon Press, Oxford.

[5] Stauss, G.H. (1980) Studies on the Reduction of Intrmodulation Generation in Communication Systems. NRL Memorandum Report 4233, 75-79.

[6] Smyth, C.P. (1995) Dielectric Behavior and Structure. McGraw-Hill, New York.

[7] Zheludev, I.S. (1971) Physics of Crystalline Dielectrcs, Volume 2, Electrical Properties. Plenum Press, New York.

[8] Bogoroditskii, N.P. and Pasynkov, V.V. (1968) Radio and Electronic Materials. Iliffe Books, London.

[9] Bottcher, C.J.F. (1973) Theory of Electric Polarization, Volume 1. Elsevier, Amsterdam, 318-323.

[10] Kumar, A. (1987) Passive IM Products Threaten High Power Satcom Systems. *Microwave & RF*, **26**, 98-103.

[11] Nishikawa, T., Ishikawa, Y. and Hattori, J. (1988) Measurement Method of Intermodulation Distortion of Dielectric Resonator. *Japanese Journal of Applied Physics*, 39-45.

20

Synthesis and Investigation of the Activity of Cu-Cr-Co/Al_2O_3/Al-Catalysts in the Microwave Radiation-Stimulated Reaction Joint Deep Oxidation of Hydrocarbons and Carbon Monoxide

Peri A. Muradova, Yuriy N. Litvishkov

Institute of Catalysis and Inorganic Chemistry Named after M. Nagiyev of Azerbaijan National Academy of Sciences, Baku, Azerbaijan

Email: muradovaperi@rambler.ru, mmanafov@gmail.com

Abstract

This article shows main principles and presents ideas described in the scientific and technical literature, on the mechanism of interaction of microwave radiation with a solid phase materials, which were used as a basis of creation of new perspective, energy efficient and environmentally safe technologies of preparation of heterogeneous catalysts for the reactions which were carried out under the influence of electro-magnetic radiation of microwave radiation. Author's research results confirm possibilities of practical use of proposed method of hydrothermal oxidation of industrial low-dispersing of aluminum powders with presence of bulk phase of $Al(OH)_3 \cdot nH_2O$, with further thermal treatment in microwave field for acquisition of armored A1/$A1_2O_3$ compositions, which effectively consume energy of microwave radiation. Due to the textured characteristics and thermo-transforming ability, synthesized components can be used as potential universal bearings of catalysts for reactions stimulated by electromagnetic radiation of (2.45 GHz) microwave frequency.

Keywords

Microwave Radiation, Heterogeneous Catalysis, γ-$A1_2O_3$/Al-Carrier, Conversion of Exhaust Gas, Carbon Monoxide, Oxides of Metals with Variable Valence

1. Introduction

While 50-60-ies of XX century, the main toxic air pollution, were made of industrial and domestic companies, the level of emissions of toxic components into the environment of internal combustion engines (ICE) of last decade has exceeded many industrial facilities as a fact of rapid growth in the number of cars and trucks (currently in the world there are over 450 million vehicles units).

While improving of motor fuel quality and design of internal combustion engines achieved by countries of the "far" and "near" abroad has significantly improved environmental characteristics of used vehicle park, however, the massive use of catalytic converters of exhaust gases should be considered the most important protection measure ensured environmental compliance of vehicles [1].

Previously, we have performed a comprehensive study on the development of effective catalysts for deep oxidation of hydrocarbons and carbon monoxide in the exhaust gas carburettor, containing the active composition which is not deficient oxides of copper, chromium, cobalt and manganese on the surface volumetrically structured Al_2O_3/A1-frame carriers [2].

It is shown that the highest activity in the oxidative conversion of carbon monoxide exhibit catalytic systems is characterized by a maximum content of secondary alumina coating phase chromate copper ($CuCrO_4$), while as for deep oxidative conversion of hydrocarbon (n-butane) responsible phase chromite cobalt ($CoCr_2O_4$). To realize the optimum profile of the macroscopic-ray distribution in the matrix phase, designated carrier has been proposed a method for separate application of opposite binary combinations of active metals, described in detail in [3].

However, heat treatment of potential catalysts under traditional heat transfer due to the presence of a temperature gradient from the surface of the carrier granules to the center, occurs uneven distribution of active components, and as a consequence, the generated samples exhibit poorly reproducible catalytic activity [4].

In this regard, it should be noted that a distinct advantage of the method of the microwave heat treatment of materials has spread in recent years in the field of high technologies such as the preparation of heterogeneous catalysts and catalytic transformations of the further stimulated by microwave radiation [5] [6].

In this paper, there was an attempt to intensify the process of preparation of Cu-Cr-Co/Al_2O_3/Al-catalysts joint deep oxidation of hydrocarbons and mono-oxide, carbon (for example, the conversion of n-butane and CO) through the formation of catalytically active phase under the action of the microwave field and holding reaction at stimulating effects of microwave radiation.

The task of selecting the component composition of the catalysts for the reaction indicated in a microwave field, thermal treatment was significantly facilitated by the development of the formulation of cooking in a conventional heating [7].

2. Methods and Apparatus

Microwave brand EM-G5593V (Panasonic) with the volume of the cavity 23 liters. functioning at an operating frequency of 2450 MHz, with a maximum input power of the radiation generator (magnetron) of 800 watts. X-ray diffractometer DRON-3 with a graphite monochromator. Laboratory microwave brand NE-1064F (Panasonic) with the volume of the resonator 14 liters.

3. Experimental

Samples of the catalysts were prepared by impregnation (incipient wetness) absorbing the microwave radiation is γ-$A1_2O_3$/Al-carrier nitrates of copper, chromium and cobalt-based content of the oxide forms of metals in the active mass (calculated as oxides CuO, Cr_2O_3 and Co_3O_4), respectively, 53%, 32% and 15% (wt.), at a total ratio of metal oxide to the weight of active γ-$A1_2O_3$/Al-carrier of 10%, 15% and 20% (wt.).

Heat treatment of the samples was carried out at the facility, constructed on the basis of microwave brand EM-G5593V (Panasonic) with the volume of the cavity 23 liters. functioning at an operating frequency of 2450 MHz with a maximum input power of the radiation generator (magnetron) of 800 watts. The technical capabilities of the microwave oven allows for heat treatment as a normal sample electrically heated coil, so and programmed to vary the ratio of power of microwave and electric heating. To avoid overheating the sample in the cavity furnace installed capacity of the circulating flow of distilled water. For X-ray diffraction patterns of the powders was applied to glass and instrumentation fixed with varnish, which has its own structural reflections.

X-ray diffraction powder samples were obtained on an automated X-ray diffractometer DRON-3 with a graphite monochromator.

The measurements were carried out on CuKα radiation in step-scan mode with a step $2\theta = 0, 1°$. The exposure time per point is 3 seconds. Treatment-togramm diffraction was carried out using software for qualitative and quantitative X-ray analysis [8].

Temperature-reduced catalysts was investigated by the method described in [9].

Experiments to assess the penetration depth of microwave radiation into the charge Ni-Co-Cr/Al_2O_3/Al-catalyst and its termotransformatsionnyh properties conducted on an apparatus constructed based on laboratory microwave brand NE-1064F (Panasonic) with the volume of the resonator 14 liters [10].

Specific surface of the samples were examined for the device "Sorbie-MS" BET.

4. Results and Discussion

Figure 1 illustrates the dynamics of changes in temperature of samples of potential catalysts in the process of heat treatment due to the absorption of microwave energy radiation with a frequency 2450 MHz varying power. For comparison, shows the dynamics of changes of sample temperature γ-Al_2O_3/Al-carrier under microwave irradiation prior to impregnation with a solution of salts of the active metal directly after the hydrothermal treatment step, the curve (2).

It can be seen that regardless of the component composition of the active mass in the first few minutes of exposure wet sample due to the high value of the dielectric loss intensively absorb microwave energy even when the power of the magnetron, to be ~30% of maximum. At the same time, a high rate of heating of samples (about 25 - 30 K/min.) The intense evaporation leads (due to heat) to reduce the temperature of the samples, and for this reason the initial phase of the microwave heating of the dynamics of change in temperature is extreme.

After completion of the drying process, the samples impregnated with the appropriate metal salts, raise the magnetron power of 300 to 600 Wt. It leads to an extreme increase in temperature with highs in the range 450 - 500 K.

The observed changes in the dynamics of the sample temperature in that range is associated with an exothermic reaction thermolysis nitrates of copper, chromium and cobalt in an air atmosphere (at the outlet of the catalyst with the charge capacity observed efflux dyed brown nitrogen oxides). Further increases in radiation power

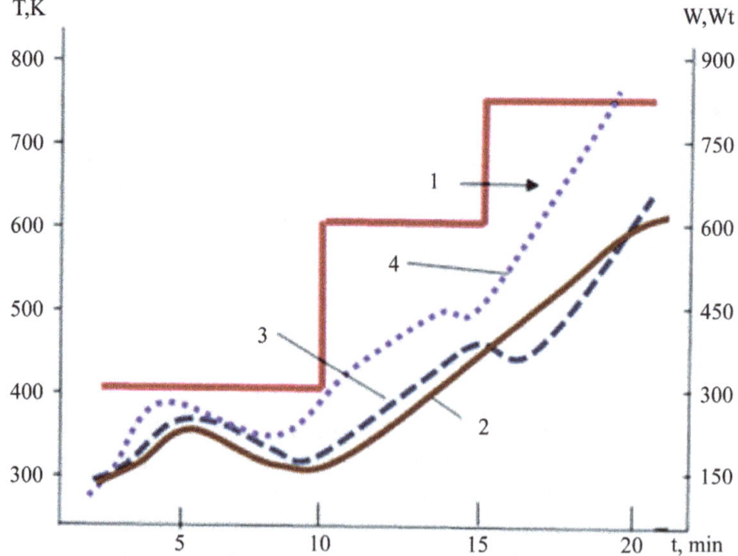

Figure 1. Influence of power microwave radiation (1); the dynamics of change in tempera tours samples γ-Al_2O_3/Al carrier in the drying process—(2); Media samples simultaneously impregnated with a solution of nitrate Cu, Cr, and Co—(3); samples prepared by the method of separate application of binary combinations (Cu and Cr) and (Cr and Co) based oxides, the amount of content in the matrix media 20% (wt.)—4.

results in an increase in temperature with entry temperature region start the formation of active metal oxide phase (623 - 653 K) [9] [10].

Exposure to microwave field patterns γ-Al_2O_3/Al-carrier subjected single-stageimpregnation of a mixture of nitrates of Cu, Cr, and Co (curve 3) results in a less intensive heating the samples when the power of magnetron 600 to 800 Wt. In the case of the samples, obtained by the application of separate components of the active mass (curve 4) observed a higher rate of temperature rise, with comparable values of the magnetron power.

This fact is probably related to an event during heat treatment of the samples the formation of the phase composition of oxide forms of the active metals are characterized by a relatively large amount of dielectric loss.

On radiographs presented in **Figure 2** there is qualitative agreement reflexes oxide phases, formed in conditions of traditional heat treatment of the samples (range A) and under microwave irradiation (spectra B and C).

Comparison of the X-ray samples synthesized by a joint (simultaneous spectrum B) and separate (consecutive spectrum C) the introduction of the matrix carrier oxides of active metals allowed to come to the conclusion that the latter are implemented more favorable conditions for the formation of catalytically active phase chromite, cobalt ($CoCr_2O_4$) and copper chromate ($CuCrO_4$), as evidenced more intensive reflections in the spectrum of the data phase, evidenced reflexes phase data in the spectrum (C) a relatively large area.

The proximity of the qualitative composition of the oxide forms of metals, are part of an asset-term surface of the Cu-Cr-Co/Al_2O_3/Al-catalysts also follows from a comparison of the spectra of the temperature-programmed reduction of samples prepared by thermal treatment of electric heating [4], and under the influence of the microwave field (**Figure 3**).

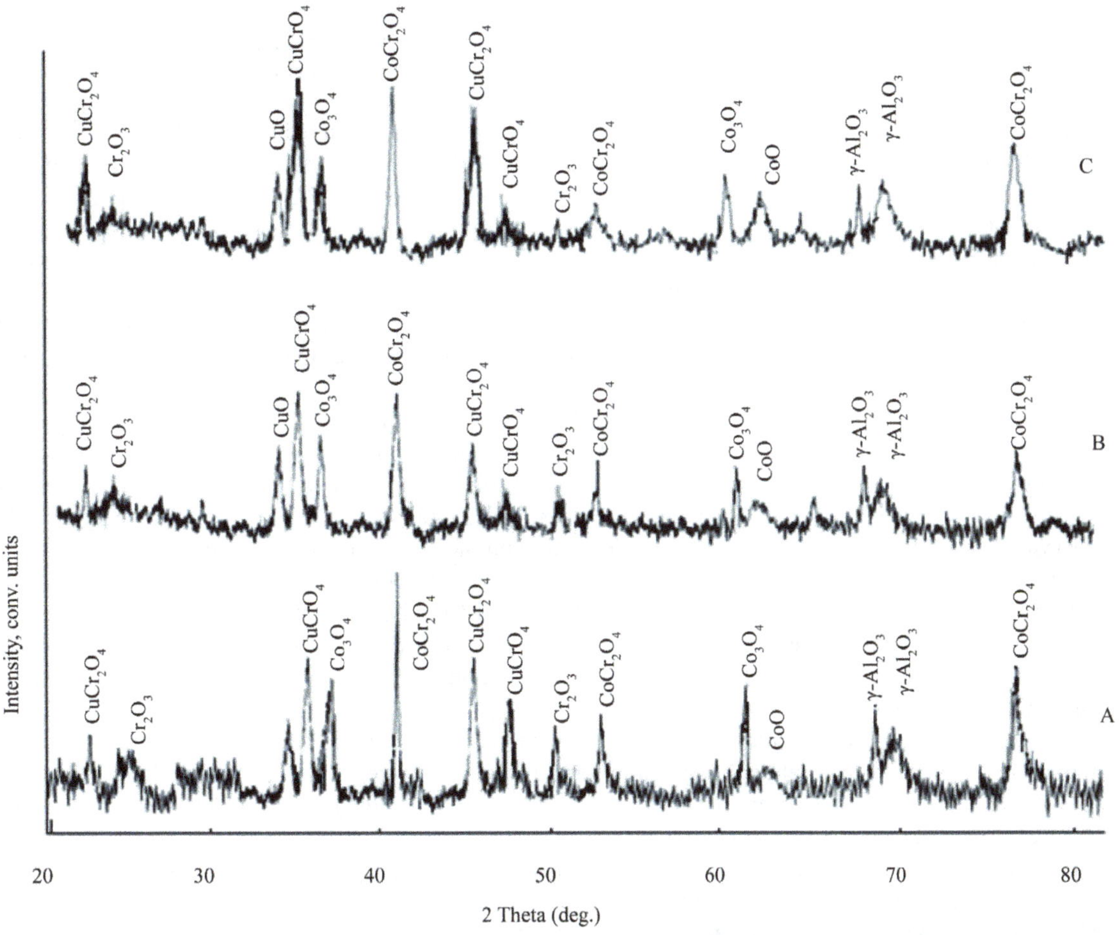

Figure 2. The XRD pattern of samples of Cu-Cr-Co/Al_2O_3/Al-catalysts obtained by impregnation step Al_2O_3/Al-Cu nitrate medium, Cr and Co in a conventional heat treatment (A); obtained by heat treatment under conditions of microwave-radiation for one-step impregnation of Al_2O_3/Al-Cu nitrate medium, Cr and Co (B); obtained under step impregnation of the support by binary combinations of nitrates of Cu-Cr and Co-Cr (C).

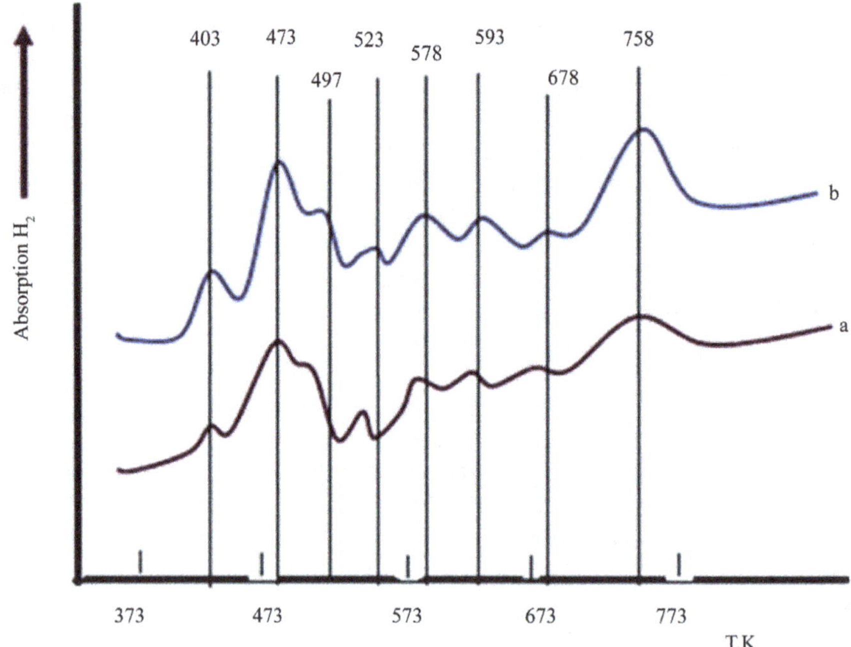

Figure 3. The curves of temperature-programmed reduction of samples of Cu-Cr-Co/Al_2O_3/Al-catalysts obtained by heat treatment electrical heating (a) and in the microwave field (b).

It is seen that the absorption maxima are observed temperature hydrogen reduction of the oxides in the Cr (VI) (403 K) and Cu (II) (523 K) in phase chromate copper ($CuCrO_4$); Cr (III) (593 K) in the phase of copper chromite ($CuCr_2O_4$); Co (II) (478 K) in the phase of cobalt chromite ($CoCr_2O_4$); Co (II) (678 K) in the oxide-cobalt oxide (Co_3O_4); Cr (III) (758) in the chromium oxide Cr_2O_3 for both types of samples are in close agreement.

It should be noted that the completion of the phase formation of chromate copper and cobalt chromium, responsible for the conversion of carbon monoxide and n-butane in a conventional heat treatment the Cu-Cr-Co/ Al_2O_3/Al-catalysts electric heating occurs at a temperature of 773 - 773 K for 6 - 7 h., while the formation of catalytically active phases $CuCrO_4$ and $CoCr_2O_4$ samples during heat treatment in the microwave field occurs within 25 to 30 minutes.

It was also found **Figure 1** that the samples of Cu-Cr-Co/Al_2O_3/Al-catalyst formed under conditions conducive to the formation of a maximum in the matrix of the secondary carrier phase $CuCrO_4$ and $CoCr_2O_4$ characterized by a sufficiently high intensity of heat due to the absorption of microwave energy field, increases with increasing content in samples of the active metal oxides. The temperature in the reaction device by transforming the absorbed electromagnetic energy reaches (depending on the amount of supported oxides) values of the order of 600 - 800 K, and stabilized by establishing heat balance with the environment.

Thus, the results suggest that the formation of the Cu-Cr-Co/Al_2O_3/Al-catalysts joint deep oxidation of hydrocarbons and CO matched component composition using a heat treatment in a microwave field is acceptable for practical implementation, and from the point of view saving time and energy cost compares favorably with the traditional processes of heat.

It is known that in order to achieve the necessary temperature to activate the reaction system under microwave irradiation requires a high level of absorption of the catalyst the charge energy microwave field and its transformation into heat. Also, a significant factor affecting the uniformity of the heating of the catalyst bed is the depth of penetration of the microwave radiation.

Figure 4 is a bar graph illustrating the thermal transformation properties of samples Cu-Cr-Co/Al_2O_3/Al-catalyst and penetration depth of the volume of the microwave radiation. For comparison, the characteristics of Al_2O_3/Al-carrier before applying the active composition (A)

It is seen that the initial capacity loss of radiation for samples prepared by single-step impregnation of Al_2O_3/ Al-medium slightly exceeds the weight loss of the carrier, in while samples prepared by separate impregnation exceed said samples by this parameter.

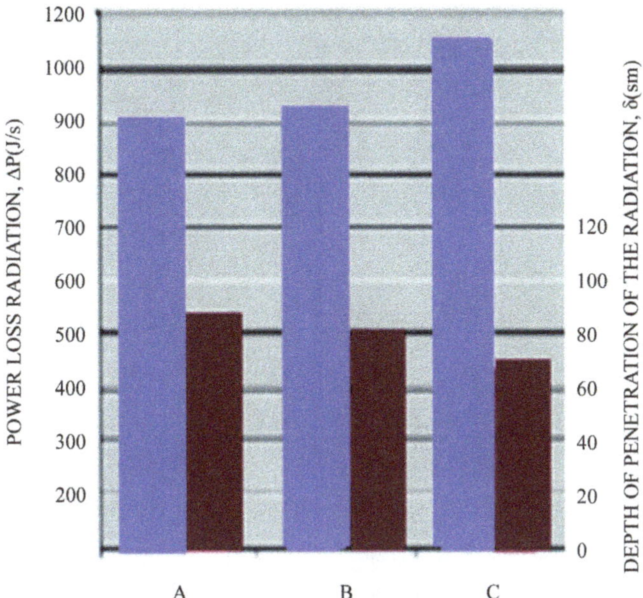

Figure 4. Thermo transformation properties and the depth of penetration of microwave radiation: in a lot of Al_2O_3/Al-carrier before applying the active ingredients (A); catalyst prepared under microwave irradiation method of the single-step impregnation of the support with nitrates Cu, Cr and Co (B); catalyst prepared by the method of time-limiting (step) impregnation with a fractional separation of the components (B). Terms: P_{BX} magnetron = 800 wt., the exposure time of 2.5 minutes.

The level of the microwave radiation power loss (ΔP) in the case of samples prepared by the method of separate application of active ingredients is sufficient, in accordance with the expression given in [11] [12]:

$$\frac{\Delta T}{\Delta \tau} = \frac{\Delta P}{c \cdot d} \tag{1}$$

where: $\Delta T/\Delta \tau$ —rate of temperature rise (K/sec.), the power loss is proportional to the radiation of the magnetron (ΔP); c—average heat capacity of the sample (kcal/deg·mol); d—density (g/sm^3), for a few minutes to ensure the temperature of the "ignition" of the catalyst—473K and above.

In the same sufficiently to ensure a uniform (no gradient), temperature distribution in the catalyst bed is the penetration depth of microwave radiation δ, ~70 sm.

As seen from the data presented in **Table 1**, samples of Cu-Cr-Co/Al_2O_3/Al-catalyst, heat treatment which takes place in the field of microwave radiation, under identical conditions, the reaction co-deep oxidation of n-butane and CO are very active.

The observed increase in activity of the samples produced using microwave heating is probably associated with the formation on the surface of the catalysts deposited type a fine crystalline phase of active oxides metal fishing variable valence resulting from uniform sample heating when exposed to microwave radiation in a shorter period of time. This indirectly indicates a relatively large area of the peaks of the absorption maxima of hydrogen at a temperature programmed reduction of metal oxides comprising the active mass of the catalyst samples.

5. Conclusions

Thus, the results suggest that the formation of the Cu-Cr-Co/Al_2O_3/Al-catalysts joint deep oxidation of hydrocarbons and CO matched component composition using a heat treatment in a microwave field is acceptable for practical implementation, and from the point of view saving time and energy cost compares favorably with the traditional processes of heat.

One important factor in improving the operating performance of heterogeneous catalytic processes, the thermal activation of catalysts is carried out at the stage of their preparation, and in the immediate environment of

Table 1. Dependence of rate and extent of conversion of n-butane at a flow rate of the gas flow at the joint oxidation of carbon monoxide in the presence of. Cu-Cr-Co/Al_2O_3/Al-catalyst. Terms and conditions of reaction: Temperature 573 ± 5 K, $P^0_{C_4H_{10}}$ = 0.02 atm.; P^0_{CO} = 0,02 atm.; $P^0_{O_2}$ = 0,15 atm. Magnetron power of 800 wt.

	V, h^{-1}	αC_4H_{10}, %	$W_{C_4H_{10}} \cdot 10^2$, mol/m^2·h	Partial pressure, $P_i \cdot 10^2$, atm.			
				P_{C4H10}	P_{CO}	P_{CO2}	P_{O2}
Heat treatment with electric heating	120,000	59.4	1.232	0.812	0.032	6.732	7.179
	100,000	63.2	1.093	0.736	0.028	7.036	5.794
	70,000	72.1	0.870	0.558	0.017	7.748	4.637
	50,000	77.9	0.671	0.442	0.008	8.212	3.883
	30,000	84.9	0.442	0.302	-	8.772	2.973
Heat treatment of microwaves	120,000	62.0	1.926	0.760	0.026	9.340	2.050
	100,000	72.4	1.611	0.560	0.022	9.428	1.907
	70,000	80.8	1.153	0.384	0.015	9.564	1.686
	50,000	86.1	0.828	0.278	0.010	9.668	1.517
	30,000	90.5	0.504	0.190	0.006	9.780	1.335

use. Thermal activation of the catalyst systems and methods of convection heat transfer methods is highly energy intensive and not very effective. With the increasing complexity of the structure and functionality of the catalysts, conventional methods of heat treatment are even less effective.

Results achieved by this study determines that the process formation of Cu-Cr-Co/Al_2O_3/Al-catalysts of deep oxidation of hydrocarbons and CO matched component composition using a heat treatment in a microwave field is acceptable for practical implementation, but in terms of saving time and energy cost it compares favorably with the traditional processes of heat.

References

[1] Heck, R. and Farrauto, R. (1998) Blocking Catalysts: Modern and Future Generations. *Kinetics and Catalysis*, **39**, 646-652.

[2] Muradova, P.A. and Litvishkov, Y.N. (2015) Steelframed Catalysts for Neutralization of Vehicle Exhaust. *Autogas Filling Complex + Alternative Fuel. Moscow*, **3**, 3-13. (In Russian)

[3] Thostenson, E.T. and Chou, T.W. (1999) Microwave Processing: Fundamentals and Applications. *Composites Part A: Applied Science and Manufacturing*, **30**, 1055-1071.

[4] Muradova, P.A. (2006) Joint Deep Oxidation of n-Butane and Carbon Monoxide with Present of Al-Frame Catalysts. PhD Tezis, Institute of Catalysis and Inorganic Chemistry Named after Academician M. Nagiyev of ANAS. Azerbaijan, Baku.

[5] Haque, K.E. (1999) Microwave Energy for Mineral Treatment Processes—A Brief Review. *International Journal of Mineral Processing*, **57**, 1-24.

[6] Bagirzade, G.A., Taghiyev, D.B. and Manafov, M.R. (2015) Vapor Phase Ammoxidation of 4-Phenylo-Xylene into 4-Phenylphthalonitrile on V-Sb-Bi-Zr/γ-Al_2O_3 Oxide Catalyst. *Modern Research in Catalysis*, **4**, 59-67. http://dx.doi.org/10.4236/mrc.2015.43008

[7] Litvishkov, Yu.N., Muradova, P.A., Godjayeva, N.S., Afandiyev, M.R., Jafarova, S.A., Zulfugarova, S.M., Shakunova, N.V., Mardanova, N.M. and Sheinin, V.E. (2005) Catalyst for the Oxidation of Carbon Monoxide. Patent of Azerbaijan Republic, No. İ 2005 0110.

[8] Wan, J.K.S., Wolf, K. and Heyding, R.D. (1984) Some Chemical Aspects of Microwave Assisted Catalytic Hydro-Assisted Processes. *Catalysis on the Energy Scene*. Elsevier, Amsterdam, 561.

[9] Litvishkov, Yu.N., Muradova, P.A., Efendiev, M.R., Godzhaeva, N.S., Guseynova, E.M. and Kulieva, L.A. (2004) Temperature-Programmed Reduction Cu-Cr-Co/Al_2O_3/Al-Frame Catalysts of Deep Oxidation of Hydrocarbons. *Azerbaijan Chemical Journal*, **4**, 88-92.

[10] Muradova, P.A., Talyshinsky, R.M., Mardanova, N.M., Litvishkov, Yu.N. and Godjayeva, N.S. (2003) Synthesis of the Multicomponent Catalyst for Reaction of Carbon Monoxide Oxidation: Kinetic Aspects of the Problem. *Process of Petrochemistry and Oil Refining*, **1**, 74-87.

[11] Dydenko, A.N. (1993) The Possibility of Using High-Power Microwave Oscillations for Texnological Purposes. *Reports of RAS*, **331**, 571-572.

[12] Bagirzade, G.A., Tagiyev, D.B. and Manafov, M.R. (2014) Synthesis of 4-Phenylphthalo-nitrile by Vapor-Phase Catalytic Ammoxidation of Intermediate 4-Phenyl-o-Tolunitrile: Reaction Kinetics. *Modern Research in Catalysis*, **3**, 6-11. http://dx.doi.org/10.4236/mrc.2014.31002

21

Microwave Plasma Enhanced Chemical Vapor Deposition of Carbon Nanotubes

Ivaylo Hinkov[1], Samir Farhat[2*], Cristian P. Lungu[3], Alix Gicquel[2], François Silva[2], Amine Mesbahi[2], Ovidiu Brinza[2], Cornel Porosnicu[3], Alexandru Anghel[3]

[1]University of Chemical Technology and Metallurgy, Sofia, Bulgaria
[2]Laboratoire des Sciences des Procédés et des Matériaux, CNRS, LSPM-UPR 3407, Université Paris 13, Villetaneuse, France
[3]National Institute for Laser, Plasma and Radiation Physics, Bucharest, Romania
Email: [*]farhat@lspm.cnrs.fr

Abstract

Multi-walled carbon nanotubes (MWCNTs) were grown by plasma-enhanced chemical vapor deposition (PECVD) in a bell jar reactor. A mixture of methane and hydrogen (CH_4/H_2) was decomposed over Ni catalyst previously deposited on Si-wafer by thermionic vacuum arc (TVA) technology. The growth parameters were optimized to obtain dense arrays of nanotubes and were found to be: hydrogen flow rate of 90 sccm; methane flow rate of 10 sccm; oxygen flow rate of 1 sccm; substrate temperature of 1123 K; total pressure of 10 mbar and microwave power of 342 Watt. Results are summarized and significant main factors and their interactions were identified. In addition a computational study of nanotubes growth rate was conducted using a gas phase reaction mechanism and surface nanotube formation model. Simulations were performed to determine the gas phase fields for temperature and species concentration as well as the surface-species coverage and carbon nanotubes growth rate. A kinetic mechanism which consists of 13 gas species, 43 gas reactions and 17 surface reactions has been used in the commercial computational fluid dynamics (CFD) software ANSYS Fluent. A comparison of simulated and experimental growth rate is presented in this paper. Simulation results agreed favorably with experimental data.

Keywords

Nanotubes, Growth, CVD, Modeling, Kinetics

[*]Corresponding author.

1. Introduction

Since their discovery by Iijima in 1991 [1], carbon nanotubes have generated much interest due to their quasi one-dimensional structure and their unique combinations of electronic, field emission, mechanical and chemical properties coupled with the new ability to grow them aligned on a substrate. This opened unlimited possibilities of applications such as field emitters, sensors, high-density energy storage devices, photonic crystals, active media for lasers, non-linear optical media etc... To grow vertically-aligned nanotubes, chemical vapor deposition (CVD) has emerged as a key technique. Indeed, contrarily to the arc, laser and HiPCO processes where the nanotubes are produced separately, purified and then manipulated for producing devices [2], CVD allows spatially controlled and highly functional components in (2D) and (3D) architecture opening the way to produce self-assembly devices with higher packing density and performances [3]. In addition, CVD offers low-temperature and large-scale production possibilities. In CVD systems, a thin catalyst layer is first deposited on silicon wafer by a separate physical vapor deposition PVD technique. When heated, the continuous catalyst layer disaggregates and forms small particles, with size controlled by the layer thickness in the range of 1 to 10 nm [4]. Then, the growth of nanotubes occurs through catalytic decomposition of a carbon gas source over the catalyst. The nanotube characteristics such as diameter, density, Single-walled SWNT versus Multi-walled MWNT depend on the size of these particles but also on the gas feedstock activation technique. Two distinct activation routes emerged, 1) thermally via an oven or hot-filament heating and 2) plasma enhanced chemical vapor deposition (PECVD) via DC, RF or microwave discharges. Plasma activation has the advantage to prevent thermal damage to the substrate allowing lower operating temperatures and better nanotubes vertical positioning due to the presence of an electric field normal to the substrate [5].

In the present work, we used thermionic vacuum arc (TVA) technology to produce uniform nickel layers of ~1 nm thickness. Then, a mixture of methane, hydrogen and oxygen ($CH_4/H_2/O_2$) was used to produce carbon nanotubes. Indeed, the addition of a controlled amount of a weak oxidizer as oxygen or water into the growth ambient of CVD was reported to significantly enhance the activity and lifetime of the catalyst resulting in efficient nanotubes growth [6] [7].

2. Experimental

Nickel films deposited on silicon substrates were prepared using thermionic vacuum arc (TVA) technology developed at NILPRP Bucharest [8]-[11]. The coating device consists of a tungsten filament surrounded by an electron focusing Wehnelt cylinder heated by an external high current source as cathode and an anode made of nickel. For ignition and maintaining the TVA arc two circuits are necessary: 1) for the heating of the cathode filament, where a relatively low voltage source (0 - 24 V) provides a 10 - 150 A current and 2) for the running up of the arc discharge, being used for this an adjustable source of high voltage (0 - 4 kV) and a current up to 3 A. The electrons coming from the cathode heats up and evaporates the anode and pure Ni plasma is ignited by applying a high dc voltage on the anode as illustrated in **Figure 1(a)**. The deposition chamber of **Figure 1(b)** has been under a residual pressure of 3×10^{-6} torr before the beginning of the coatings. For plasma ignition, the TVA gun filament has been heated with a 60 A current and at an alternative voltage of 20 V. The continuous voltage has been applied on the anode with an increasing rate of approximately 1000 V/min, being followed by the focusing process of the electron beam by the Wehnelt cylinder on the anode crucible. When the powder in the crucible has melt, the applied voltage was adjusted in order to ensure the ignition of the discharge in the Ni

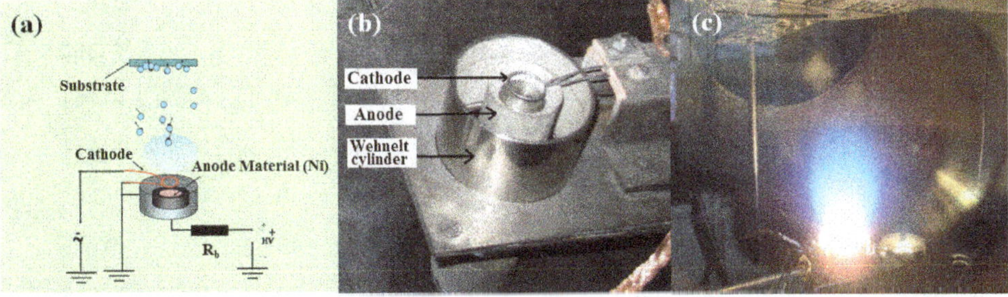

Figure 1. Thermionic vacuum arc TVA set-up for catalyst deposition. (a) Principle, (b) deposition chamber and (c) plasma running in Ni vapors.

vapors. A stable discharge shown in **Figure 1(c)**, is obtained and the film thickness measured during all the duration of the deposition process with a quartz balance equipment. The deposition has been interrupted when the thickness of 1 nm is reached. At this step, the anode voltage and the applied current to the TVA gun filament have been reduced to zero and the sample kept in the deposition chamber, under high vacuum for about 120 minutes to slowly cool down. Finally, Ni/Si substrates with ~1 nm nickel thickness were obtained in 18 - 20 s, due to a fine control of the deposition rate of ~0.05 nm/s.

For nanotube growth, we used a 10 cm diameter silica bell jar low pressure reactor activated by a microwave electric field (**Figure 2**) and developed originally to CVD diamond growth [12] [13]. The input gases ($CH_4/H_2/O_2$) with mass flow rates controlled electronically were injected in the reactor and exit via the reactor pumping system. The Ni/Si substrate is held in a resistance boat made in molybdenum and electrically heated to a temperature ranging from 973 to 1123 K. During all the experiment, substrate temperature was monitored by an optical pyrometer. The reactor utilizes 1.2 kW SAIREM microwave generator operating at 2.45 Ghz. The electromagnetic waves are generated, guided in a rectangular wave guide and applied inside the cavity delimited by Faraday cage (**Figure 2(a)**). The short-circuit piston at the end of the wave guide helped to create stationary waves and to situate the maximum of the electric field near the substrate. Input power was varied with the pressure simultaneously in order to hold plasma volume constant (**Figure 2(b)**). Efficient operation is assumed with good microwave coupling, and minimal radial diffusion to the quartz enclosure thereby leading to greater discharge stability and better plasma uniformity. As shown in **Figure 2(b)**, a quasi hemispherical active plasma zone of radius of 2.5 cm is created near the substrate. The function of this zone is to produce the charged and radical species that diffuse to the catalyst particle and contribute to nanotube growth. For microwave PECVD nanotube synthesis we developed an experimental protocol composed by three steps: 1) thermal annealing of Ni/Si substrates, 2) hydrogenation of Ni catalyst, 3) nanotube growth.

For this protocol, the essential parameters were optimized using Taguchi design method of **Table 1**, with 3 factors and 2 levels. These parameters are namely, substrate temperature, hydrogen flow rate, and total pressure. This last parameter was coupled with the total input microwave power to hold the plasma volume constant. For all the 8 experiments, the silicon substrate covered by 1 nm thick nickel was first annealed in vacuum at specified temperature, then nickel catalyst particles was reduced using hydrogen plasma for 10 minutes. Finally, 10 sccm of methane were introduced to grow nanotubes during 20 minutes.

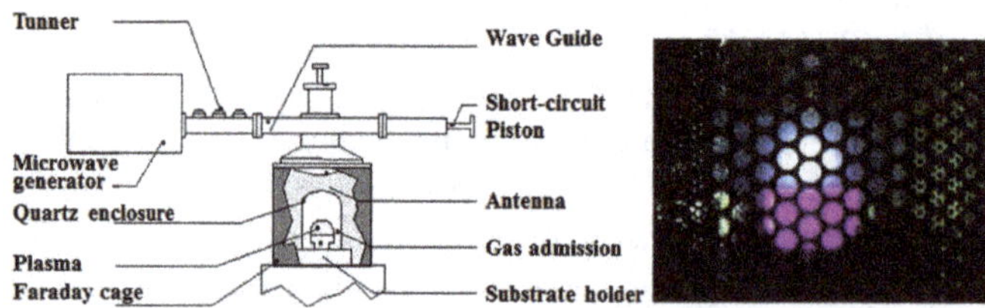

Figure 2. PECVD Bell jar reactor. (a) Scheme and (b) plasma picture through the Faraday cage during nanotubes growth.

Table 1. Factors and levels for the PECVD synthesis of nanotubes. Pressure and microwave power are coupled.

Experiment No.	*Substrate Temperature* T_{sub} (K)	H_2 *Flow rate* Q_{H_2} (sccm)	CH_4 *Flow rate* Q_{CH_4} (sccm)	*Total pressure* P (mbar)	*Microwave power* P_{mw} (Watt)
E1	973	10	10	10	342.82
E2	1123	10	10	10	342.82
E3	973	10	10	40	814.54
E4	1123	10	10	40	814.54
E5	973	100	10	10	342.82
E6	1123	100	10	10	342.82
E7	973	100	10	40	814.54
E8	1123	100	10	40	814.54

After removing samples from the reactor, they were analyzed by scanning electron microscopy SEM LEO 440. In addition, surface morphology of the substrates, before and after thermal annealing was examined by atomic force microscopy AFM D3100, Nanoscope NS3.

3. Modeling Approach

A two-dimensional simulation of the PECVD process was performed in order to compare the theoretical prediction and the experimental measurements. Computational fluid dynamics (CFD) modeling evaluations were made for the temperature and species concentrations profiles as well as for the carbon nanotube growth rate in the reactor. Simulations were performed by using the commercial software ANSYS Fluent version 12.

3.1. Geometry and Assumptions

The 2D computational domain shown in **Figure 3** includes the PECVD reactor quartz enclosure, containing the substrate holder and the plasma zone. From experimental observations, plasmas are most intense along with the top edge of the substrate holder. The microwave plasma volume used in the present simulations was estimated from visual observations under experimental growth conditions in the reactor. The geometry was created using ANSYS Design Modeler and the mesh is generated using ANSYS Meshing application. The chosen dimensions of the reactor refer to the experimental setup. The grid is composed of an unstructured triangular mesh. Because of the strong temperature and concentrations gradients near the substrate where the CNT are grown, a condensed quadrilateral mesh refinement was applied on this region. After numerous checks for grid sensitivity and mesh constraints, the total number of elements is 4583, the final grid has 2487 nodes with a grid skewness maximum of 0.78. To understand the effect of the macroscopic process parameters on carbon nanotubes growth rate, a 2D model has been developed based on the following assumptions:

- The plasma is in Local Thermodynamic Equilibrium (LTE). This assumption allows us to define a unique temperature of all plasma species in localized areas in the plasma;
- Laminar flow: it is characterized by relatively low values of the Reynolds number caused by small inlet flow rate;
- The plasma is modeled using a steady state time formulation;
- Axisymmetric physical domain;
- The radiative losses are neglected;
- Only neutral species are involved in the gas-phase and surface chemistry.

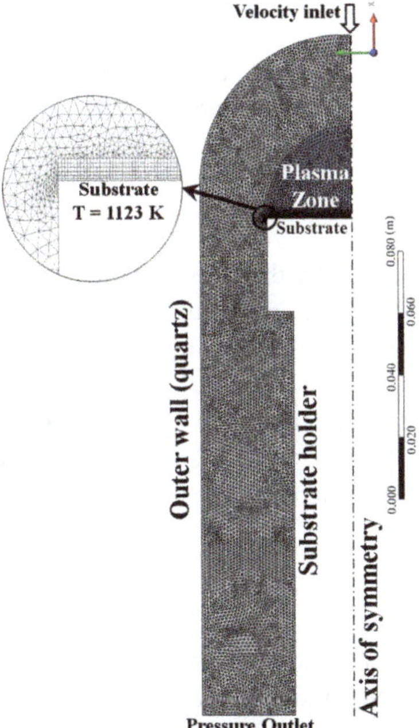

Figure 3. Two-dimensional computational domain.

These assumptions lead to the following limitations: The existence of only neutral species affects the accuracy of the calculated concentrations of considered species. We expect the influence of the charged species on carbon nanotubes growth rate to be minor because their molar fraction is not great. The major species formed in the plasma are H, H_2, CH_4, CH_3, C_2H_2, and C_2H_4 and the gas phase chemistry in the plasma is dominated by the neutral species [14]. Furthermore, assuming LTE reduces the complexity of the mixture considerably.

3.2. Gas-Phase Chemistry

In the methane/hydrogen plasma, different species are created due to chemical reactions. The considered gas-phase chemistry involves 13 neutral species and consists of 43 reactions leading to the conversion of CH_4. This reaction set is given in **Table 2**. The gas phase chemistry model describes homogeneous reactions that influence the species concentration distribution near the deposition surface through the production/destruction of chemical species in the gas phase. Each reaction is assumed to be reversible. The temperature dependence of the forward rate constants is usually described through a modified Arrhenius type of expression.

3.3. Surface Chemistry

The surface chemistry model used in the present study describes the reactions and other processes that take place at the substrate surface, involving both gaseous species impinging on the surface, adsorbed molecules, atoms and free sites. These surface processes lead to the growth of solid carbon nanotubes. Actually, surface reactions in the PECVD process are not fully understood. The proposed surface reaction mechanism consists of 18 heterogeneous reactions involving vacant surface sites S_{Ni} on a nickel catalyst particle, 7 surface species (CH_4(s), CH_3(s), C_1H_2(s), CH(s), C(s), H(s), CNT) and 5 gaseous species (C_2H_2, CH_4, CH_3, H_2, H). These reactions include surface site adsorption/desorption, hydrogen abstraction/addition, and carbon diffusion toward a carbon nanotube growth edge (**Table 3**).

The rate of deposition is governed by both chemical kinetics and the diffusion rate from gas to the surface. The reactions create sources of chemical species in the bulk phase and determine the rate of deposition of surface species.

Following the above considerations, the growth of MWNT on the substrate is expected to occur as follows. The plasma generates vapors and provides carbon contamination to the nickel particles. These particles having suitable temperature and size will be the sites of CNT growth.

The surface structure is associated with a surface site density Γ (given in $mol \cdot cm^{-2}$) required to evaluate the surface growth rate of MWNT. Since experimental determination of site density is difficult, we used the value for the reconstructed diamond (100) surface $\Gamma = 2.61 \times 10^{-9}$ as an upper limit. In the specific case of nanotube growth in PECVD reactor, the surface site density is certainly much lower than this value. It could be estimated from the substrate density of catalytic nickel particles and the concentration of the surface sites occupied by C atoms on each nanoparticle.

3.4. Initial and Boundary Conditions

The temperature at the outer boundary walls and the substrate holder except for the substrate surface was fixed to 400 K. The temperature of the substrate surface was much higher, at 1123 K.

The inlet conditions of the simulations were derived from the experimental conditions *i.e.*, volume %: 10% CH_4 and 90% H_2. The gas mixture was initialized to a uniform temperature of 298 K. The gas velocity was specified as a uniform inflow condition with vertical upward velocity of 0.1326 m/s at a temperature of 298 K. The initial inlet mole fractions for all species were calculated using a thermochemical model based on Chemkin software in 0D [31]. The gas outlet was specified as a pressure outlet. The initial pressure inside the reactor was fixed at 10 mbar.

3.5. Computational Procedure

2D reactor simulation including coupled momentum, heat and species transfer was performed by using the CFD code ANSYS Fluent. It utilizes the finite volume method to solve the governing equations, *i.e.*, conservation of total mass, momentum, and energy, and the individual species conservation equations. The reactive flow is modeled using the 2D axisymmetric laminar finite-rate model, including the above-mentioned volumetric and

Table 2. Gas-phase reactions.

No.	*Gas Phase Reactions*	A^*	β (-)	E_a (cal/mol)	*Ref.*
1.	$H + H + M = H_2 + M$	1.00×10^{18}	−1.0	0.0	[15]
	H_2 Enhanced by 2.0				
2.	$H + H + H_2 = H_2 + H_2$	9.20×10^{16}	−0.6	0.0	[15]
3.	$CH_3 + CH_3 (+M) = C_2H_6 (+M)$	9.22×10^{16}	−1.174	636.0	[16]
	Low pressure limit: 1.14×10^{36} −5.246 1705.0				
	TROE centering: 0.405 1120.0 69.6 1.0×10^{15}				
	H_2 Enhanced by 2.0				
4.	$CH_3 + H (+M) = CH_4 (+M)$	2.14×10^{15}	−0.4	0.0	[17]
	Low pressure limit: 3.31×10^{30} −4.0 2108.0				
	TROE centering: 0.0 1.00×10^{-15} 1.0×10^{-15} 40				
	H_2 Enhanced by 2.0				
5.	$CH_4 + H = CH_3 + H_2$	2.20×10^{4}	3.0	8750.0	[18]
6.	$CH_3 + H = CH_2 + H_2$	9.00×10^{13}	0.0	15100.0	[18]
7.	$CH_3 + M = CH + H_2 + M$	6.90×10^{14}	0.0	82469.0	[19]
8.	$CH_3 + M = CH_2 + H + M$	1.90×10^{16}	0.0	91411.0	[19]
9.	$CH_2 + H = CH + H_2$	1.00×10^{18}	−1.56	0.0	[18]
10.	$CH_2 + CH_3 = C_2H_4 + H$	4.00×10^{13}	0.0	0.0	[18]
11.	$CH_2 + CH_2 = C_2H_2 + H + H$	4.00×10^{13}	0.0	0.0	[18]
12.	$CH_2(S) + M = CH_2 + M$	1.00×10^{13}	0.0	0.0	[18]
	H_2 Enhanced by 12.0				
	C_2H_2 Enhanced by 4.0				
13.	$CH_2(S) + CH_4 = CH_3 + CH_3$	4.00×10^{13}	0.0	0.0	[18]
14.	$CH_2(S) + C_2H_6 = CH_3 + C_2H_5$	1.20×10^{14}	0.0	0.0	[18]
15.	$CH_2(S) + H_2 = CH_3 + H$	7.00×10^{13}	0.0	0.0	[18]
16.	$CH_2(S) + H = CH + H_2$	3.00×10^{13}	0.0	0.0	[18]
17.	$CH_2(S) + CH_3 = C_2H_4 + H$	2.00×10^{13}	0.0	0.0	[18]
18.	$CH + H = C + H_2$	1.50×10^{14}	0.0	0.0	[18]
18.	$CH + CH_2 = C_2H_2 + H$	4.00×10^{13}	0.0	0.0	[18]
20.	$CH + CH_3 = C_2H_3 + H$	3.00×10^{13}	0.0	0.0	[18]
21.	$CH + CH_4 = C_2H_4 + H$	6.00×10^{13}	0.0	0.0	[18]
22.	$C + CH_3 = C_2H_2 + H$	5.00×10^{13}	0.0	0.0	[18]
23.	$C + CH_2 = C_2H + H$	5.00×10^{13}	0.0	0.0	[18]
24.	$C_2H_6 + CH_3 = C_2H_5 + CH_4$	5.50×10^{-1}	4.0	8300.0	[18]
25.	$C_2H_6 + H = C_2H_5 + H_2$	5.40×10^{2}	3.5	5210.0	[18]
26.	$C_2H_5 + H = C_2H_4 + H_2$	1.25×10^{14}	0.0	8000.0	[15]
27.	$C_2H_5 + H = CH_3 + CH_3$	3.00×10^{13}	0.0	0.0	[20]
28.	$C_2H_5 + H = C_2H_6$	1.00×10^{14}	0.0	0.0	[15]
29.	$C_2H_4 + H = C_2H_3 + H_2$	3.36×10^{-7}	6.0	1692.0	[21]
30.	$C_2H_4 + CH_3 = C_2H_3 + CH_4$	6.62	3.7	9500.0	[15]
31.	$C_2H_4 + H (+M) = C_2H_5 (+M)$	1.08×10^{12}	0.454	1822.0	[22]
	Low pressure limit: 1.112×10^{34} −5.0 4448.0				
	TROE centering: 1.0 1.00×10^{-15} 95.0 200.0				
	H_2 Enhanced by 2.0				

Continued

32.	C_2H_4 (+M) = C_2H_2 + H_2 (+M)	1.80×10^{13}	0.0	76000.0	[23]
	Low pressure limit: 1.50×10^{15} 0.0 55443.0				[24]
33.	C_2H_4 (+M) = C_2H_3 + H (+M)	2.00×10^{16}	0.0	110000.0	[25]
	Low pressure limit: 1.40×10^{15} 0.0 81833.0				[24]
34.	C_2H_3 + H = C_2H_2 + H_2	4.00×10^{13}	0.0	0.0	[18]
35.	C_2H_3 + C_2H = C_2H_2 + C_2H_2	3.00×10^{13}	0.0	0.0	[18]
36.	C_2H_3 + CH = CH_2 + C_2H_2	5.00×10^{13}	0.0	0.0	[18]
37.	C_2H_3 + CH_3 = C_2H_2 + CH_4	2.00×10^{13}	0.0	0.0	[26]
38.	C_2H_3 + C_2H_3 = C_2H_4 + C_2H_2	1.45×10^{13}	0.0	0.0	[26]
39.	C_2H_2 + CH_3 = C_2H + CH_4	1.81×10^{11}	0.0	17289.0	[17]
40.	C_2H_2 + M = C_2H + H + M	4.20×10^{16}	0.0	107000.0	[18]
41.	C_2H_2 +H (+M) = C_2H_3 (+M)	3.11×10^{11}	0.58	2589.0	[27]
	Low pressure limit: 2.25×10^{40} −7.269 6577.0				
	TROE centering: 1.0 1.00×10^{-15} 675.0 1.0×10^{15}				
	H_2 Enhanced by 2.0				
42.	C_2H + H_2 = C_2H_2 + H	4.09×10^5	2.39	864.3	[18]
43.	C_2 + H_2 = C_2H + H	4.00×10^5	2.4	1000.0	[18]

*Units for A depend on the reaction order but are defined in terms of mol, cm^3 and s.

Table 3. Surface reactions.

No.	*Surface Reactions*	*A**	*B* (-)	E_a (cal/mol)	*Ref.*
S1.	H_2 + S_{Ni} + S_{Ni} → H(s) + H(s)	0.01**	0.0	0.0	[28]
S2.	H(s) + H(s) → H_2 + S_{Ni} + S_{Ni}	2.545×10^{19}	0.0	19379.0	[28]
S3.	CH_4 + S_{Ni} → CH_4(s)	0.008**	0.0	0.0	[28]
S4.	CH_4(s) → CH_4 + S_{Ni}	8.705×10^{15}	0.0	8962.0	[28]
S5.	CH_4(s) + S_{Ni} → CH_3(s) + H(s)	3.700×10^{21}	0.0	13770.8	[28]
S6.	CH_3(s) + H(s) → CH_4(s) + S_{Ni}	6.034×10^{21}	0.0	14701.6	[28]
S7.	S_{Ni} + CH_3 → CH_3(s)	5.000×10^{12}	0.0	0.0	[29]
S8.	CH_3(s) + H = C_1H_2(s) + H_2	2.800×10^7	2.0	7700.0	[29]
S9.	C_1H_2(s) + H = CH(s) + H_2	2.800×10^7	2.0	7700.0	[29]
S10.	CH(s) + H = C(s) + H_2	2.800×10^7	2.0	7700.0	[29]
S11.	C_1H_2(S) + H → S_{Ni} + CH_3	3.000×10^{13}	0.0	0.0	[29]
S12.	CH(s) + S_{Ni} → H(s) + C(s)	3.700×10^{21}	0.0	4486.8	[28]
S13.	C(s) + H(s) → CH(s) + S_{Ni}	4.562×10^{22}	0.0	38448.5	[28]
S14.	S_{Ni} + H → H(s)	1.000×10^{13}	0.0	0.0	[29]
S15.	H(s) + H → S_{Ni} + H_2	1.300×10^{14}	0.0	7.3	[29]
S16.	$2S_{Ni}$ + C_2H_2 → 2C(s) + H_2	7.700×10^{10}	0.0	67160.0	[30]
S17.	C(s) = CNT + S_{Ni}	1.300×10^{12}	0.0	31104.0	[29]
S18.	C(s) + CNT = 2CNT + S_{Ni}	1.300×10^{12}	0.0	31104.0	[29]

*Units for A depend on the reaction order but are defined in terms of mol, cm^3 and s. **Sticking coefficient.

surface reactions. The Simple method for pressure-velocity coupling and the second order upwind scheme to interpolate the variables on the surface of the control volume were selected.

The model requires knowledge of the thermo-chemical and transport properties of the gases in the reactor chamber. Thermo-chemical properties of the gas species as a function of temperature have been taken from CHEMKIN thermodynamic database [32]-[33]. Temperature and species dependence was imposed in calculations of thermodynamic and transport properties. The required Lennard-Jones parameters for many CVD gases can be found in e.g. Ref. [34]. Viscosity of the individual species was calculated by using kinetic theory.

The mixture viscosity was calculated using ideal gas mixing law. Thermal conductivity for individual species was calculated using kinetic theory. Specific heat capacity of individual species was calculated using piece-wise-polynomial approximation.

To account for the plasma heating from the microwave power, a constant volumetric heat source, applied in the plasma zone, was included in the governing energy equation. It was calculated from the input plasma power and the estimated plasma volume. For input plasma power of 342.8 W, the calculated power density was 0.634×10^7 W/m^3. The heat source using CH_4 and H_2 as a medium creates plasma field with a temperature over 2000 K.

The solution was initialized from the inlet. It was monitored approximately up to 20,000 iterations with residual convergence fixed between 1×10^{-3} and 1×10^{-5}.

The calculated rate of production s_{CR} expressed in moles/cm^2/s is converted to linear nanotube growth rate G in m/s by using nanotube bulk mass density ρ_{CNT} = 2.20 g/cm^3 and molecular weight M_{CNT} = 12.01 g/mol using the equation:

$$G = s_{CR} \cdot M_{CNT} / \rho_{CNT} \quad (1)$$

4. Results and Discussion

4.1. Experimental Results

Atomic force microscopy (AFM) of the as produced by TVA Ni/Si substrates and annealed at 850°C during 20 minutes then hydrogenated with pure hydrogen plasma during 10 minutes was carried out to determine the surface morphology. **Figure 4** shows AFM images of the substrate before (a and b) and after (c and d) annealing.

The change in the root mean square roughness RMS was from 1.864 nm before annealing to 3.485 nm after annealing. This higher rough surface clearly indicates an agglomeration of individual nickel clusters. After thermal annealing of Ni/Si substrates at the consigned temperature during 20 minutes, we followed the same experimental protocol for all the 8 experiments of **Table 1**.

First, substrates are treated with pure hydrogen plasma during 10 minutes then 10 sccm of methane was added into the mixture for 20 minutes. Scanning Electron Microscope indicates that no or few nanotubes were found

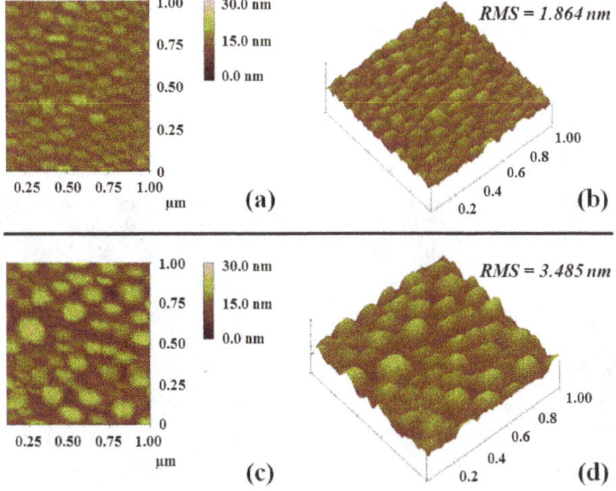

Figure 4. AFM images of the substrate before ((a) and (b)) and after ((c) and (d)) annealing.

for samples at 700°C E1, E3, E5 and E5 suggesting that 700°C is the lower growth temperature. Long spiral or helical nanotubes were observed for sample E2 and shorter nanotubes at different densities were found in samples E4, E6 and E8. Since carbon diffusion in bulk Ni is characterized by a large energy, namely, activation energy of 1.4 eV (33 kcal/mole) [35], we can estimate the diffusion coefficient of carbon in nickel using the following equation at lower and upper temperatures.

$$D_{C-Ni} = 0.1\times\exp\left(-33000/RT\right) \quad (2)$$

At 700°C, $D_{C-Ni} \sim 3.9 \times 10^{-9}$ $m^2 \cdot s^{-1}$ and at 850°C, $D_{C-Ni} \sim 3.8 \times 10^{-8}$ $m^2 \cdot s^{-1}$, hence carbon diffusion in the catalyst is increased by one order of magnitude.

To explain the role of the combined parameter on growth, we calculated the plasma kinetics using a thermochemical model based on Chemkin software in 0D [36]. This model actually includes 119 species in C/H system with atoms and molecules including Polycyclic aromatic hydrocarbons (PAHs), ions and electrons and 336 chemical reactions [13] [37]. **Figure 5**, shows the calculated hydrogen atom mole fraction in the plasma for the different conditions of **Table 1**. It clearly indicates that hydrogen atoms excess is not suitable for nanotube growth in microwave plasma systems.

In extreme case E3, SEM pictures revealed a strong etching of the substrate. Based on these observations, the best combination of parameters are those of experiment E2 which will be retained to explore the effect of oxygen. It was reported by several workers that addition of a controlled amount of oxygen or water ranging from 500 ppm to 2%, significantly enhances nanotubes growth.

Since the lower limit of our oxygen mass flow rate is 1 sccm, we increased the H_2 flow rate of experiment E2 from 10 to 90 sccm in order to avoid the limit of explosion of the hydrogen/oxygen mixture. The conditions of this experiment called OPTI are summarized in **Table 4**.

In **Figure 6** are showed the Scanning Electron Microscope images of sample OPTI. All the silicon surface is regularly covered by multiwalled carbon nanotubes of about 40 nm outer diameter and 1 μm long. Each nanotube is terminated by a catalyst particle on its top suggesting a top-growth mechanism.

In **Figure 7**, we can see the limit between the nickel covered and intentionally non covered witness area during the PVD step. Results demonstrate that the combination of TVA and PECVD is a powerful tool to uniform cover a large surface area with nanotubes.

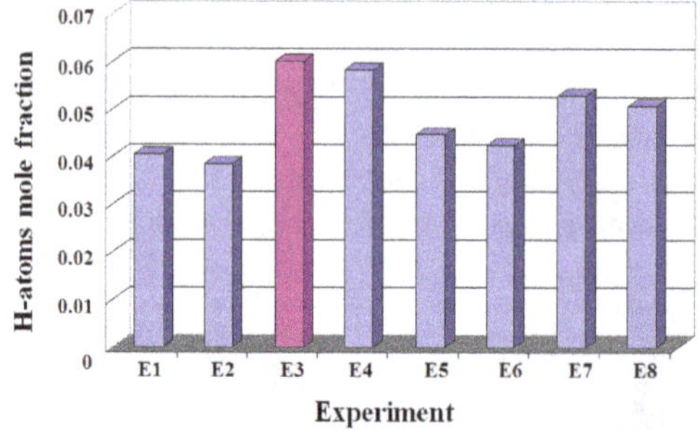

Figure 5. Calculated H-atom mole fraction for the 8 experimental conditions of **Table 1**.

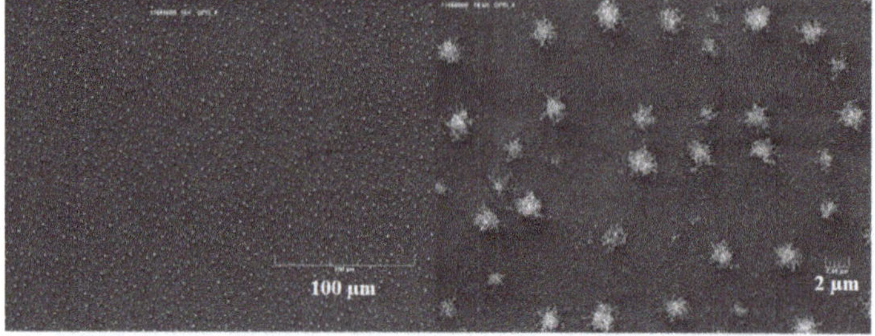

Figure 6. Scanning Electron Microscope images of samples OPTI showing a regular coverage of the substrate by nanotubes at two magnifications levels. Tool bars are100 μm for left image and 2 μm for right image.

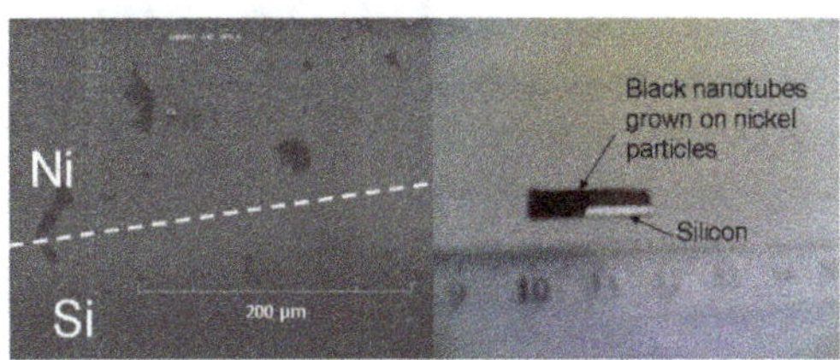

Figure 7. Left: Scanning Electron Microscope images of samples OPTI showing the border (dashed line) between nickel covered and non-covered silicon. Right: macroscopic image of the substrate, the black part is covered by nanotube and the shiny part is silicon.

Table 4. Optimal conditions, experiment OPTI.

Experiment No.	*Substrate Temperature T_{sub}* (K)	*Hydrogen Flow rate Q_{H_2}* (sccm)	*Methane Flow rate Q_{CH_4}* (sccm)	*Oxygen Flow rate Q_{O_2}* (sccm)	*Total pressure P* (mbar)	*Microwave power P_{mw}* (Watt)
OPTI	1123	90	10	1	10	342.82

4.2. Modeling Results

A two dimensional steady-state reactor simulations performed by CFD code ANSYS Fluent 12 provided information about the temperature and species distribution in the reactor otherwise difficult to characterize. The simulations conditions are summarized in **Table 5**.

As first results we present the simulated temperature contours inside the reactor (**Figure 8(a)**). The maximal temperature in the center of the plasma at a distance of ~17 mm to the substrate is 2163 K. Such temperatures are expected to yield a complete dissociation of the carbon precursor and the availability of atomic carbon. Thermal balance of plasma heating and substrate cooling determines the substrate temperature. Hence, there is a steep temperature gradient between the substrate and the region where temperature is highest as shown in the 1D profile along the centerline of the reactor (**Figure 8(b)**). Also, temperature near the substrate is within the range for the appropriate MWCNT-synthesis temperature condition (950 - 1150 K). Thus, the experimentally observed, stably CNT-synthesizing conditions correspond to the appropriate conditions to produce MWCNTs, in terms of both supplied carbon concentrations and temperature.

The gas temperature decreases when the gas reaches around the quartz walls and the gas goes down.

The inlet gas is introduced in the reactor from the top side of the quartz enclosure. It splits into two components: one flows to the outlet and other flows to the substrate. **Figure 9** shows the gas flow around the substrate. The flow and trajectories of gas species are visualized by using path lines. The length of arrows corresponds to the gas flow velocity. On the top surface of the substrate, gas flows from the center to the substrate edges. The gas stagnates and the magnitude of flow is relatively smaller than other regions. The presence of vortices in the region above the substrate is suspected to increase the mass flux of carbon species from the plasma zone to the cold substrate region, and to enhance the nanotube growth.

The simulated species profiles presented in **Figure 10** show that C_2H_2, CH_3 and CH_4 are important species that may significantly contribute to carbon nanotubes growth. There is a region of uniform C_2H_2, CH_3 and CH_4 distribution where CNTs were synthesized for the experiments, showing a relatively broader region close to the substrate surface. Other species such as atomic carbon and hydrogen have also a rather important contribution. Finally, simulation shows that large amounts of H_2 are produced in the gas phase, but H_2 production arises also from the surface desorption (**Figure 11**).

First simulation results were obtained by adjusting the surface site density in order to reproduce the experimental deposition rate values. For $\Gamma = 5.0 \times 10^{-10}$ the calculated nanotube growth rate was 8.3 µm/h. We checked the influence of the substrate temperature by varying it from 873 K to 1273 K and confirmed that increasing the temperature leads to increasing of the nanotubes growth rate from 0.5 to 45 µm/h. To calculate the activation energy of the surface reactions, we plotted the logarithm of the calculated nanotubes growth rates against $10{,}000/T_{sub}$ where T_{sub} is the temperature of the substrate (**Figure 12**). The plotted points follow an Arrhenius type law that allows us to calculate activation energy of 1.2 kJ/mol.

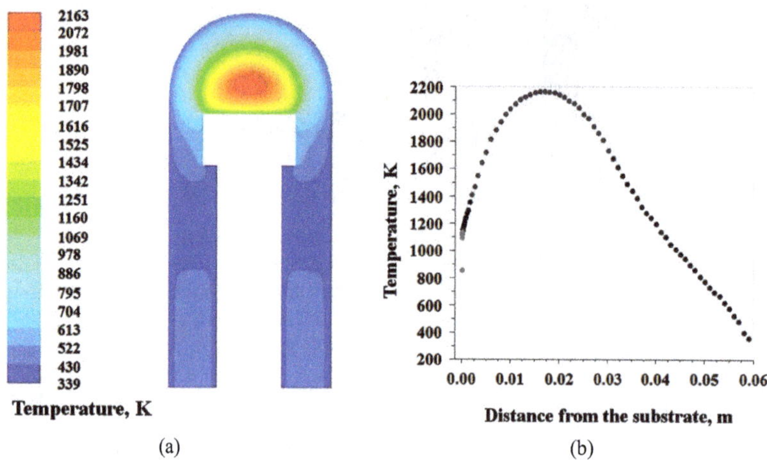

Figure 8. (a) Simulated temperature profile inside the reactor; (b) 1D temperature profile along the centerline of the reactor.

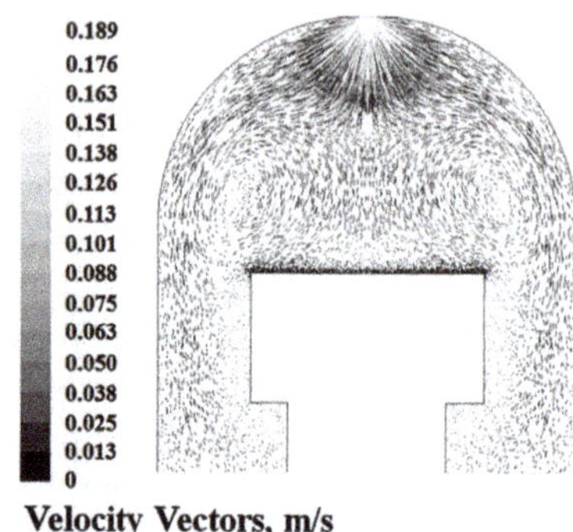

Figure 9. Velocity vectors around the substrate.

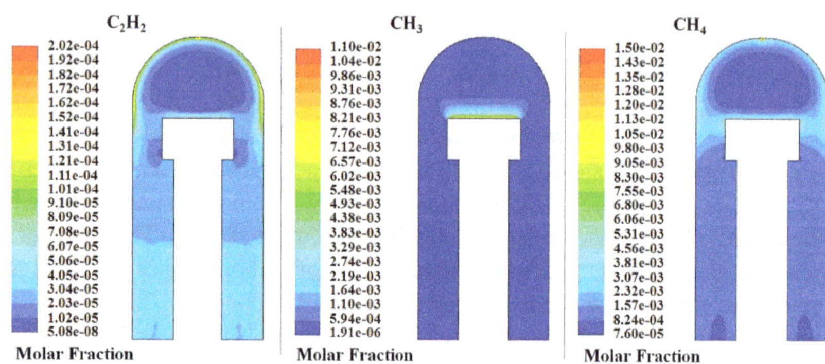

Figure 10. Simulated molar fractions of C_2H_2, CH_3 and CH_4.

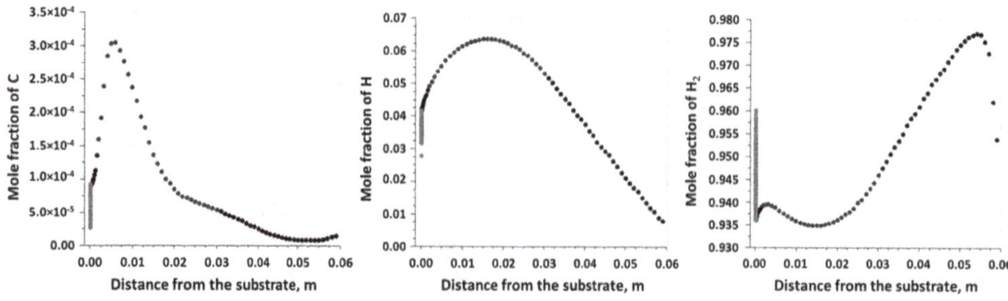

Figure 11. 1D temperature profile of the C, H and H_2 mole fractions along the centerline of the reactor.

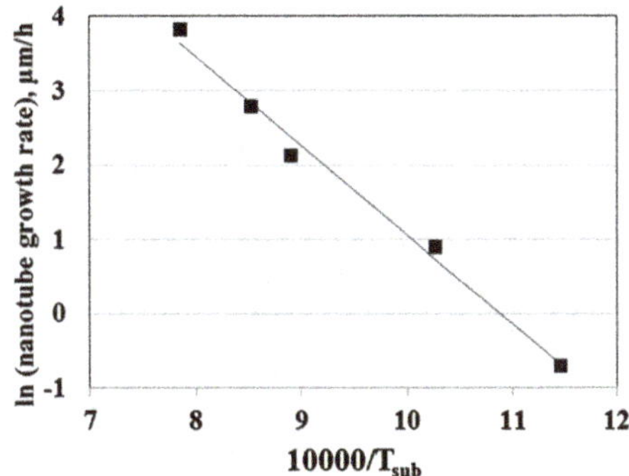

Figure 12. Logarithm of the calculated nanotubes growth rates against 10,000/T_{sub}.

Table 5. Simulations conditions.

%vol. CH_4	10
%vol. H_2	90
Inlet Velocity, m/s	0.1326
Substrate Temperature, K	1123
Wall Temperature, K	400
Plasma Heat Source, W/m^3	0.634×10^7
Pressure, mbar	10
Site density, mol/cm^2	5.0×10^{-10}

5. Conclusions

In this work, we have successfully grown multi-walled carbon nanotubes on Ni/Si substrates using a combination of two methods: 1) thermionic vacuum arc (TVA) to catalyst 1 nm ultra-thin films deposition and 2) microwave plasma PECVD with a mixture of methane and hydrogen to CNT's growth. By using an experimental factor plan, substrate temperature and plasma power density were observed to significantly influence nanotube growth. Substrate temperature affects carbon diffusion into the catalyst particle while plasma power controls the atomic hydrogen in the plasma. Based on SEM observations, higher substrate temperature and lower hydrogen atom concentration are favorable to nanotube growth. In addition, a limited fraction of oxygen added to the plasma enhances the catalytic activity improving nanotube growth. Further work is underway in order to explore the number of walls and alignment of the CNTs by controlling the catalyst size.

Plasma reactor simulation results confirmed these experimental trends. Hydrocarbon species such as C_2H_2, CH_3 and C are likely to be key deposition species influencing CNT growth rate. The reaction mechanism used in these simulations will be improved to further confirm these preliminary results.

References

[1] Iijima, S. (1991) Helical Microtubules of Graphitic Carbon. *Nature*, **354**, 56-58. http://dx.doi.org/10.1038/354056a0

[2] Farhat, S. and Scott, C. (2006) Review of the Arc Process Modeling for Fullerene and Nanotube Production. *Journal of Nanoscience and Nanotechnology*, **6**, 1189-1210. http://dx.doi.org/10.1166/jnn.2006.331

[3] Yu, B. and Meyyappan, M. (2006) Nanotechnology: Role in Emerging Nanoelectronics. *Solid State Electronics*, **50**, 536-544. http://dx.doi.org/10.1016/j.sse.2006.03.028

[4] Schäffel, F., Schünemann, C., Rümmeli, M.H., Täschner, C., Pohl, D., Kramberger, C., Gemming, T., Leonhardt, A., Pichler, T., Rellinghaus, B., Büchner, B. and Schultz, L. (2008) Comparative Study on Thermal and Plasma Enhanced CVD Grown Carbon Nanotubes from Gas Phase Prepared Elemental and Binary Catalyst Particles. *Physica Status Solidi* (*b*), **245**, 1919-1922. http://dx.doi.org/10.1002/pssb.200879605

[5] Delzeit, L., Nguyen, C.V., Stevens, R.M., Han, J. and Meyyappan, M. (2002) Growth of Carbon Nanotubes by Thermal

and Plasma Chemical Vapour Deposition Processes and Applications in Microscopy. *Nanotechnology*, **13**, 280-284. http://dx.doi.org/10.1088/0957-4484/13/3/308

[6] Hata, K., Futaba, D.N., Mizuno, K., Namai, T., Yumura M. and Ijima, S. (2004) Water-Assisted Highly Efficient Synthesis of Impurity-Free Single-Walled Carbon Nanotubes. *Science*, **306**, 1362-1364. http://dx.doi.org/10.1126/science.1104962

[7] Zhang, G., Mann, D., Zhang, L., Javey, A., Li, Y., Yenilmez, E., Wang, Q., McVittie, J.P., Nishi, Y., Gibbons, J. and Dai, H. (2005) Ultra-High-Yield Growth of Vertical Single-Walled Carbon Nanotubes: Hidden Roles of Hydrogen and Oxygen. *PNAS*, **102**, 16141-16145. http://dx.doi.org/10.1073/pnas.0507064102

[8] Lungu, C.P., Mustata, I., Zaroschi, V., Lungu, A.M., Anghel, A., Chiru, P., Rubel, M., Coad, P. and Matthews, G.F. (2007) Beryllium Coatings on Metals for Marker Tiles at JET: Development of Process and Characterization of Layers. *Physyca Scripta*, **T128**, 157-161. http://dx.doi.org/10.1088/0031-8949/2007/T128/030

[9] Lungu, C.P., Mustata, I., Musa, G., Lungu, A.M., Zaroschi, V., Iwasaki, K., Tanaka, R., Matsumura, Y., Iwanaga, I., Tanaka, H., Oi, T. and Fujita, K. (2005) Formation of Nanostructured Re–Cr–Ni Diffusion Barrier Coatings on Nb Superalloys by TVA Method. *Surface and Coating Technology*, **200**, 399-402. http://dx.doi.org/10.1016/j.surfcoat.2005.02.172

[10] Lungu, C.P. (2005) Nanostructure Influence on DLC-Ag Tribological Coatings. *Surface and Coating Technology*, **200**, 198-202. http://dx.doi.org/10.1016/j.surfcoat.2005.02.103

[11] Lungu, C.P., Mustata, I., Zaroschi, V., Lungu, A.M., Chiru, P., Anghel, A., Burcea, G., Bailescu, V., Dinuta, G. and Din, F. (2007) Spectroscopic Study of Beryllium Plasma Produced by Thermionic Vacuum Arc. *Journal of Optoelectronics and Advanced Materials*, **9**, 884-886.

[12] Silva, F., Gicquel, A., Chiron, A. and Achard, J. (2000) Low Roughness Diamond Films Produced at Temperatures Less than 600˚C. *Diamond and Related Materials*, **9**, 1965-1970. http://dx.doi.org/10.1016/S0925-9635(00)00347-2

[13] Scott, C.D., Farhat, S., Gicquel, A., Hassouni, K. and Lefebvre, M. (1996) Determining Electron Temperature and Density in a Hydrogen Microwave Plasma. *Journal of Thermophysics and Heat Transfer*, **10**, 426-435. http://dx.doi.org/10.2514/3.807

[14] Garg, R.K., Kim, S.S., Hash, D.B., Gore, J.P. and Fisher, T. (2008) Effects of Feed Gas Composition and Catalyst Thickness on Carbon Nanotube and Nanofiber Synthesis by Plasma Enhanced Chemical Vapor Deposition. *Journal of Nanoscience and Nanotechnology*, **8**, 3068-3076. http://dx.doi.org/10.1166/jnn.2008.082

[15] Marinov, N.M. and Malte, P.C. (1995) Ethylene Oxidation in a Well-Stirred Reactor. *International Journal of Chemical Kinetics*, **27**, 957-986. http://dx.doi.org/10.1002/kin.550271003

[16] Walter, D., Grotheer, H.H., Davies, J.W., Pilling, M.J. and Wagner, A.F. (1990) Experimental and Theoretical Study of the Recombination Reaction $CH_3 + CH_3 - C_2H_6$. *Symposium (International) on Combustion*, **23**, 107-114. http://dx.doi.org/10.1016/S0082-0784(06)80248-1

[17] Tsang, W. and Hampson, R.F. (1986) Chemical Kinetic Data Base for Combustion Chemistry. Part 1. Methane and Related Compounds. *Journal of Physical and Chemical Reference Data*, **15**, 1087-1279. http://dx.doi.org/10.1063/1.555759

[18] Miller, J.A. and Melius, C.F. (1992) Kinetics and Thermodynamic Issues in the Formation of Aromatic Compounds in Flames of Aliphatic Fuels. *Combustion and Flame*, **91**, 21-39. http://dx.doi.org/10.1016/0010-2180(92)90124-8

[19] Markus, M.W., Woiki, D. and Roth, P. (1992) Two-Channel Thermal Decomposition of CH_3. *Symposium (International) on Combustion*, **24**, 581-588. http://dx.doi.org/10.1016/S0082-0784(06)80071-8

[20] Warnatz, J. (1984) Rate Coefficients in the C/H/O System. In: Gardiner Jr., W.C., Ed., *Combustion Chemistry*, Book Chapter, Springer-Verlag, New York.

[21] Dagaut, P., Cathonnet, M., Aboussi, B. and Boettner, J.-C. (1990) Allene Oxidation: A Kinetic Modeling Study. *Journal de Chimie Physique et de Physico-Chimie Biologique*, **87**, 1159-1172.

[22] Feng, Y., Niiranen, J.T., Bencsura, A., Knyazev, V.D., Gutman, D. and Tsang, W. (1993) Weak Collision Effects in the Reaction C_2H_5=C_2H_4+H. *Journal of Physical Chemistry*, **97**, 871-880. http://dx.doi.org/10.1021/j100106a012

[23] Towell, G.D. and Martin, J.J. (1961) Kinetic Data from Nonisothermal Experiments: Thermal Decomposition of Ethane, Ethylene, and Acetylene. *AIChE Journal*, **7**, 693-698. http://dx.doi.org/10.1002/aic.690070432

[24] Kiefer, J.H., Kapsalis, S.A., MAlami, M.Z. and Budach, K.A. (1983) The Very High Temperature Pyrolysis of Ethylene and the Subsequent Reactions of Product Acetylene. *Combustion and Flame*, **51**, 79-93. http://dx.doi.org/10.1016/0010-2180(83)90085-8

[25] Dean, A.M. (1985) Predictions of Pressure and Temperature Effects upon Radical Addition and Recombination Reactions. *Journal of Physical Chemistry*, **89**, 4600-4608. http://dx.doi.org/10.1021/j100267a038

[26] Fahr, A., Laufer, A., Klein, R. and Braun, W. (1991) Reaction Rate Determinations of Vinyl Radical Reactions with

Vinyl, Methyl, and Hydrogen Atoms. *Journal of Physical Chemistry*, **95**, 3218-3224. http://dx.doi.org/10.1021/j100161a047

[27] Knyazev, V.D., Bencsura, A., Stoliarov, S.I. and Slagle, I.R. (1996) Kinetics of the C_2H_3+H_2=H+C_2H_4 and CH_3+ H_2=H+CH_4 Reactions. *Journal of Physical Chemistry*, **100**, 11346-11354. http://dx.doi.org/10.1021/jp9606568

[28] Janardhanan, V.M. and Deutschmann, O. (2006) CFD Analysis of a Solid Oxide Fuel Cell with Internal Reforming: Coupled Interactions of Transport, Heterogeneous Catalysis and Electrochemical Processes. *Journal of Power Sources*, **162**, 1192-1202. http://dx.doi.org/10.1016/j.jpowsour.2006.08.017

[29] Lysaght, A.C. and Chiu, W.K.S. (2008) Modeling of the Carbon Nanotube Chemical Vapor Deposition Process Using Methane and Acetylene Precursor Gases. *Nanotechnology* **19**, 165607. http://dx.doi.org/10.1088/0957-4484/19/16/165607

[30] Lacroix, R., Fournet, R., Ziegler-Devin, I. and Marquaire, P.-M. (2010) Kinetic Modeling of Surface Reactions Involved in CVI of Pyrocarbon Obtained by Propane Pyrolysis. *Carbon*, **48**, 132–144. http://dx.doi.org/10.1016/j.carbon.2009.08.041

[31] Farhat, S., Panham, S., Gicquel, A., Silva, F., Brinza, O. and Lungu, C.P. (2010) Synthèse de Nanotubes Orientés par PECVD. Matériaux 2010, 18-22 October 2010, Nantes.

[32] Kee, R.J., Rupley, F.M., Miller, J.A., Coltrin, M.E., *et al.* (2001) CHEMKIN Collection, Release 3.6, Reaction Design, Inc., San Diego.

[33] JANAF (1965) "Thermochemical tables, National Standards Reference Data Series" Report NSRDS-NBS: Dow Chemikal Company, distributed by Clearinghouse for federal Scientific and Technical Information, PB168370.

[34] Kee, R.J., Dixon-Lewis, G., Warnatz, J. and Miller, J.A. (1986) A FORTRAN Computer Code Package for Evaluation of Gas-Phase, Multicomponent Transport Properties. Technical Report SAND86-8426, Sandia National Laboratories, Albuquerque.

[35] Sickafus, E.N. (1970) Sulfur and Carbon on the (110) Surface of Nickel. *Surface Science*, **19**, 181-197. http://dx.doi.org/10.1016/0039-6028(70)90117-2

[36] Kee, R.J., Miller, J.A. and Jefferson, T.H. (1980) CHEMKIN: A General-Purpose, Problem-Independent, Transportable, Fortran Chemical Kinetics Code Package. Technical Report SAND80-8003, Sandia National Laboratories, Albuquerque.

[37] Farhat, S., Findeling, C., Silva, F., Hassouni, K. and Gicquel, A. (1997) Third Edition of the International Workshop Microwave Discharges: Fundamentals and Applications. Abbaye de Fontevraud, Fontevraud-l'Abbaye.

22

Behaviors of Crystallization for Osmotic Pressure under Microwave Irradiation

Ryosuke Nakata, Yusuke Asakuma*

Department of Mechanical and System Engineering, University of Hyogo, Kobe, Japan
Email: *asakuma@eng.u-hyogo.ac.jp

Abstract

We studied chemical garden in order to investigate precipitation behavior for osmotic pressure under microwave irradiation. The salt concentration and microwave irradiation power were varied. Microwave irradiation induced release of osmotic pressure and change of precipitation pattern because polar molecules vibrate and rotate in an electromagnetic field. For example, the width of precipitation increased and the number of rapture of the membrane decreased due to the release of osmotic pressure by the irradiation. Accordingly, microwave irradiation accelerated the diffusion of ionic molecules through the membrane.

Keywords

Microwave, *In-Situ* Observation, Osmotic Pressure

1. Introduction

Chemical garden for osmotic pressure forms attractive and unique tree-like patterns. The hollow tubes of basic chemical garden structure are formed when the precipitation reaction takes place at the interface between two different solutions. Precipitation from a concentrated salt solution becomes semipermeable colloidal membrane. Moreover, the growth rate pattern depends on the salt kinds and the concentration of sodium silicate (Na_2SiO_3; water glass) in aqueous solution. Salt continues to dissolve into the solution inside the membrane and the higher osmic pressure causes the rupture of the membrane wall and jet fluid into the surrounding solution. This tubular fiber develops at the point. The morphological pattern is scientifically interesting because such pattern formation frequently inspires microstructural design in materials science [1]-[4]. However, the mechanism of chemical garden has not been clear because of the nonlinearity. In this study, we tried to control such complex pattern by outside fields.

*Corresponding author.

Microwave irradiation may be helpful in this context. Microwave irradiation has recently been a very active research due to the many advantages of nonthermal effects such as reaction promotion, nucleation induction, and diffusion facilitated by the molecular vibration of polar molecules [5] [6]. We hypothesized that the precipitation reaction and diffusion characteristics through semipermeable membrane of chemical garden could be modulated or explained via various microwave properties. In addition, precipitation of chemical garden may suggest a new application of microwave irradiation and suitable operating conditions for highly functional inorganic materials of hollow tubes. The concentration of aqueous solution is essential for solving the diffusion and precipitation challenges of chemical garden phenomena [1]. In this study, we varied the concentration and microwave power to understand the microwave effect. We used a microwave reactor equipped with an *in-situ* observation system to capture the dynamic growth behavior of the tree-like pattern.

2. Experimental

Chemical garden behavior, which is crystal growth like a plant by release of osmotic pressure, is performed by addition of copper sulfate ($CuSO_4$) to an aqueous solution of sodium silicate ($NaSiO_3$; water glass) because copper silicates are insoluble in water and are colored.

First, grain of copper sulfate is placed on bottom of thin cell, which consists of three acryl resin layers as shown in **Figure 1(a)**. Thickness of center and sides of the cell, are 2 and 1 mm, respectively. The grain size is smaller than thickness of the width. Sodium silicate aqueous solution listed in **Table 1**, is poured into the cell. Examples of chemical garden are shown in **Figure 1(b)** and **Figure 1(c)**. Tree like pattern is formed from the bottom to the top.

Temperature is important factor for the precipitation. We used two different methods, microwave irradiation (MW) and conventional heating (thermostatic bath, CH) for maintaining solution temperature. We measured the pattern for two heating methods.

A. Microwave heating

Figure 2 shows a tube-guide microwave reactor (Shikoku Instrumentation), which is specially designed for preventing microwave leakage and equipped with a microscope camera (Sigma Koki, Model SK-TC202USB-AT) for *in-situ* observation [5]-[7]. Cell containing the solution of chemical garden is placed at the center of the reactor. Direction of thin cell is parallel to the side and the pattern is captured through the side of the reactor, with the fixed microscope. Immediately after putting on the cell, microwave was irradiated for 20 min. We measured the growth rate of the pattern, which is defines as time for the growth of 3 mm length, the width and the number of ruptures through the movie. And width is measured by every 1 mm. Moreover, the rupture of membrane for 30 min is counted. Average values of six crystals for growth rate, width and the number of rup-

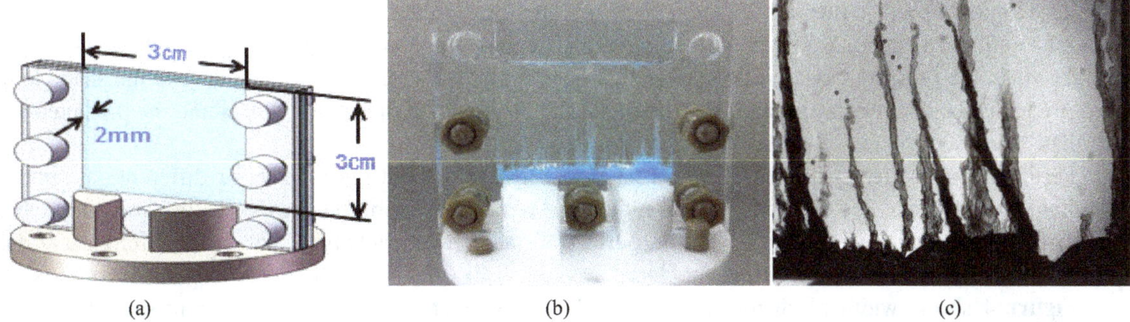

Figure 1. Cell for observation of chemical garden. (a) Cell; (b) Chemical garden in cell; (c) Enlarge view of chemial garden (6 mm × 8 mm).

Table 1. Experimental conditions.

Na_2SiO_3 conc. [w%]	MW output [W]	Temperature of bath [˚C] of CH
18		
20	21	20
22	22.5	40
24		

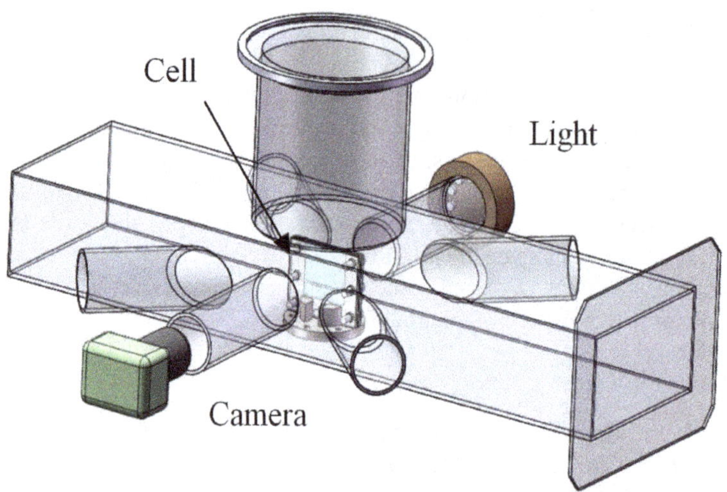

Figure 2. Microwave reactor for observation of chemical garden.

ture are calculated. These approaches enable dynamic evaluation of chemical garden with respect to microwave irradiation. After 20 min, we measured the final temperature with an optical fiber (Anritsu Meter) because the solution temperature increases under irradiation.

B. Conventional heating

We used thermostatic bath to maintain the solution temperature. The bath temperature was same with the final temperature observed in the microwave experiment. The cell is immersed into the bath, and movie of the pattern is captured after 20 min with the microscope camera.

3. Results and Discussion

$CuSO_4$ starts to dissolve in the aqueous solution after the seed crystals contact with the solution. Insoluble copper silicate immediately forms by a double decomposition reaction (anion metathesis reaction). This copper silicate is a semipermeable membrane. Because the ionic strength of the copper solution inside the membrane is higher than the sodium silicate solution, the osmotic pressure increases within the membrane. Water in the sodium silicate solution is drawn osmotically from outside into the membrane, in the meanwhile dissolution of seed crystal continues. Balance between the dissolution and the entry of water under osmotic pressure causes a rupture around tip of the membrane, and new membrane forms at the rupture points by the reaction of the copper cations and silicate anions. The crystals relatively grow upwards since the pressure at the bottom of the cell is higher than the pressure closer to the top. In this way, the growth continues in the cell and the shape looks like colored plants as shown in **Figure 1(b)**. Generally, the position and time of rupture cannot be predicted and controlled.

Figure 3 shows growth rate of chemical garden with and without microwave for different concentration of sodium silicate. Solid and unfilled symbols indicate MW and CH conditions, respectively. Growth rate becomes higher when temperature is higher. However, effect of microwave irradiation, that is, the difference at the same temperature between MW and CH, is not clear.

Figure 4 shows width of chemical garden with and without microwave for different sodium silicate. The width increased with the concentration because much sodium near the interface is supplied. Moreover, the width became wider when temperature becomes higher. On the other hand, microwave effect was observed for the width at the same temperature condition. It became a little wider by microwave irradiation. The diffusion of water molecule thorough the membrane is accelerated by molecular rotation of microwave irradiation [8] [9]. Accordingly, the diffusion causes the lower concentration of copper sulfate near the interface and slower release of osmotic pressure.

Figure 5 shows the number of ruptures for chemical garden with and without microwave for different sodium silicate. The number of ruptures decreases by microwave irradiation. In case of MW, rupture happened only around the time after pouring the solution into the cell. It means that osmotic pressure is gradually released due

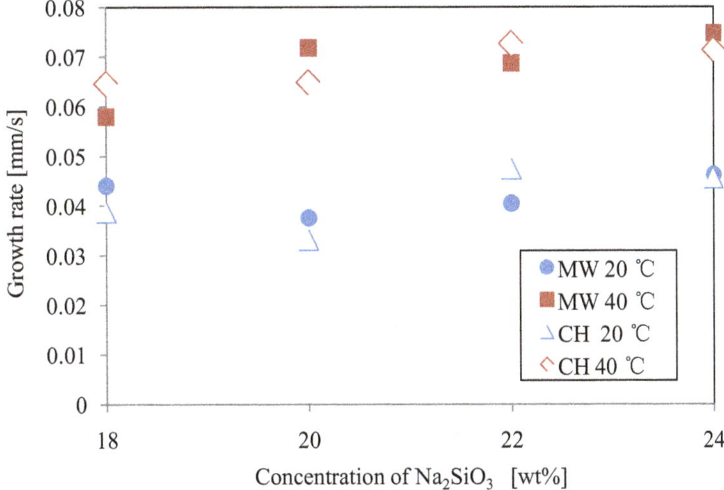

Figure 3. Growth rate of chemical garden for Na_2SiO_3 concentration with and without microwave.

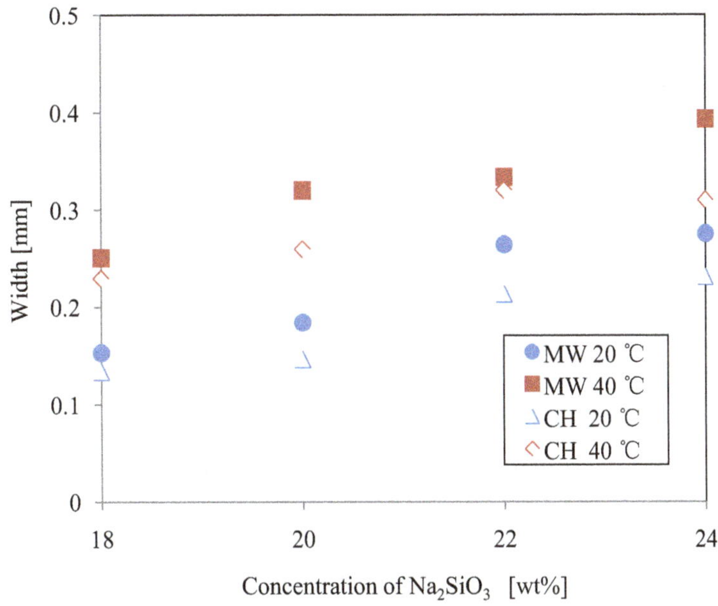

Figure 4. Width of chemical garden for Na_2SiO_3 concentration with and without microwave.

to the water molecule entry into the membrane by the irradiation, and the rupture does not happen anymore after initial rupture. Particularly, condition of higher irradiation, single rupture is observed.

Figure 6 show examples of microscope photos around tip of chemical garden with and without microwave. There are several characteristics between two heating methods. In case of no microwave, the membrane is formed perfectly, and the interface between two liquids is clear. Pattern by multiple ruptures is disordered like tree because jet flows of the solution are ejected. Consequently, the higher osmic pressure by dissolution of seed crystal and the lower pressure after the rupture repeat irregularly for the time and the position. Crystal grows up when osmotic pressure is over the limit values. The rupture will happen near thin wall of the membrane, which is pale colored in photos. According, border of membrane obtained as quick precipitation becomes darker.

On the other hand, behavior under the irradiation is different with conventional heating. For example, the border of straight membrane is not clear by microwave irradiation. Moreover, weak jet flow continues without closing the tip due to higher diffusion of water through the membrane. Accordingly, osmotic pressure is gradu-

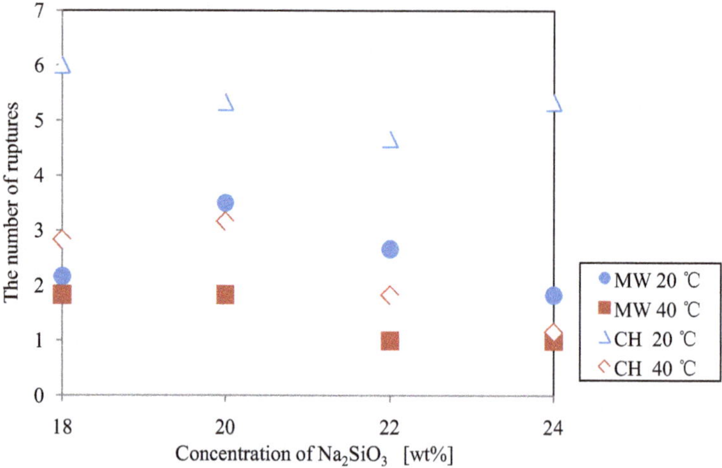

Figure 5. Number of ruptures for chemical garden for Na_2SiO_3 concentration with and without microwave.

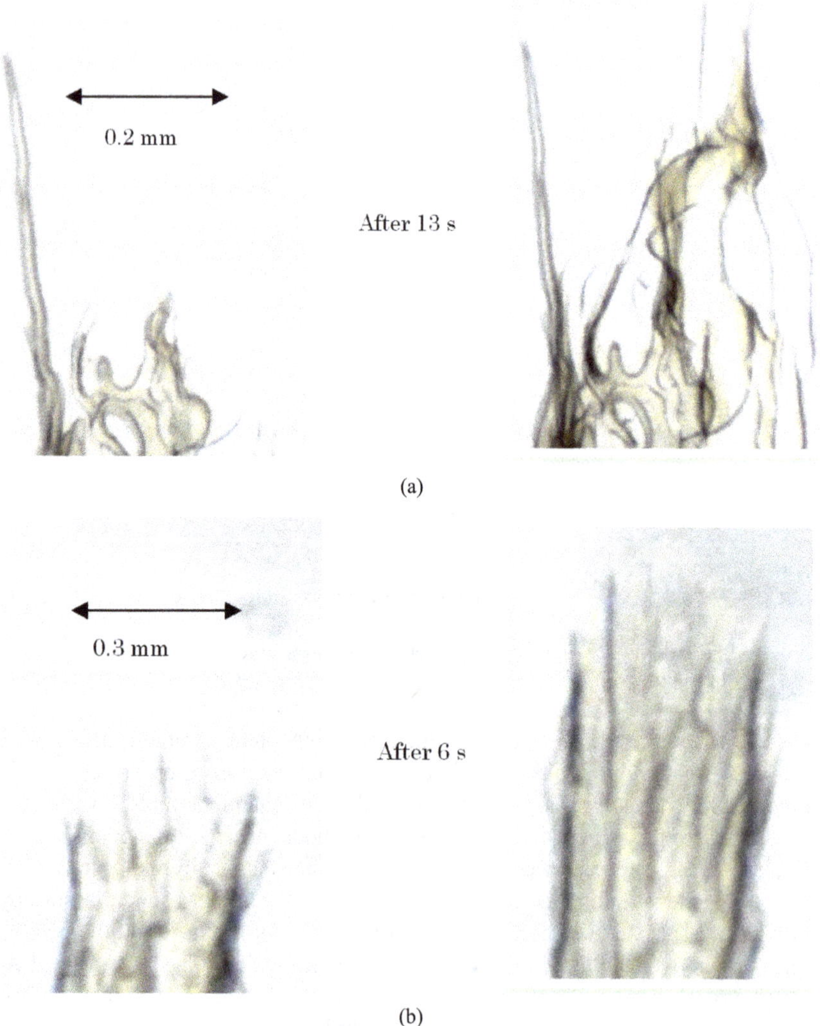

Figure 6. Microscope photos around tip of crystal with and without microwave (18 w%). (a) Without microwave; (b) With microwave.

ally released and condition of slower precipitation is achieved due to the smaller number of the rupture. The microwave irradiation, which prevents the rupture for osmotic pressure, can modulate the molecular diffusion and manifestation of complex patterns for chemical garden. Finally, the irradiation may be a useful tool in other fine-scale, nonequilibrium self-organization processes for nonlinear precipitation.

4. Conclusion

In this study, the mechanism of the crystal growth for chemical garden was investigated to clarify the osmotic pressure and the diffusion problem through the membrane under microwave irradiation. Effect of microwave on the number of the ruptures for the $NaSiO_3$ concentration, which is the release of the osmotic pressure, is interesting. Although crystal growth rates with and without microwave irradiation at the same temperature condition are almost same, the number of ruptures under microwave irradiation becomes smaller. On the other hand, the osmotic pressure becomes gradually smaller due to the higher diffusion of water molecule by the irradiation. Although the rupture behavior is decided by the balance between water molecule entry through the membrane and dissolution of seed crystal, unstable and explosive release of the osmotic pressure was avoidable by the irradiation.

References

[1] Cartwright, J.H.E., García-Ruiz, J.M., Novella, M.L. and Otálora, F. (2002) Formation of Chemical Gardens. *Journal of Colloid and Interface Science*, **256**, 351-359. http://dx.doi.org/10.1006/jcis.2002.8620

[2] Pratama, F.S., Robinson, H.F. and Pagano, J.J. (2011) Spatially Resolved Analysis of Calcium-Silica Tubes in Reverse Chemical Gardens. *Colloids and Surfaces A: Physicochemical and Engineering Aspects*, **389**, 127-133. http://dx.doi.org/10.1016/j.colsurfa.2011.08.041

[3] Bormashenko, E., Bormashenko, Y., Stanevsky, O. and Pogreb, R. (2006) Evolution of Chemical Gardens in Aqueous Solutions of Polymers. *Chemical Physics Letters*, **417**, 341-344. http://dx.doi.org/10.1016/j.cplett.2005.10.049

[4] Barge, L.M., Doloboff, I.J., White, L.M., Stucky, G.D., Russell, M.J. and Kanik, I. (2011) Characterization of Iron-Phosphate-Silicate Chemical Garden Structures. *Langmuir*, **28**, 3714-3721. http://dx.doi.org/10.1021/la203727g

[5] Aaskuma, Y., Murakami, Y. and Konishi, M. (2014) Anti-Solvent Effect of Crystallization by Feeding Ethanol under Microwave Radiation. *Crystal Research and Technology*, **49**, 129-134. http://dx.doi.org/10.1002/crat.201300327

[6] Asakuma, Y. and Miura, M. (2014) Effect of Microwave Radiation on Diffusion Behavior of Anti-Solvent during Crystallization. *Journal of Crystal Growth*, **402**, 32-36. http://dx.doi.org/10.1016/j.jcrysgro.2014.04.031

[7] Parmar, H., Kanazawa, Y., Asada, M., Asakuma, Y., Phan, C., Pareek, V. and Evans, G. (2014) Influence of Microwave on Water Surface Tension. *Langmuir*, **30**, 9875-9879. http://dx.doi.org/10.1021/la5019218

[8] Nakai, Y., Tsujita, Y. and Yoshimizu, H. (2002) Control of Gas Permeability for Cellulose Acetate Membrane by Microwave Irradiation. *Desalination*, **145**, 375-377. http://dx.doi.org/10.1016/S0011-9164(02)00439-3

[9] Nakai, Y., Yoshimizu, H. and Tsujita, Y. (2005) Enhanced Gas Permeability of Cellulose Acetate Membranes under Microwave Irradiation. *Journal of Membrane Science*, **256**, 72-77.

23

Design and Implementation of Hybrid Light and Microwave Switches Based on Wavelength Selective Switch for Future Satellite Networks

Bin Wu[1], Hongxi Yin[1*], Anliang Liu[1], Chang Liu[1], Jingchao Wang[2]

[1]Laboratory of Optical Communications and Photonic Technology, School of Information and Communication Engineering, Dalian University of Technology, Dalian, China
[2]Institute of China Electronic System Engineering Company, Beijing, China
Email: *hxyin@dlut.edu.cn

Abstract

A hybrid switching node structure with light and microwave links is proposed, which is applicable to the future data relay satellite systems, aiming at the development trend of coexistence of light-link and microwave-link in the future. An experimental system for the light and microwave hybrid switching node based on wavelength selective optical switches (WSS) and optical transceiver modules, is established. It is shown by our experiment that this hybrid switching node can realize the dynamic bandwidth allocation and wavelength routing while the bit error rate of light link is less than 10^{-12}, which provides a method for solving the hybrid switching problem of light-link and microwave-link on the future data relay satellite systems.

Keywords

Light-Link and Microwave-Link, Data Relay Satellite, WSS, Hybrid Switching

1. Introduction

With the continuous developments of high-speed data communication, navigation and positioning, high resolution image acquisition and deep space exploration, the transmission demand of large capacity and high-speed data relay satellite services for inter-satellites is rapidly increasing. Therefore, the inter-satellite light-link has become an important development trend for data relay satellites. In recent years, many developed countries, such as Europe, America and Japan, have successively carried out theoretical research and spaceborne demonstration about the data relay satellite system with inter-satellite light-links, and launched out into corresponding engineering verification and satellite networking project [1] [2]. With the improving performance of inter-satellite light link, it has been a trend for satellite communication (SATCOM) to construct an satellite optical networks based on light link and achieve broadband satellite data communication, which is a principal way to solve contradiction between the exponentially increasing information and the limited bandwidth in existing satellite communication. However, the matured satellite largely depends on microwave link for now, because communication based on microwave can realize multi-beam coverage and get the merits of widespread applications, accessing rapidly and high degree of technological maturity. Moreover, microwave link in space-terrestrial communication system can partially overcome the influence of atmospheric turbulence to ensure the reliability in the satellite communication link. Therefore, microwave link will still be the main body of satellite communication system and space information network system in the future.

The inter-satellite light link possesses many advantages such as high bandwidth, small size, low power consumption, excellent transmission performance and etc., while microwave link in satellite-ground communication could make full use of the mature technology of ground station and avoid the severe influence of atmospheric turbulence on the light link. Hence, the satellite information system with high bandwidth and flexibility will definitely be a heterogeneous hybrid network in the future where light-link and microwave-link can achieve coexisting and complementary advantages. When the information from different satellites or ground stations converges to a data relay satellite node, there will be a spaceborne convergence of microwave link and light link. In this situation, data relay satellite node should have the ability to switch the data of backbone light link as well as to accomplish data access from light-link and microwave-link to build a bridge for both light and microwave network in the inter-satellite and space-terrestrial communications. In recent years, with the continuous development of satellite light communication and the trend of multi-satellite networking in many countries, to solve the problem of light and microwave hybrid switching on data relay satellite is becoming increasingly urgent.

In this paper, a structure of spaceborne light-link and microwave-link hybrid switch is proposed based on wavelength selective optical switch (WSS). A hybrid switching system is designed and experimentally implemented, which can achieve dynamical reconfiguration, optical cross connection and add/drop multiplexing. It is shown by the experimental result of light and microwave hybrid switching that the proposed hybrid switching system can be available for the future satellite networks.

2. Structure for Spaceborne Light-Link and Microwave-Link Hybrid Node

The data relay satellite system based on light and microwave link hybrid switching is shown in **Figure 1**. The satellite backbone network is formed between Geosynchronous Satellites (GSO) by the light link, which takes full advantage of high speed, large capacity, high bandwidth, low power consumption, small volume and anti-electromagnetic interference. Backbone network and access network are connected with light-links or microwave-links, while satellite and ground station are mainly connected by relative mature microwave links [3]. Therefore, the data relay node in backbone network is a hybrid satellite node, which should have the ability of both microwave access and optical switching.

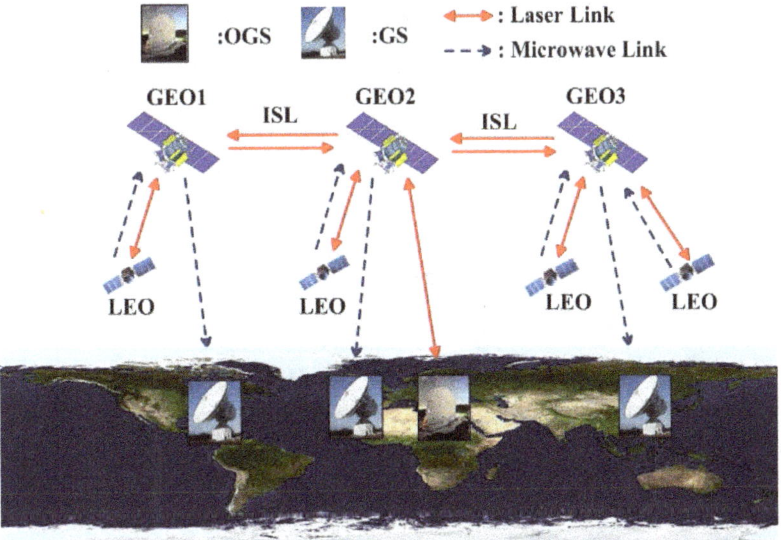

Figure 1. Data relay satellite system with hybrid light-links and microwave-links.

The structure of spaceborne light and microwave hybrid switches based on WSS is shown in **Figure 2**. It can interconnect light links between the backbone network, realize optical cross-connect and wavelength routing and simultaneously access light and microwave signals. The accessing microwave signal from low earth orbit (LEO) satellites or ground station (GS) is modulated into optical domain to implement transparent processing of microwave signal and then it passes the switching matrix before sent to the next satellite node. The hybrid node also has the ability of optical cross connection and add-drop multiplexing [4] [5].

The switching fabric is made up of 1 × N wavelength selective switching matrix, shown in **Figure 3**. WSS has the ability to independently switch any wavelength to any port without any restrictions imposed by the switching of other wavelengths. WSS is completely reconfigurable by allocating channel spacing and bandwidth dynamically. This property improves the spectrum efficiency and makes it applicable to dynamically changeable network with limited spectrum resources [6] [7].

3. Experimental Results of Hybrid Switching Node

The test system of light and microwave hybrid switching node is connected as shown in **Figure 4**, which includes the Agilent E4438 vector signal generator, the DWDM optical transceiver, the optical modulator, the polarization controller, the wavelength division multiplexer and de-multiplexer, the 1 × 4 wavelength selective switch, the photodetector, the Anritsu MS9740A optical spectrum analyzer, the Agilent N9020A MXA signal analyzer and the Anritsu MP1800A BERT. The operating wavelength of the optical transmitter module is 1554.94 nm and 1553.33 nm respectively, which corresponds to Channel 28 and Channel 30 in dense wavelength division multiplexing standard (ITU-T). The insertion loss of WSS is approximately 7 dB, and the isolation is higher than 34 dB.

Firstly, the channel spacing and bandwidth allocation for 1 × 4 WSS will be tested. The wide-spectrum light from a light diode enters the input "com" port of the WSS through a variable optical attenuator. A personal computer (PC) is employed to send control commands through its serial port and after the channel spacing, wavelength routing and attenuation for each channel are allocated the spectra is observed by optical spectrum analyzer. The bandwidths assigned to each channel are 25 GHz, 50 GHz, 100 GHz and 200 GHz, respectively and the attenuation for each channel is 0 dB. As shown in **Figure 5**, the −3 dB bandwidths of the four channels are

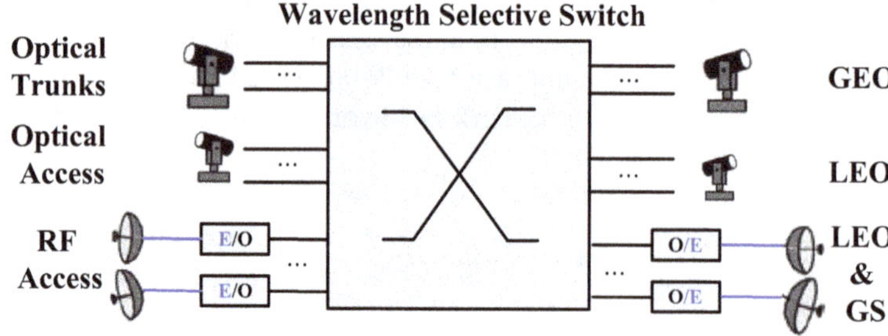

Figure 2. Light and microwave hybrid switching node for data relay satellite.

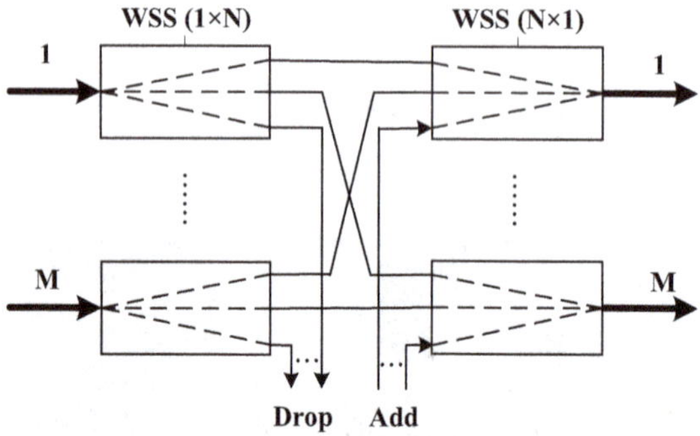

Figure 3. Schematic of the internal structure of the wavelength selective switching matrix.

0.19 nm, 0.39 nm, 0.78 nm and 1.59 nm, respectively.

Secondly, let's verify the hybrid switching ability between light-links and microwave-links. The experiment is based on the designed testbed shown in **Figure 4**. The microwave link signal is generated by E4438 vector signal generator. The rate of the digital baseband signal is 50 Mbps and the carrier frequency is 5 GHz, and the modulation mode is 16QAM. The microwave signal is transformed into an optical signal and then, it passes through the hybrid node (B→D) and outputs from the third port of the wavelength division demultiplexer at the bottom right. Its spectra are shown in **Figure 6** and **Figure 6(a)** shows the spectrum of input 5 GHz signal modulated by 50 Mbps RF signal, and **Figure 6(b)** displays the RF signal spectrum of hybrid light-microwave switching node. It can be seen by comparison of two figures that the central frequency of this signal is 5 GHz and the total loss of the switching system is 29 dB. The light link data is generated by the MP1800A BERT and an optical transmitter. The test pattern is NRZ $2^{31}-1$ of 10Gbps data rate. The signal outputs from the first port of the wavelength division demultiplexer through the hybrid node (A→C) and the bit error rate of light link in the loopback test is less than 10^{-12}.

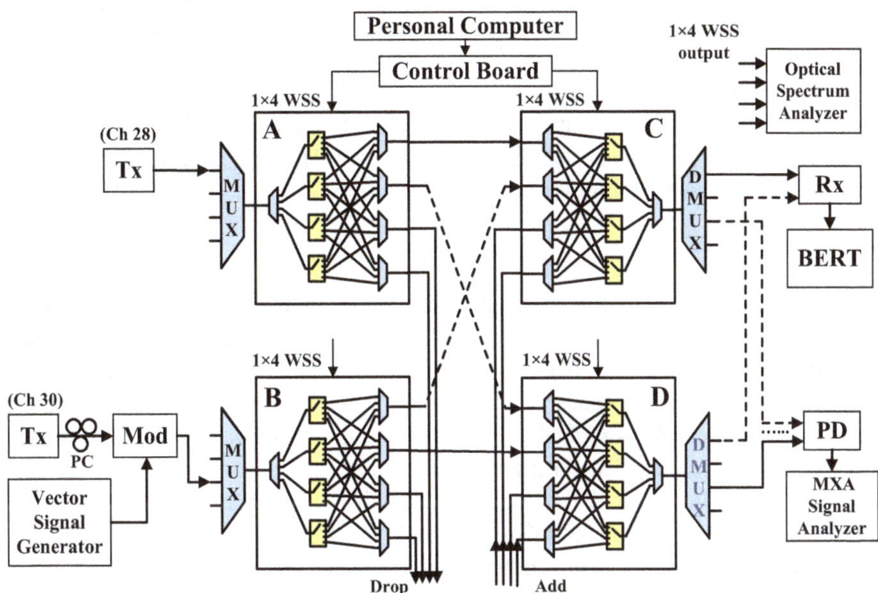

Figure 4. Experimental system for the light and microwave hybrid switching node.

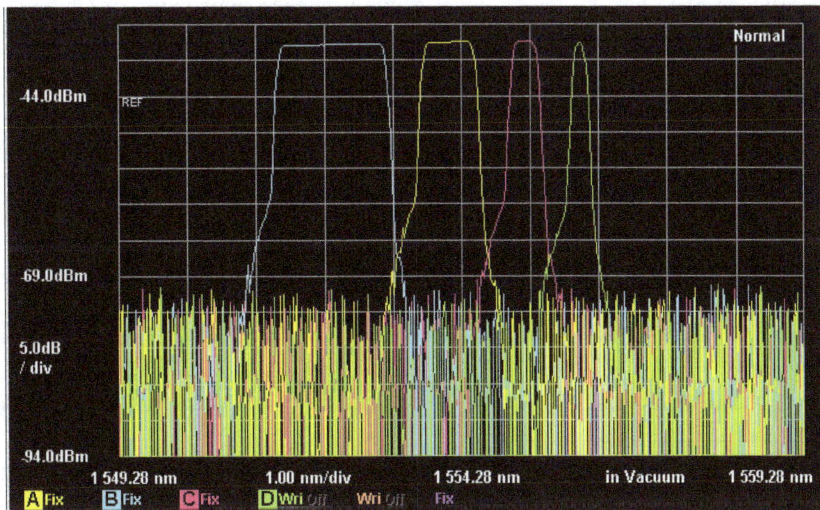

Figure 5. Spectra of the four channels (the SLICE number assigned to each channel is 16, 8, 4, 2 respectively).

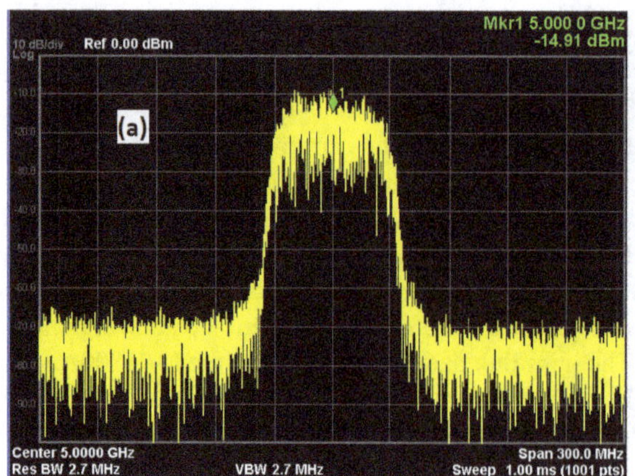

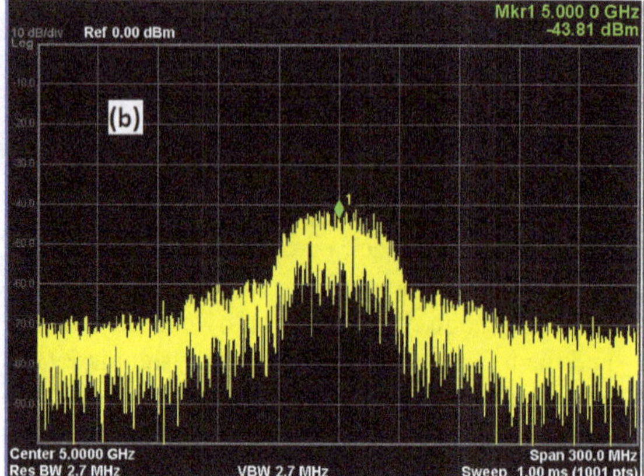

Figure 6. Spectra of the input and output 5 GHz signals modulated by 50 Mbps RF signal. (a) Input RF signal after 16QAM modulation; (b) Output RF signal of light and microwave hybrid switching node.

The dotted line shown in **Figure 4** indicates the signal flow of between the light-links and microwave-links after switching controlled by the computer (A→D, B→C). The experimental results are consistent with those of previous tests. The hybrid switching node designed can accomplish data exchange between light-links and microwave-links to achieve the function of the data relay satellite.

4. Conclusion

In this paper, a light-microwave hybrid switching node applicable to data relay satellites, is designed and an experimental platform based on 1 × 4 WSS and optical transceiver module, is set up. The experiment results indicate that the hybrid node can realize dynamic bandwidth allocation and wavelength routing, which highly improves the spectrum efficiency of spaceborne communication system. It reveals through experimental testing that this node can implement the light and microwave hybrid switching. The signal of microwave link is transformed into optical domain to implement the transparent processing using optical transceiver module, while the bit error rate of light link in the loopback test is less than 10^{-12}. The hybrid node designed is applicable for the situation of converge for light-links and microwave-links on a data relay satellite, which has many advantages such as high spectrum efficiency, transparent data processing and etc.

Acknowledgements

This work was supported in part by the National Natural Science Foundation of China (NSFC) under Grants 61231011 and 61071123, and The First HAEPC Science and Technology Project in 2015(5217Q014006U).

References

[1] Arruego, I., Guerrero, H., Rodrıguez, S., Martınez-Oter, J., Jimenez, J.J., *et al.* (2009) OWLS: A Ten-Year History in Optical Wireless Links for Intra-Satellite Communications. *IEEE Journal on Selected Areas in Communications*, **27**, 1599-1610. http://dx.doi.org/10.1109/JSAC.2009.091210

[2] Hanada, T., Yamakawa, S. and Kohata, H. (2011) Study of Optical Inter-orbit Communication Technology for Next Generation Space Data-relay Satellite. *Proceedings of Free-Space Laser Communication Technologies Conference of SPIE*, San Francisco, 26-27 January 2011, 7923.

[3] Hossein, S. and Murat, U. (2013) End-to-End Performance of Mixed RF/FSO Transmission Systems. *Journal of Optical Communications and Networking*, **5**, 1139-1144. http://dx.doi.org/10.1364/JOCN.5.001139

[4] Daneshmand, M. and Mansour, R.R. (2011) RF MEMS Satellite Switch Matrices. *Microwave Magazine*, **12**, 92-109. http://dx.doi.org/10.1109/MMM.2011.941417

[5] Hwan, S.C., Sun, H.C., Sang, S.L. and Kwangjoon, K. (2008) Experimental Demonstration of Optical Multicast Using WSS Based Multi-Degree ROADM. *Proceedings of Optical Fiber Communication Conference and Exposition (OFC/NFOEC) and the National Fiber Optic Engineers Conference*, San Diego, 24-28 February 2008, 88-91.

[6] Winston, I.W., Philip, N.J. and Ankitkumar, N.P. (2013) Wavelength Contention-Free via Optical Bypass within a

Colorless and Directionless ROADM [Invited]. *Journal of Optical Communications and Networking*, **5**, 220-229. http://dx.doi.org/10.1364/JOCN.5.00A220

[7] Strasser, T.A. and Wagener, J.L. (2010) Wavelength-Selective Switches for ROADM Applications [Invited]. *IEEE Journal of Selected Topics in Quantum Electronics*, **16**, 1150-1157. http://dx.doi.org/10.1109/JSTQE.2010.2049345

24

Preparation and Optical Patterning of Organic-Inorganic Hybrid Color-Filter Films Using Latent Pigments by Utilizing Photo-Acid-Generator and Microwave Irradiation

T. Ohishi*, S. Sugawara

Department of Applied Chemistry, Faculty of Engineering, Shibaura Institute of Technology, Tokyo, Japan
Email: *tooishi@sic.shibaura-it.ac.jp

Abstract

Preparation and photo-patterning characteristics of organic-inorganic hybrid thin film containing latent pigment by using photo-acid-generator (PAG) and microwave irradiation have been investigated. The acrylic thin film modified with methoxysilane containing PAG was formed on a glass substrate and irradiated with ultraviolet rays to promote sol-gel reaction by catalytic action of acid which was generated from PAG. And then the film was hardened with microwave irradiation, yielding organic-inorganic hybrid polymer film having hardness, highly transparency and strong adhesion with a glass substrate. Since this reaction only occurred in the optically (UV) irradiated regions, by exploiting the difference between the adhesivenesses of these regions photo-irradiated through photomask with a glass substrate, it was possible to form a patterned film with pitch of 100 to 50 μm by a simple lift-off method. A pigment-containing film using latent pigments (with subtractive three primary colors of coloring materials) and a patterned film were prepared, and it was possible to make these films multi-colored by varying the mixing ratio of the pigments. This multi-colored film-preparation method is effective for simply and efficiently forming a color-filter film by applying optical and microwave irradiation.

Keywords

Organic-Inorganic Hybrid Film, Color Filter, Photo-Patterning, Latent Pigment,

---

*Corresponding author.

Photo-Acid-Generator, Microwave, Lift-Off Method

1. Introduction

In recent years, organic-inorganic hybrid films—which combine the flexibility and responsiveness of organic substances with the hardness of inorganic substances—are being applied in fields like integrated circuits and optical devices [1]-[10]. Among organic-inorganic hybrid films, those including functional organic dyes combine the expression of new optical properties with improvement in durability. Accordingly, it is expected that they will be applied in a wide range of fields—starting from recording elements and display devices [11]-[15]. Color filters for display devices are an important field in which these organic-inorganic hybrid films are being applied. Present mainstream color filters consist of an organic pigment dispersed in a photosensitive organic polymer [16]. However, degradation of their optical characteristics by the onset of agglomeration of the organic pigment in a solvent, low scratch resistance (derived from the organic polymer), and degradation of adhesion with the film's substrate are some of the common problems with these filters. Moreover, the process for manufacturing the color filter involves optical patterning by photolithography, which is beset by problems such as occurrence of liquid residue during post-light-exposure etching and shortening of film-hardening time [17] [18]. In light of the above-described problems, it is required to develop a new material for forming a color filter with higher performance and an easy-to-use optical-patterning method for manufacturing the filter.

The aim of the present study was to develop an easy-to-use optical patterning process and improve the optical characteristics and strength of a color-filter film. By doping a latent pigment into an organic-inorganic hybrid film by utilizing a siloxane-group-linked acrylic resin, a high-performance color-filter film with outstanding transparency, hardness, and durability (such as high adhesiveness) was prepared, and its properties were investigated. Furthermore, as an easy-to-use patterning approach, optical patterning based on a "lift-off" method and formation of a sol-gel film by using a photo-acid generator was investigated. As for an efficient method for curing the film, microwave irradiation was used instead of the conventionally used heat treatment.

As a method for optical patterning of the organic-inorganic hybrid film, irradiation by ultraviolet (UV) light onto a sol-gel film (composed of a chemically modified alkoxide) is known to be available. As for a film prepared by this method, variation of its solubility with regard to the solvent of the UV-irradiated gel film is exploited, and patterns in the film are formed by etching [19]-[23]. In this study, an easy-to-use optical patterning method—which uses a lift-off technique and does not generate solvent residue due to the etching—is applied.

2. Experimental

2.1. Preparation of Latent Pigments

Synthetic method of quinacridon (Qn) latent pigment is described below as an example.

Qn pigment 0.30 g (0.96 mmol), di-t-buthyl dicarbonate 0.83 g (3.80 mmol: t-BOC) and 4-dimethylamino-pyridine 0.25 g (2.10 mmol: 4-DMAP) were mixed in N,N-dimethylformamide (70 ml: DMF). The mixed solution was stirred for 24 hours at room temperature. This solution was dropped into water, and the resulting precipitates were filtered and dried. Yellow precipitates were recrystallized from chloroholm and hexane to give Qn latent pigment (Qn-BOC). Yield: 0.190 g (39%).

Indigo latent pigment (Indigo-BOC) and Pigment Yellow 93 latent pigment (PY-BOC) were prepared by a similar method to that of Qn-BOC.

2.2. Preparation of PMPTMS

Poly-methacryloxypropyltrimethoxysilane (PMPTMS) was synthesized by block polymerization.

3-methacry-loxypropyltrimethoxysilane (MPTMS) 2.96 g (11.9 mmol) and 1-Hydroxycyclohexylphenylketone (HCPK: Photo-polymerization initiator) 0.06 g (0.29 mmol) were mixed and stirred for 5 min at rotating speed 500 rpm. This solution was irradiated with ultraviolet ray (low pressure mercury lamp, emitting wavelength: 254 nm, 185 nm, photo-intensity: 5 mW/cm^2) under nitrogen. After 30 min., a stir bar stopped due to high viscosity of the solution because of polymerization. This polymer was diluted with THF and then poured into hexane. After standing in a refrigerator for 24 hours, white precipitate was decanted with hexane and filtered and dried in N_2. Yield: 1.586 g (54%).

2.3. Thin Film Formation

PMPTMS 20 wt%, triphenylsulfoniumtriflate (PAG) 1.0% and H_2O 0.5% were dissolved in THF 58.5%. And then cyclohexanone 20 wt% was added to this solution, resulting in the spin-coating solution for thin film formation. The solution was spin-coated on the glass substrate (30 × 30 × 1.2 mmt) at rotating speed 3000 rpm for 60 sec. to form the thin film. Photomask with 100 μm or 50 μm slit was attached on the film, following to be irradiated with ultraviolet ray using low pressure mercury lamp at photo intensity of 5.0 mW/cm^2 for 10 sec. After UV irradiation the film was peeled off by a lift-off method using scotch tape. And then the patterned film was irradiated with microwave (MW, 2.45 GHz, 100 W) for 5 to 10 min. to harden the film.

2.4. Thin Film Formation of PMPTMS Containing Latent Pigments

Latent pigment Qn-BOC 0.70 wt% (Indigo-BOC 1.80 wt%, PY-BOC 1.50 wt%) was added to the PMPTMS solution for thin film formation described in 2.3, respectively. By using this solution PMPTMS thin films containing latent pigment were prepared in the same way to the formation method of PMPTMS thin film.

To make multi-coloring thin film the appropriate ratios of Qn-BOC:Indigo-BOC, Qn-BOC:PY-BOC and Indigo-BOC:PY-BOC were mixed and added to the coating solution. The multi-coloring thin films were prepared by using this solution.

2.5. Evaluation

Infrared spectroscopy (Shimadzu FT-IR 8400S) and ^{1}H nuclear magnetic resonance spectroscopy (JEOL FT-NMR JNM-ECS400) were used to analyze the structure of latent pigments and organic-inorganic hybrid polymer (PMPTMS). Ultraviolet-Visible spectroscopy (Shimadzu UV2450) was measured to analyze a transparency and absorption characteristics of the films containing latent pigment. The positions of the films containing pigments on chromaticity coordinates were obtained from the values of absorption spectra. Surface morphology of the films was measured with optical microscope (Kyowakogyo ME-LUX2) and laser microscope (Keyence VK-X200). For the evaluation of hardness of the coating films, the dynamic hardness was measured with a dynamic ultra-microhardness tester (Shimadzu DUH 211S) and pencil hardness was evaluated by a tape test conforming to Japan Industrial Standard (JIS K5600).

3. Results and Discussion

3.1. Fine Patterning Using Photo-Acid Generator and Organic-Inorganic-Hybrid Polymer

An outline of the synthesis route for the organic-inorganic-hybrid polymer and formation of a fine pattern are shown in **Figure 1** and **Figure 2**, respectively. As the starting material for the organic-inorganic-hybrid polymer, 3-methacryloxypropyltrimethoxysilane (MPTMS) was used. MPTMS consists of methacrylate organic regions and methoxysilane inorganic regions existing in a molecular framework. First, the organic regions are subjected to photo radical polymerization by UV irradiation. The C=C bonds of the organic framework are cleaved by radical polymerization, and poly-MPTMS (PMPTMS) is synthesized with the progression of organic polymerization. Next, the methoxysilane regions are subjected to inorganic polymerization by a sol-gel reaction. After a photo-acid generator (PAG) is added to the solvent for forming the PMPTMS film, a film is formed on a glass substrate. When that film is irradiated with UV light, acid that catalyzes the sol-gel reaction is generated from the PAG in the film, and the sol-gel reaction progresses. Via the sol-gel reaction, methoxy groups are then inorganically polymerized and converted into siloxane bonds. Finally, a condensation reaction is initiated by MW irradiation, and the film is hardened. The organic-inorganic hybrid resin synthesized by this process consists of organic regions and inorganic regions within its molecular framework; consequently, it is expected that its compatibility with organic pigment will be improved by the organic constituents, resulting in high chromogenicity, and its durability will be high due to the inorganic constituents.

The chemical structure of the PAG used (triphenylsulfoniumtriflate) and the acid-generation mechanism are shown in **Figure 3**. The PAG generates acid only in the irradiated regions, so when the UV irradiation is limited

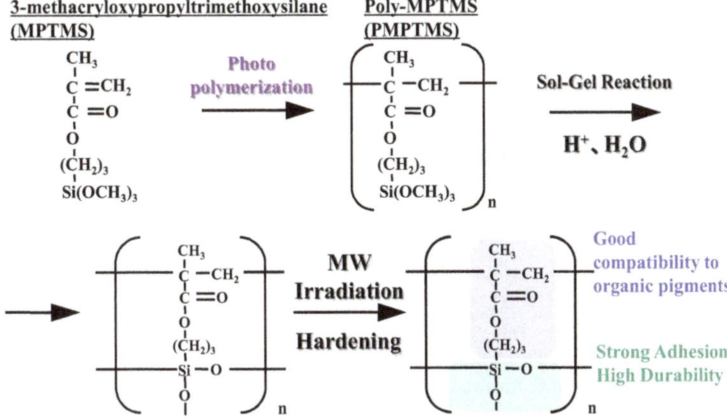

Figure 1. Preparation of photo-sensitive organic-inorganic hybrid polymer.

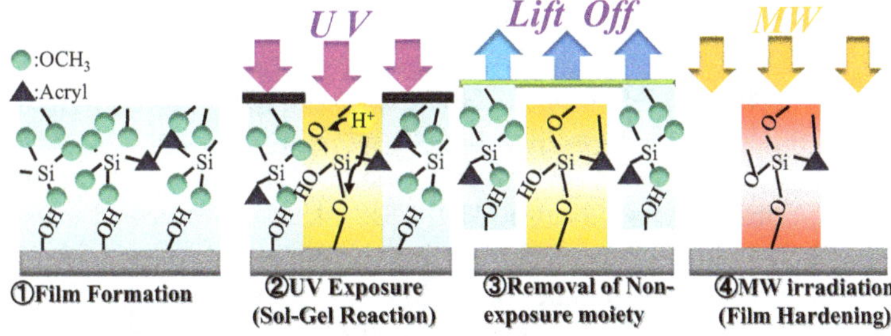

Figure 2. Schematic drawing of thin film patterning process using PAG.

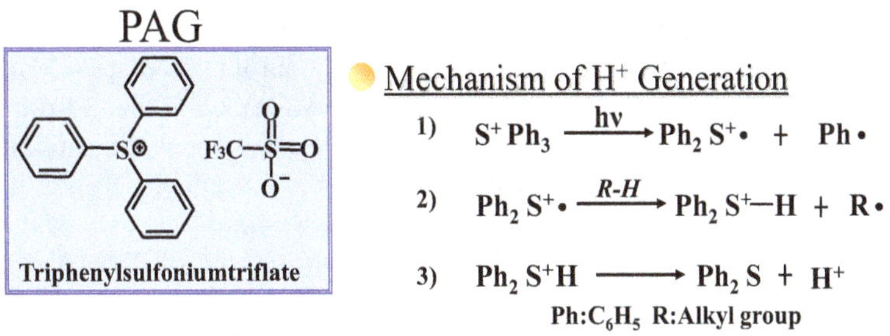

PAG

Triphenylsulfoniumtriflate

Mechanism of H^+ Generation

1) $S^+Ph_3 \xrightarrow{h\nu} Ph_2S^{+}\bullet + Ph\bullet$

2) $Ph_2S^{+}\bullet \xrightarrow{R\text{-}H} Ph_2S^{+}\text{—}H + R\bullet$

3) $Ph_2S^+H \longrightarrow Ph_2S + H^+$

$Ph: C_6H_5$ R:Alkyl group

Figure 3. Chemical structure of photo acid generator (PAG) and mechanism of H^+ Generation.

through a photomask, the sol-gel reaction is promoted by the acid catalyst in the UV-irradiated regions only. As a result, as the inorganic polymerization proceeds, the Si-OH groups of the glass surface react with the PMPTMS film, and chemical bonds are formed. The adhesiveness of the PMPTMS film with the glass substrate is thereby improved by this bonding. Exploiting the difference between the adhesiveness of the irradiated regions and that of the non-irradiated regions with the glass substrate makes it possible to form a fine patterning film by means of a simple lift-off method. The MWs (used instead of heat treatment for hardening the film) can act directly on the polarity of molecules in a different manner to heating by thermal convection and heat the interior of the film. As a result, a thin film can be heated efficiently, and processing time can be shortened.

3.2. NMR and IR Spectra of Organic-Inorganic Hybrid Polymer

Infrared spectra of MPTMS and photo-polymerized MPTMS (*i.e.*, PMPTMS) are shown in **Figure 4**. MPTMS (namely, the starting material) shows stretching vibration of the C=C bond derived from methacrylate of around 1600 cm^{-1}, bending vibration of $\text{-}OCH_3$ derived from the methoxy group of about 800 cm^{-1}, stretching vibration of C-H bond due to $\text{-}CH_2\text{-}$ around 2950 cm^{-1}, and stretching vibration of the C-H bond due to $C\text{-}CH_3$ around

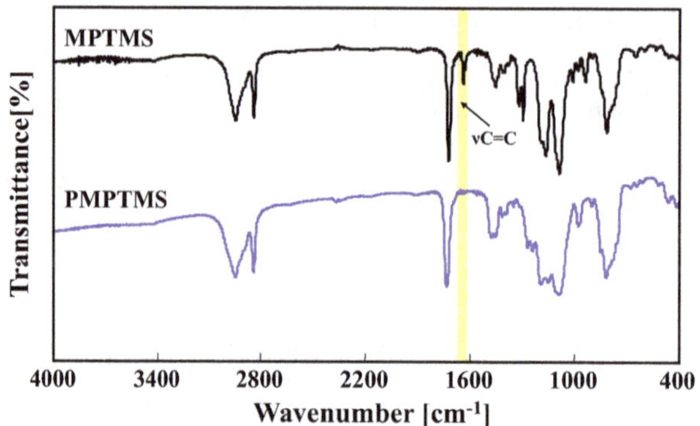

Figure 4. Infrared spectra of MPTMS and PMPTMS.

2830 cm^{-1}. As for photo-radical-polymerized MPTMS (*i.e.*, PMPTMS), the stretching vibration of the C=C bond has disappeared. This result confirms the polymerization of MPTMS.

^{1}H-NMR spectra of PMPTMS (synthesized from MPTMS) and MPTMS are shown in **Figure 5**. MPTMS shows peaks ascribed to C=C-double-bond protons at 6.10 ppm and 5.56 ppm, -CH_2- protons adjoining a -COO-group at 4.13 ppm, -OCH_3 protons at 3.58 ppm, -CH_3 protons at 1.95 ppm, -CH_2- protons in the range 1.76 to 1.83 ppm, and silicon-adjoining -CH_2- protons in the range 0.68 to 0.72 ppm. Moreover, the integration ratio of hydrogen, namely, 1:1:2:9:3:2:2 (from the low-magnetic-field side), agrees with the proton number of MPTMS.

As for the ^{1}H-NMR spectra of the synthesized PMPTMS, the peaks ascribed to the double-bond protons (at 6.10 ppm and 5.56 ppm) have disappeared, inferring that the double bond has been cleaved. As for the integration ratio of proton number, the proton number of methoxy group with a peak at 3.58 ppm is nine, while the integration ratio of the peaks at 0.65 ppm, 1.73 ppm and 3.90 ppm is 2:2:2, indicating that these peaks are methylene group. That means the three peaks can be ascribed to the alkyl chains of the methacryloxy group. From the peak positions, it is inferred that the 0.65 ppm is peak represents silicon-adjoining -CH_2- protons, 1.73 ppm represents -CH_2- protons, and 3.90 ppm represents -CH_2- protons adjoining a -COO-group. However, a weak peak at 1.10 ppm can also be seen. Although it is likely to be an intermediate -CH_2- peak, its proton number does not agree. Moreover, as for PMMA (polymethylmethacrylate) synthesized from MMA (methylmethacrylate), after the C=C (*i.e.*, the main chain) bond is cleaved, the -CH_2- peak appears near 1.84 ppm [24]. In the case of the synthesized PMPTMS as well, since a peak appears close to 1.85 ppm, it is inferred that it comes from the protons of the polymerized main chain of the PMPTMS.

The ^{1}H-NMR peaks of the synthesized PMPTMS are similar to those of PMMA with a syndiotactic configuration [25]. Peaks also appear at 1.27 ppm and 3.75 ppm, however. It is thus considered that the framework of PMPTMS is partly atactic. In light of that fact, in regard to the stereoregularity of PMPTMS, although it mainly involves a syndiotactic configuration, it also partly involves an atactic configuration.

3.3. Optical Patterning of PMPTMS Film by Sol-Gel Process Using Photo-Acid Generator

The state of a PMPTMS patterned thin film (width: 100 μm; no organic pigment) prepared by the proposed method is shown in **Figure 6** by a) an optical-micrograph and b) 3D and cross-sectional laser micrographs of the pattern shape in the thickness direction of the film. The images show that an organic-inorganic hybrid film with a satisfactory pattern (with width of 100 μm) was formed and that the pattern is formed with an edge (with thickness of about 1 μm) cut perpendicularly into it.

In the optically irradiated regions, a condensation reaction between the Si-OH group of the glass-substrate surface and the alkoxy group of the PMPTMS is promoted. As a result, adhesion of the film with the glass substrate by formation of chemical bonds becomes extremely strong. In the meantime, in the non-optically irradiated regions, inorganic polymerization does not proceed, chemical bonding with the glass substrate does not occur, and the adhesion with the glass substrate is weak. By exploiting this difference between the adhesivenesses of the respective regions with glass, it is conceivable that a lift-off method could be used for precise patterning. A lift-off method, which forms extremely simple patterns by "peeling off" a tape, does not produce liquid residue or need exclusive-use equipment in the manner of methods like etching; accordingly, it is a low-cost means of forming patterned films.

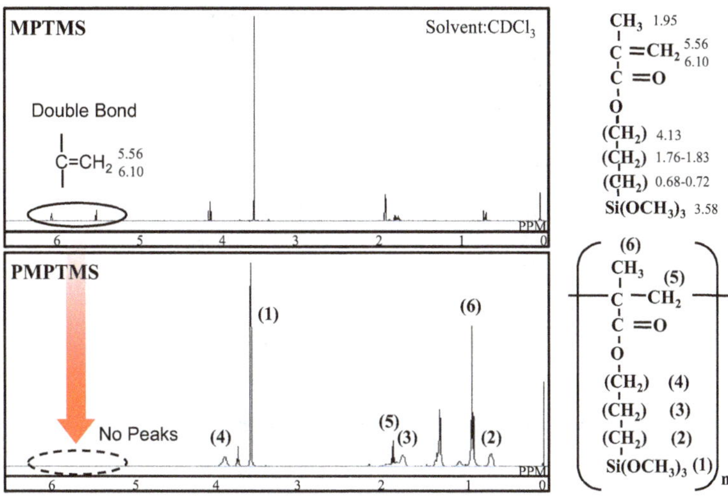

Figure 5. NMR spectra of MPTMS and PMPTMS.

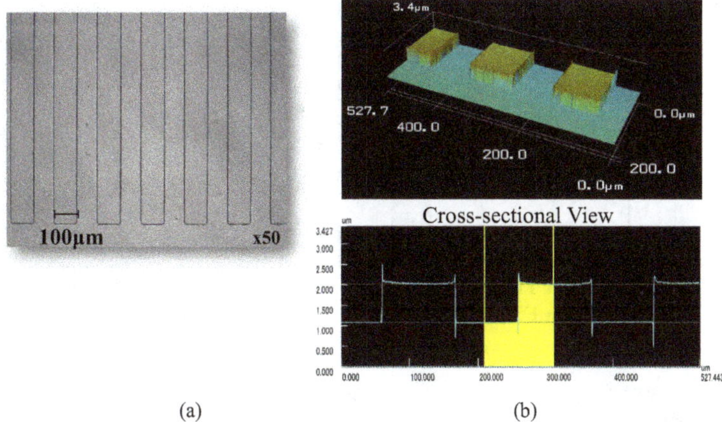

Figure 6. Surface photographs of patterned organic-inorganic hybrid thin film. (a) Optical microscope; (b) Laser microscope.

3.4. Properties of PMPTMS Film

The absorption spectrum of a thin film formed on glass substrate is shown in **Figure 7**. The transparency of the film is extremely high, indicating no absorption in the visible-light region (380 - 800 nm). In fact, it overlaps with the transmission line of the glass substrate. The measured hardness of the organic-inorganic hybrid thin film is plotted in **Figure 8**. The graph compares the PMPTMS film and an equivalent acrylic film (PMMA). It is clear that the pencil hardness of the hybrid film is considerably increased compared to that of the PMMA film (namely, from 6B to 4H). When the hardness of the thin films was measured by a thin-film microhardness tester, a significant difference between the films—in terms of the load behavior of the indenter under loading—was revealed. The load-displacement curves of the films (for a loading depth of 0.1 μm) are shown in **Figure 8(b)**. The load on the PMMA film is lower than that on the PMPTMS film (0.15 mN compared to 0.25 mN). Moreover, in contrast to the load behavior of the PMPTMS film (namely, the load-deformation line returns to the origin), that of the PMMA film shows that deformation damage remains (namely, the line does not return to the origin). While the PMPTMS film plastically deforms to a certain extent, it shows high elasticity during the loading-unloading process. The hardness of the PMPTMS film is significantly improved in comparison to that of the PMMA film; namely, the Martens' hardness (determined by loading-unloading test) of the former is 456.5 N/mm^2, and that of the latter is 258.6 N/mm^2. It is supposed that as a result of the sol-gel reaction (which is promoted by applying the photo-acid generator as previously described), siloxane bonds (Si-O-Si) are formed in the film, and a hard inorganic network is formed, and that hard-network formation explains the superior hard

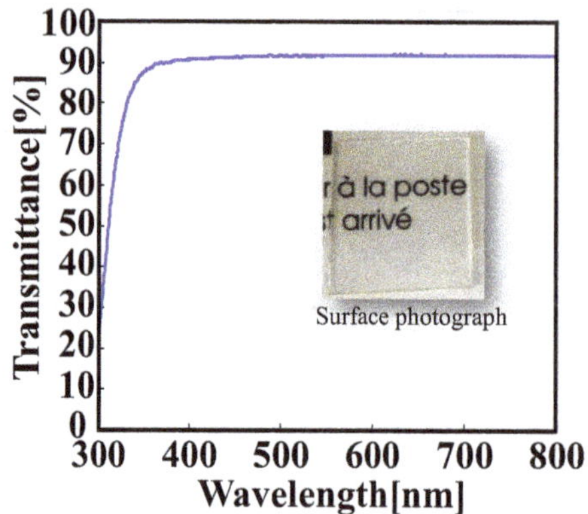

Figure 7. Absorption spectrum of organic-inorganic hybrid thin film on glass substrate.

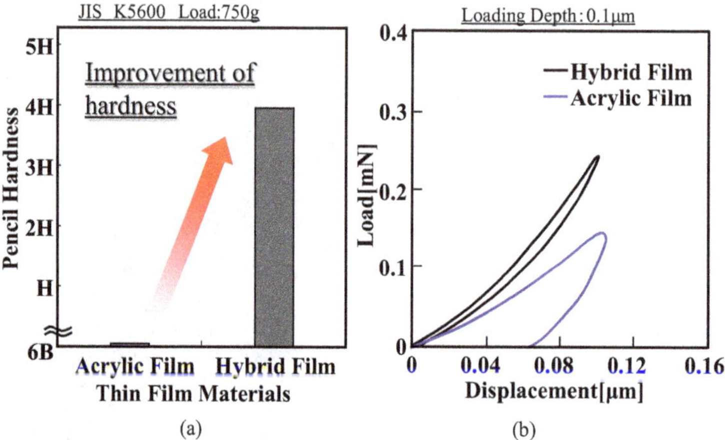

Figure 8. Hardness of organic-inorganic hybrid thin film. (a) Pencil test; (b) Microhardness.

ness of the PMPTMS film. Cross-cut adhesion test (ISO Cross-cut Test) was carried out to evaluate adhesiveness of PMPTMS film on a glass substrate. PMPTMS film showed the value of 25/25 (the number of residual films/the number of cross-cut films), indicating strong adhesiveness. In stark contrast, for acrylic film, which has been used for color filter, both the MW-irradiated film and the heat-treated film at 150°C showed the values of 0/25, indicating weak adhesiveness. The strong adhesiveness of PMPTMS film would be attributed to the chemical bonding between PMPTMS film and a glass substrate.

3.5. Preparation of PMPTMS Film Containing Latent Pigments

"Multi-coloring" by mixing coloring materials, namely, latent pigments (with three "primary colors" of subtractive mixture) into the synthesized PMPTMS film was investigated. The latent pigments used to form the three subtractive primary colors were cyan (*i.e.*, indigo), yellow (*i.e.*, PY93), and magenta (*i.e.*, quinacridone)). The chemical structures of the pigments are shown in **Figure 9**. The latent pigments are compounds in which the hydrogen of the N-H group in the pigment molecules is substituted by t-BOC ($(CH_3)_3COCO$) groups, and they are soluble in a solvent by means of blocking hydrogen bonding between the pigment molecules [26]. The latent pigments are soluble in various solvents, and they can dissolve in solutions used for forming films. Moreover, as a result of converting them to pigments in which t-BOC groups introduced by heating are eliminated, it is possible to precipitate extremely small, uniform pigment particles in the film. It is thus possible to prepare a color film with excellent optical characteristics.

Changes in the surface color of latent-pigment-containing films (prepared on a glass substrate) after MW irradiation (500 W for 10 min) or heat treatment are compared in **Figure 10**. After MW irradiation, it is clear that

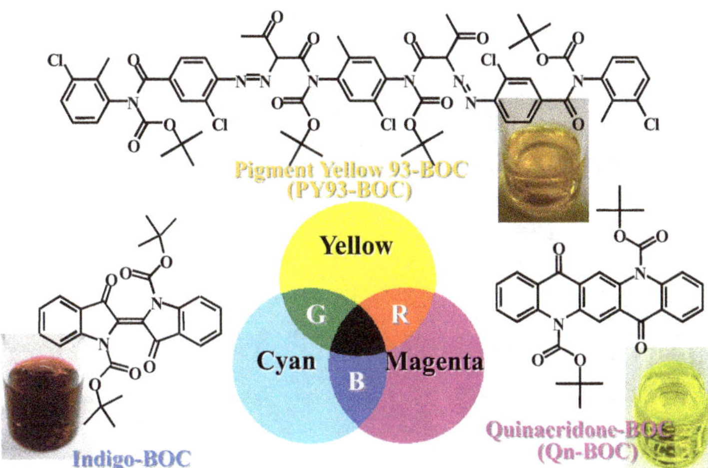

Figure 9. Chemical structure of latent pigments of subtractive three primary colors.

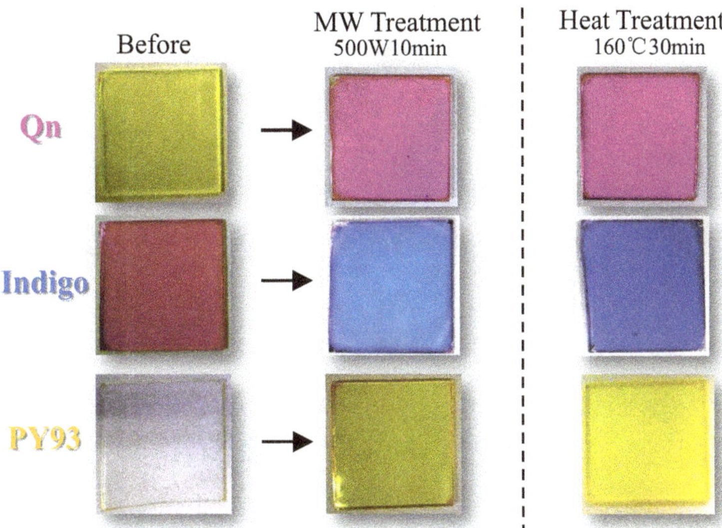

Figure 10. Changes of surface color after MW treatment or heat treatment.

the film produces the three subtractive "primary colors" of the coloring materials (magenta, cyan, and yellow). As for the heat-treated film, heating at 180°C for 30 min is necessary. The latent pigment is transformed by the MW irradiation (500 W for 5 to 10 min) to active pigment, and a pigment-containing film is formed efficiently by MW irradiation. While hardening the PMPTMS film, the MW irradiation also contributes to forming active pigment from the latent pigment. Moreover, by varying the blend ratio of the latent pigments with the three subtractive "primary colors," it is possible to produce the three primary colors of light (namely, red, green, and blue) by subtractive mixture. Photographs of the surfaces of films with systematically varied color mixtures of latent pigments are shown in **Figure 11**. It can be clearly seen that just about all colors were successfully produced; in other words, a technique of colorization by applying latent pigments is feasible. The positions of the chromaticity coordinates of the prepared films are shown in **Figure 12**. The coordinates of the film showing the three primary colors of light mostly coincide with the a* and b* axes of the chromaticity-coordinate system, indicating that it properly expresses red, green, and blue.

Changes in absorption spectra when the mixing ratio of latent pigments is changed are shown in **Figure 13**. When the mixing ratios of the pigment mixtures (namely, indigo and Qn, PY93 and indigo, and Qn and PY93) are changed, absorption spectra with isosbestic points are obtained. This result shows that the two constituents are uniformly mixed in the films. Patterned films (prepared by the proposed method) containing the pigments

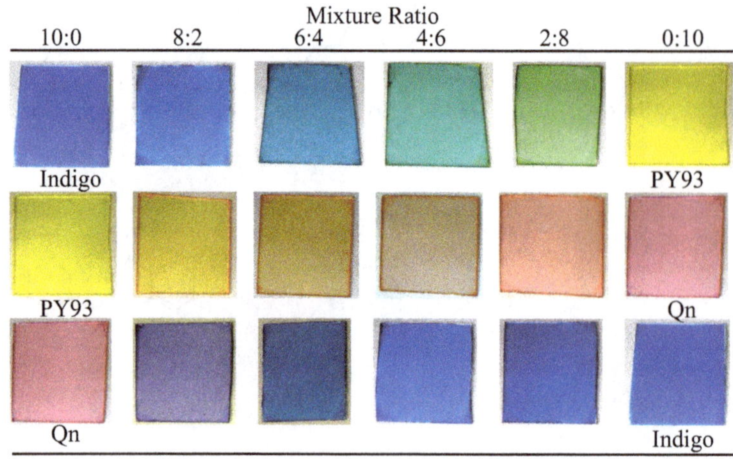

Figure 11. Surface photographs of color mixture films of latent pigments.

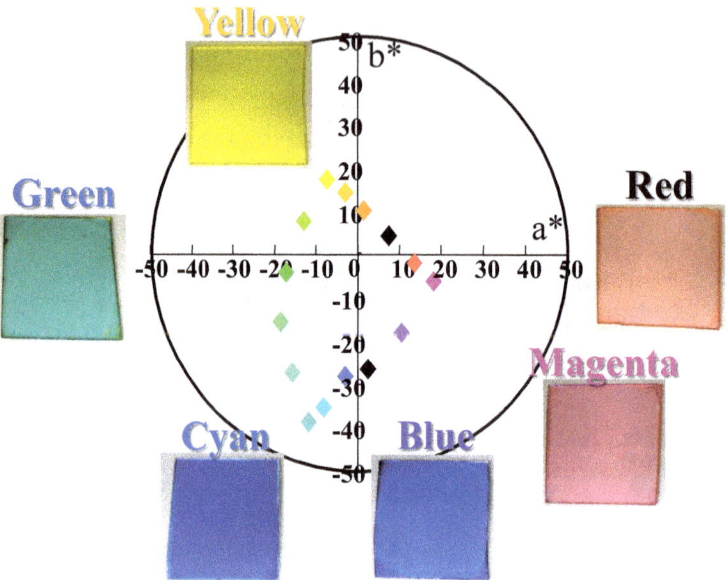

Figure 12. Chromaticity coordinates of the films prepared by using latent pigments.

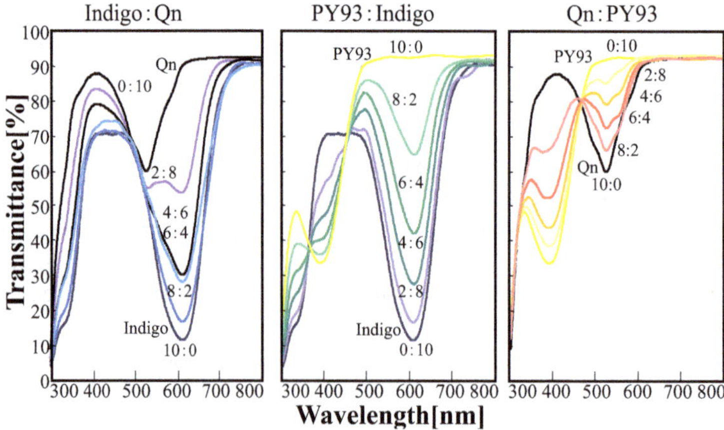

Figure 13. Absorption spectra of color mixture films of latent pigments.

with the three subtractive primary colors of the coloring materials are shown in **Figure 14**. The photographs clearly show fine patterns with widths of 50 μm and 100 μm were uniformly formed. Optical micrographs and laser-microscope photographs of a 100-μm patterned film containing Qn pigment are shown in **Figure 15**. The images show that in the case of the film containing pigment, a satisfactory pattern is formed, and the edges of the patterns are cut precisely. Compared to the PMPTMS film without pigment, the film with pigment is thicker (about 2.3 μm). It is presumed that this additional thickness is due to the inclusion of particles of organic pigments in the film. Furthermore, the pencil hardness and adhesiveness of the film including pigments equal that (namely, pencil hardness 4H, cross-cut test 25/25) of PMPTMS without organic pigments. The other pigment-containing films (namely, the ones containing indigo and PY93) gave similar results. The developed PMPTMS film has high hardness and strong adhesiveness, indicating that the film would be applicable to color filter for displays. And microwave irradiation process is useful to harden the film and shorten processing time (5 - 10 min) in comparison with conventional heat-treatment process (30 - 60 min).

4. Conclusion

An organic-inorganic hybrid film containing latent pigments was prepared by using a photo-acid generator and

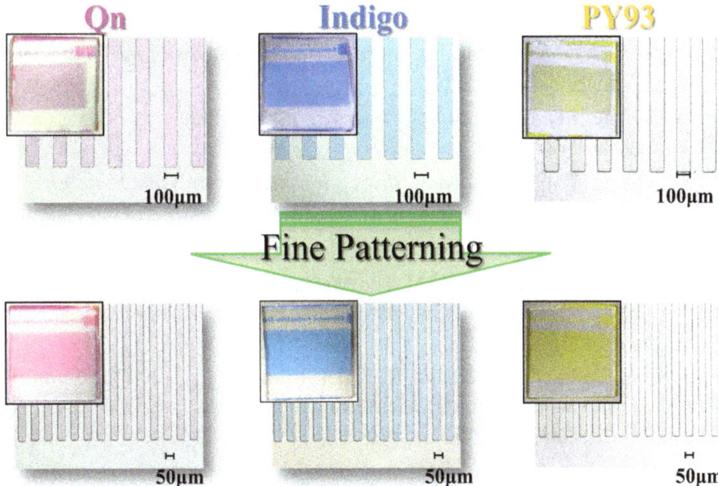

Figure 14. Surface photographs of fine patterned films containing three subtractive primary color pigments.

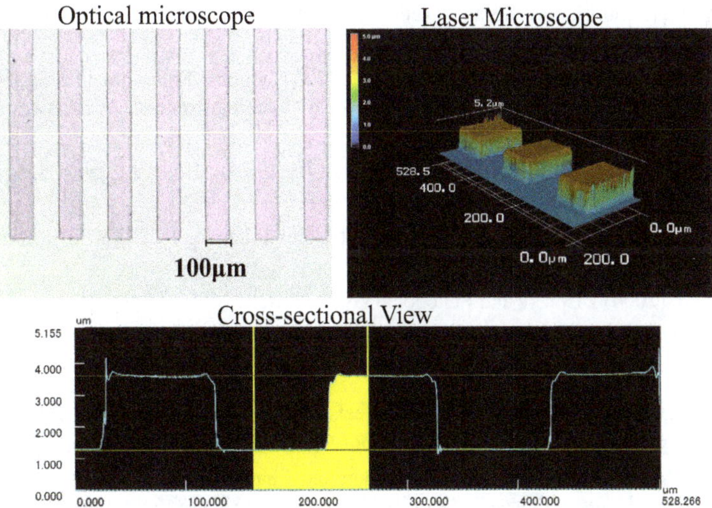

Figure 15. Surface photographs of 100 μm patterned film containing Qn Pigment.

MW irradiation and optically patterned by UV light. Acrylic resin combining methoxysilane groups forms a hard and highly transparent thin film as a result of a sol-gel reaction being promoted by an acid generated by a photo-acid generator. Since this reaction only occurs in the optically (UV) irradiated regions, by exploiting the difference between the adhesivenesses of these regions with a glass substrate, it is possible to form a patterned film with pitch of 100 to 50 μm by a simple lift-off method. Moreover, a pigment-containing film using latent pigments (with three primary colors of coloring materials) and a patterned film were prepared, and it was possible to make these films multi-colored by varying the mixing ratio of the pigments. This multi-colored film-preparation method is effective for simply and efficiently forming a color-filter film by applying optical and MW irradiation.

References

[1] Sakka, S., Ed. (2004) Handbook of Sol-Gel Technology, Sol-Gel Processing. Vol. 1, Kluwer Anademic Publishers, Dordrecht.

[2] Sakka, S., Ed. (2004) Handbook of Sol-Gel Technology, Application of Sol-Gel Technology. Vol. 3, Kluwer Anademic Publishers, Dordrecht.

[3] Sakka, S., Ed. (2005) Application of Sol-Gel Processing to Nanotechnology. CMC Publishers.

[4] Schmidt, H., Jonscher, G., Goedicke, S. and Mennig, M. (2000) Sol-Gel Process as a Basic Technology of Nanoparticle-Dispersed Inorganic-Organic Composites. *Journal of Sol-Gel Science and Technology*, **19**, 39. http://dx.doi.org/10.1023/A:1008706003996

[5] Bescher, E. and Mackenzie, J.D. (2003) Sol-Gel Coatings for the Protection of Brass and Bronze. *Journal of Sol-Gel Science and Technology*, **26**, 1223.

[6] Ro, H.W. and Soles, C.L. (2011) Mater. *Today, Silsesquioxanes in Nanoscale Patterning Applications*, **14**, 20.

[7] Cordes, D.B., Lickiss, P.D. and Rataboul, F. (2010) Recent Developments in the Chemistry of Cubic Polyhedral Oligosilsesquioxane. *Chemical Reviews*, **110**, 2081. http://dx.doi.org/10.1021/cr900201r

[8] Ainuddin, A.R., Hakiri, N., Muto, H., Sakai, M. and Matsuda, A. (2011) Mechanical Properties of Phenylsilsesquioxane-Methylsilsesquioxane Hybrid Films by Indentation. *Journal of the Ceramic Society of Japan*, **119**, 490. http://dx.doi.org/10.2109/jcersj2.119.490

[9] Matsuda, A., Ogawa, R., Daiko, Y., Tadanaga, K. and Tatsumisago, A. (2007) Micropatterning for Vinylsilsesquioxane-Titania hybrid Gel Films with Ultraviolet Light Irradiation. *Journal of Photopolymer Science and Technology*, **20**, 101. http://dx.doi.org/10.2494/photopolymer.20.101

[10] Ainuddin, A.R., Ishigaki, T., Hakiri, N., Muto, H., Sakai, M. and Matsuda, A. (2012) Influence of UV Irradiation on Mechanical Properties and Structures of Sol-Gel-Derived Vinylsilsesquioxane Films. *Journal of the Ceramic Society of Japan*, **120**, 442. http://dx.doi.org/10.2109/jcersj2.120.442

[11] Nakazumi, H., Ed. (2004) Displays and Functional Dyes. CMC Publishers.

[12] Nonomura, K., Higashino, H. and Murai, R. (2002) Plasma Display Materials. *MRS Bulletin*, **27**, 898-902. http://dx.doi.org/10.1557/mrs2002.280

[13] Okamura, T., Kitagawa, T., Koike, K. and Fukuda, S. (2004) Optical Filters for Plasma-Display Panels Using Organic Dyes and Conductive Meshlayers by Using a Roll-to-Roll Etching Process. *Proceedings of the* 8*th Asian Symposium on Information Display*, Nanjing, 15-17 February 2004, 23.

[14] Yang, Y.C., Song, K., Rho, S.G., Rho, N.S., Hong, S.J., Beul, K.B., Hong, M. and Chung, K. (2005) Society of Information Display (SID) '05 Digest, 1210.

[15] Minato, K., Itoi, T. and Ito, H. (2005) High Contrast Color Filter for LCD-TV. *Proceedings of Information Display Workshop* (*IDW*) '05, 339.

[16] Nakazumi, H., Ed. (2004) Displays and Functional Dyes. CMC.

[17] Hatajima, M. (2007) NIKKEI FPD, 126.

[18] Masuda, J. (2008) *Monthly Display*, **9**, 18.

[19] Shinmou, K., Tohge, N. and Minami, T. (1994) Effects of UV-Irradiation on the Formation of Oxide Thin Films from Chemically Modified Metal-Alkoxides. *Journal of Sol-Gel Science and Technology*, **2**, 581-585.

[20] Zhao, G., Tohge, N. and Nishii, J. (1998) Fabrication and Characterization of Diffraction Gratings Using Photosensitive Al_2O_3 Gel Films. *Japanese Journal of Applied Physics*, **37**, 1842-1846. http://dx.doi.org/10.1143/JJAP.37.1842

[21] Ono, S. and Hirano, S. (1997) Patterning of Lithium Niobate Thin Films Derived from Aqueous Solution. *Journal of the American Ceramic Society*, **80**, 2533-2540. http://dx.doi.org/10.1111/j.1151-2916.1997.tb03155.x

[22] Tadanaga, K., Owan, T., Morionaga, J., Urbanek, S. and Minami, T. (2000) Fine Patterning of Transparent, Conductive SnO_2 Thin Films by UV-Irradiation. *Journal of Sol-Gel Science and Technology*, **19**, 791-794. http://dx.doi.org/10.1023/A:1008764217509

[23] Tadanaga, K., Ueyama, K., Sueki, T., Matsuda, A. and Minami, T. (2003) Micropatterning of Inorganic-Organic Hybrid Coating Films from Various Trifunctional Silicon Alkoxides with a Double Bond in Their Organic Components. *Journal of Sol-Gel Science and Technology*, **26**, 431-434. http://dx.doi.org/10.1023/A:1020745820563

[24] Higashimura, T., *et al.* (2007) Introduction to New Polymer Chemistry. Kagaku Doujin, Tokyo.

[25] NIMS NMR Database. http://polymer.nims.go.jo/NMR/

[26] Zambounis, Z., Hao, H. and Iqbal, A. (1997) Latent Pigments Activated by Heat. *Nature*, **388**, 131-132.

25

Application Concept of Zero Method Measurement in Microwave Radiometers

Alexander V. Filatov

Tomsk State University of Control Systems and Radio Engineering, Tomsk, Russia
Email: filsash@mail.ru

Abstract

This article examined in detail microwave radiometer functioning algorithm with synchronously using of the two types of pulse modulation: amplitude pulse modulation and pulse-width modulation. This allows a zero-radiometer measurement method to realize when the fluctuation effect of the receiver gain and the influence of its own noise changes are minimized. A zero balance automatically maintains in radiometer. The antenna signal is indirectly determined through the signal duration that controls the pulse-width modulation. An analytical expression of the fluctuation sensitivity was obtained in a general form. From its analysis gain in sensitivity, conditions were defined by the optimizing of the radiometer input knot's construction. Three modifications of the radiometer input knot were researched. Fluctuation sensitivity at different measurement range was determined for modification of the radiometer input knot.

Keywords

Microwave Radiometer, Null Method of Measurement, Fluctuation Sensitivity

1. Introduction

It is well known that study of microwave appearance of various objects of earth's surface provides fundamentally different physical self-descriptiveness than using only optical and infrared earth remote sensing [1] [2]. This fact encourages continuous development of measurement systems aimed at improving the accuracy and the increasing of informative sensing saturation. A development of microwave remote sensing systems with low energy consumption intended for working in natural conditions and especially on the side refers to a difficult problem. Traditional approaches often lead to mutually exclusive solutions; therefore, the implementation of stringent requirements for the microwave radiometers is impossible without searching for new approaches, methods and solutions.

Methods for stabilizing or accounting of radiometer technical parameters' changes consist in application of different functioning methods. Among various schemes, modulation radiometers have been widely used [3] [4]. These radiometers are based on method of differential measurements. At the entrance, before radiometric receiver, modulation is generated with a certain frequency antenna signal and a stable reference noise generator signal—the antenna simulator connected to the input of the receiver, instead of the antenna. Because measurements are transferred to a higher frequency with which signals are modulated, the influence of two major destabilizing factors reduces: the influence of anomalous fluctuations of the gain near zero frequency significantly decreases; constant and quasi constant of self-noise components of the radiometer decrease. Wide application of modulation radiometers is associated with a satisfactory accuracy of measurements, which is achieved essentially by simple methods (modulation at the entrance and demodulation—synchronous detection—at the output) and a simple circuit implementation. Modulation radiometers are attracted by simple design; therefore, they are promising for repeating. This is manifested in their massive using. However, the full minimizations of effect of changes of amplifier gain and self-noise of receiver in modulation scheme do not happen. It is useful noticeably to minimize these changes by using the zero method of measurement in modulation radiometer. This method has the highest potential facilities to create precision radiometers.

Ryle and Troitsky are the authors of the conception and interpretation of zero method applying to radiometers [5] [6]. A number of successful researches have associated with the creation of zero-radiometer. These researches are proof of the Nyquist formula for spectral, fluctuating noise density of various materials' resistance validity, and discovery of the recombination rf spectral line emitted by highly excited atoms in radio astronomy, measuring abyssal temperature of the biological objects, etc. The output power of the noise reference generator is regulated before achieving the so-called zero balance in the measuring path. The so-called zero balance is being considered established if the same signals in different half-periods of a symmetric pulse of the measuring path are passing. Therefore, feedback following network and regulated reference noise source are presenting in zero radiometer. The first zero radiometer had analog regulation principle of zero balance. In this radiometer, in input knots precision operated microwave devices, adjustable attenuators or random-noise generators with regulated input power being applied. The demands of the high linear adjusting characteristic, large dynamic range, elevated response speed of regulation for bringing the measuring system in the zero balance method were established to current knots. Errors arising from using the given elements in input blocks were not allowed to fully realize the zero measurement method advantages and zero radiometers were not widely spreading.

The creation of the pulse added noise operation in the radiometer appeared as a successful development of zero method application. This operation works on pulse-width-law [7]-[9]. In this case, the average of the half-cycle modulation power of invariable reference noise signal generator is regulated by changing of the duration of its action. This led to the simplification of the input receiving block scheme (microwave switch and noise generator), improving of the linearity of calibration radiometer equation by the simple adjustment of the reference signal.

This method of mixing the reference and measured signals (in different ways pulse-modulated) led to a considerable simplification of input block design. But this method complicated modulated signals conditioning after square-law detector. It led to an increase in measurement error. Thus, schemes of the zero radiometer with pulse added noise turned out more difficult than schemes of the ordinary modulation radiometers, and therefore they were not widely used.

In this article, a new modification of the zero functioning principle of microwave radiometers by weak signal changing was considered, and the possibility of fluctuation sensitivity and stability of these systems were analyzed.

2. The Zero Method Modification of the Signal Reception

The equality of low-energy signals at the radiometric receiver input at different half-periods of a symmetric pulse modulation is achieved by the pulse duration changing. This impulse controls the introduction of additional noise signal into the supporting or the antenna paths of modulation radiometer. Time diagrams which explain the combine modulation principle in radiometer are shown in **Figure 1**. The control pulse-width signal t_{pws} changes in the range from zero to the half-period length t_{mod} of main modulator work.

The signals energy equality is the condition of specified balance in radiometer. These signals enter to the receivers input at different half-periods of symmetrical modulation. In **Figure 1(a)**, these energies are proportional to the corresponding shaded areas $Q_1(t)$ and $Q_2(t)$. The signals energy equality is continuously maintained

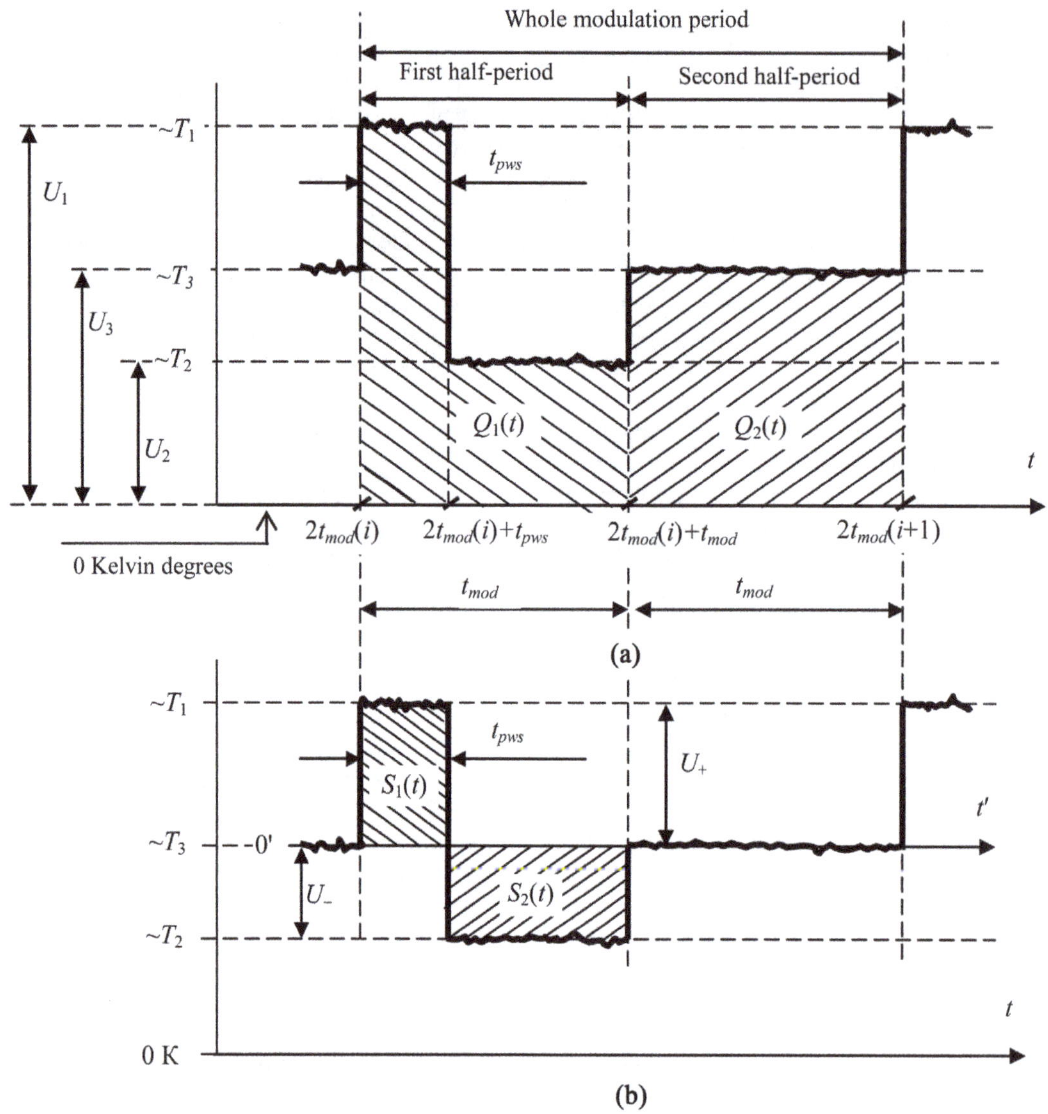

Figure 1. Signal time diagrams at the receiver output of zero radiometer which uses pulse-amplitude and pulse-width modulations.

by automatic pulse-width signal duration t_{pws} changing of radiometer controlling system. On the diagram pulse voltage amplitudes U_1, U_2, U_3 at the receiver output are proportional to appropriate signal noise temperatures T_1, T_2, T_3 at the receiver input. Proportionality constant $G(t)$ is equal to product of gain coefficient at high and low frequencies and square-law detector transmission coefficient.

Introduce the value $\Delta Q(t)$ which considers the inequality of $Q_1(t)$ and $Q_2(t)$ and includes both a constant component ΔQ_0 and the fluctuating part of $\Delta q(t)$

$$\begin{aligned}\Delta Q(t) = Q_1(t) - Q_2(t) = \sum_{i=0}^{\infty} \int_{2t_{\text{mod}}(i)}^{2t_{\text{mod}}(i)+t_{pws}} \left[T_1 G(t-\theta) + n_1(t-\theta)\right] H_1(\theta) \mathrm{d}\theta \\ + \sum_{i=0}^{\infty} \int_{2t_{\text{mod}}(i)+t_{pws}}^{2t_{\text{mod}}(i)+t_{\text{mod}}} \left[T_2 G(t-\theta) + n_2(t-\theta)\right] H_2(\theta) \mathrm{d}\theta \\ - \sum_{i=0}^{\infty} \int_{2t_{\text{mod}}(i)+t_{\text{mod}}}^{2t_{\text{mod}}(i+1)} \left[T_3 G(t-\theta) + n_3(t-\theta)\right] H_3(\theta) \mathrm{d}\theta,\end{aligned} \tag{1}$$

where $G(t) = G_0 + g(t)$, G_0—constant component, $g(t)$—transmission coefficient fluctuation, $n_1(t)$, $n_2(t)$, $n_3(t)$—signal T_1, T_2, T_3 noise entries, which take into account their noise character; $H_1(\theta)$, $H_2(\theta)$, $H_3(\theta)$—pulse response characteristics of accumulative receiver filters for each of the corresponding signals.

For a long time interval it is possible to use $n_1(t) = n_2(t) = n_3(t) = 0$ for mean value t_{pws} considering even deviations of noise signal components which represent the normal static ergodic processes. Also for the transmission coefficient $G(t)$, taking into account the statistical invariance of $g(t)$ for a long time interval its fluctuating component can be considered equal to zero. The accumulation of signals occurs by using of three integrating *RC*-circuits of the first order. Pulse characteristics of these circuits are determined by the well-known relation $H(\theta) = \exp(-\theta/\tau)/\tau$, where τ is circuit time constant; $\tau = RC$.

Then for the constant component of the parameter $\Delta Q(t)$, using (1) write

$$\Delta Q_0 = \sum_{i=0}^{\infty} \int_{2t_{\text{mod}}(i)}^{2t_{\text{mod}}(i)+\overline{t_{pws}}} T_1 G_0 \frac{\exp\left(-\frac{\theta}{\tau_1}\right)}{\tau_1} \mathrm{d}\theta + \sum_{i=0}^{\infty} \int_{2t_{\text{mod}}(i)+\overline{t_{pws}}}^{2t_{\text{mod}}(i)+t_{\text{mod}}} T_2 G_0 \frac{\exp\left(-\frac{\theta}{\tau_2}\right)}{\tau_2} \mathrm{d}\theta - \sum_{i=0}^{\infty} \int_{2t_{\text{mod}}(i)+t_{\text{mod}}}^{2t_{\text{mod}}(i+1)} T_3 G_0 \frac{\exp\left(-\frac{\theta}{\tau_3}\right)}{\tau_3} \mathrm{d}\theta. \quad (2)$$

It is possible to use the decomposition of exponential functions in a Maclaurin series with the two members approaching, because pulse durations t_{pws} and t_{mod} are much less than the accumulation signals time (these signals are defined by low-frequency filters constants τ_1, τ_2, τ_3). The solution of Equation (2) is

$$\Delta Q_0 = G_0 \frac{T_1 \overline{t_{pws}} + T_2\left(t_{\text{mod}} - \overline{t_{pws}}\right) - T_3 t_{\text{mod}}}{2t_{\text{mod}}}. \quad (3)$$

Taking into account signal noise character and availability of the receiver amplifier gain fluctuation $Q_1(t)$ and $Q_2(t)$ are not equal for a modulation period. But for a large time interval in the limit for an endless number of modulation periods $\Delta Q_0 = 0$ can be written due to automatic signals energy tracking system operation for uninterrupted leveling.

It results from (3) that

$$T_1 \overline{t_{pws}} + T_2\left(t_{\text{mod}} - \overline{t_{pws}}\right) - T_3 t_{\text{mod}} = 0. \quad (4)$$

Solving Equation (4) relatively $\overline{t_{pws}}$ obtain

$$\overline{t_{pws}} = \frac{T_3 - T_2}{T_1 - T_2} \times t_{\text{mod}} \quad (5)$$

Equation (5) is a mathematic of the implementation of the proposed modification of the zero method reception. It follows that the antenna signal can be determined indirectly through the added noise pulse duration without signals changing in the low frequency path. Equation (5) does not include the transfer constant of the measuring path. This indicates of zero method of radiometer working.

3. Zero Method Implementation Course in Microwave Radiometers

The comparison of $Q_1(t)$ and $Q_2(t)$ is replaced by equivalent comparison of volt-second areas (**Figure 1(b)**) of positive $S_1(t)$ and negative $S_2(t)$ pulses at the receiver output. The pulse amplitudes are proportional to the signal differences $T_1 - T_3$, $T_3 - T_2$ at the receiver input. If the voltage of second half-period modulation is equal to zero and time zero axis passes through the T_3 level signal, the volt-second pulse areas in the modulation first half-period are equal, $S_1(t) = S_2(t)$ (periodic sequence). As

$$S_1(t) = U_+ \overline{t_{pws}} \text{ and } S_2(t) = U_-\left(t_{\text{mod}} - \overline{t_{pws}}\right)$$

where $U_+ = G_0 kdf(T_1 - T_3)$ and $U_- = G_0 kdf(T_3 - T_2)$, then

$$G_0 kdf\left(T_1 - T_3\right)\overline{t_{pws}} = G_0 kdf\left(T_3 - T_2\right)\left(t_{\text{mod}} - \overline{t_{pws}}\right)$$

where k—Boltzmann constant, df—frequency range band. Solving the last equation relatively $\overline{t_{pws}}$ obtain (5).

A sequence of simple but necessary operations follows from this reasoning. These operations must be done to transform the signals after square-law detector and low-frequency amplification. These operations are constant component exclusion in signals and voltage sign determination in the second half period of modulation. The direct component exception reduces to the pulse signal T_3 top deflection to the zero time axis ($t \Rightarrow t'$). The zero balance condition (the voltage lack in the modulation second half-period on the output of the measuring tract) settles by the t_{pws} duration regulation. Thus antenna signal tracking is realized by the t_{pws} duration changing. This leads to the signal periodic sequence displacement up or down relatively the time zero axes.

4. The Fluctuation Sensitivity Analysis

$\Delta Q(t)$ chaotic changes in (1) are associated with the parameter $\Delta q(t)$ with nonzero components $n_1(t)$, $n_2(t)$, $n_3(t)$, $g(t)$. Compute the parameter $\Delta q(t)$ dispersion by correlation function method. Meanwhile take into account the statistical independence of the transmission coefficient fluctuations of the radiometer measuring path $g(t)$ and signals T_1, T_2, T_3 noise components $n_1(t)$, $n_2(t)$, $n_3(t)$. The total dispersion is the sum of two dispersions. First of them Δq_g^2 allows for the receiver transmission coefficient fluctuations, the second one Δq_n^2 is caused by the noise signal nature.

The noise correlation time of function $n(t)$ is determined by signal receiving bandwidth df and for radiometers is considerable less than modulation period $2t_{mod}$. Consequently, these signals T_1, T_2, T_3 noise components $n_1(t)$, $n_2(t)$, $n_3(t)$ are uncorrelated with each other. Then the dispersion Δq_n^2 is determined from (1)

$$\begin{aligned}\overline{\Delta q_n^2} &= \sum_{i=0}^{\infty}\sum_{j=0}^{\infty}\int\limits_{2t_{\text{mod}}(i)}^{2t_{\text{mod}}(i)+t_{pws}}\int\limits_{2t_{\text{mod}}(j)}^{2t_{\text{mod}}(j)+t_{pws}} \overline{n_1(t-\theta)n_1(t-\theta')}H_1(\theta)H_1(\theta')\mathrm{d}\theta\mathrm{d}\theta' \\ &+\sum_{i=0}^{\infty}\sum_{j=0}^{\infty}\int\limits_{2t_{\text{mod}}(i)+t_{pws}}^{2t_{\text{mod}}(i)+t_{\text{mod}}}\int\limits_{2t_{\text{mod}}(j)+t_{pws}}^{2t_{\text{mod}}(j)+t_{\text{mod}}} \overline{n_2(t-\theta)n_2(t-\theta')}H_2(\theta)H_2(\theta')\mathrm{d}\theta\mathrm{d}\theta' \\ &+\sum_{i=0}^{\infty}\sum_{j=0}^{\infty}\int\limits_{2t_{\text{mod}}(i)+t_{\text{mod}}}^{2t_{\text{mod}}(i+1)}\int\limits_{2t_{\text{mod}}(j)+t_{\text{mod}}}^{2t_{\text{mod}}(j+1)} \overline{n_3(t-\theta)n_3(t-\theta')}H_3(\theta)H_3(\theta')\mathrm{d}\theta\mathrm{d}\theta' \\ &= J_{1n} + J_{2n} + J_{3n}\end{aligned} \tag{6}$$

Noise $n(t)$ autocorrelation function in comparison with pulse filters characteristics $H(\theta)$ and the modulation period $2t_{mod}$ can be considered a delta function $\delta(t)$ with integral value $2G_0^2T^2$ and correlation time $1/df$. Accordingly, have

$$\overline{n_1^2} = \frac{2G_0^2T_1^2}{df}; \quad \overline{n_2^2} = \frac{2G_0^2T_2^2}{df}; \quad \overline{n_3^2} = \frac{2G_0^2T_3^2}{df}. \tag{7}$$

Exploiting (7), (6), obtain the following equation

$$\overline{\Delta q_n^2} = \frac{G_0^2}{2t_{\text{mod}}\,df}\left(\frac{T_1^2\overline{t_{pws}}}{\tau_1} + \frac{T_2^2\left(t_{\text{mod}} - \overline{t_{pws}}\right)}{\tau_2} + \frac{T_3^2 t_{\text{mod}}}{\tau_3}\right). \tag{8}$$

After the replacement of (5) into (8) finally obtain

$$\overline{\Delta q_n^2} = \frac{G_0^2}{2df}\left[\frac{T_1^2}{\tau_1}\times\frac{T_3 - T_2}{T_1 - T_2} + \frac{T_2^2}{\tau_2}\left(1 - \frac{T_3 - T_2}{T_1 - T_2}\right) + \frac{T_3^2}{\tau_3}\right]. \tag{9}$$

In case of equal low-frequency filters $\tau_1 = \tau_2 = \tau_3 = \tau$

$$\overline{\Delta q_n^2} = \frac{G_0^2}{2df\,\tau}\left[T_3\left(T_1 + T_2 + T_3\right) - T_1T_2\right]. \tag{10}$$

Next determine the dispersion caused by the fluctuations of the radiometer measuring path transmission coefficient.

Write down the formula for dispersion $\overline{\Delta q_g^2}$ calculating using the fundamental relation (1)

$$\begin{aligned}\overline{\Delta q_g^2} &= \sum_{i=0}^{\infty}\sum_{j=0}^{\infty}\int_{2t_{\text{mod}}(i)}^{2t_{\text{mod}}(i)+t_{pws}}\int_{2t_{\text{mod}}(j)}^{2t_{\text{mod}}(j)+t_{pws}} T_1^2\overline{g(t-\theta)g(t-\theta')}H_1(\theta)H_1(\theta')\mathrm{d}\theta\mathrm{d}\theta' \\ &+\sum_{i=0}^{\infty}\sum_{j=0}^{\infty}\int_{2t_{\text{mod}}(i)+t_{pws}}^{2t_{\text{mod}}(i)+t_{\text{mod}}}\int_{2t_{\text{mod}}(j)+t_{pws}}^{2t_{\text{mod}}(j)+t_{\text{mod}}} T_2^2\overline{g(t-\theta)g(t-\theta')}H_2(\theta)H_2(\theta')\mathrm{d}\theta\mathrm{d}\theta' \\ &+\sum_{i=0}^{\infty}\sum_{j=0}^{\infty}\int_{2t_{\text{mod}}(i)+t_{\text{mod}}}^{2t_{\text{mod}}(i+1)}\int_{2t_{\text{mod}}(j)+t_{\text{mod}}}^{2t_{\text{mod}}(j+1)} T_3^2\overline{g(t-\theta)g(t-\theta')}H_3(\theta)H_3(\theta')\mathrm{d}\theta\mathrm{d}\theta' \\ &+2\sum_{i=0}^{\infty}\sum_{j=0}^{\infty}\int_{2t_{\text{mod}}(i)}^{2t_{\text{mod}}(i)+t_{pws}}\int_{2t_{\text{mod}}(j)+t_{pws}}^{2t_{\text{mod}}(j)+t_{\text{mod}}} T_1T_2\overline{g(t-\theta)g(t-\theta')}H_1(\theta)H_2(\theta')\mathrm{d}\theta\mathrm{d}\theta' \\ &-2\sum_{i=0}^{\infty}\sum_{j=0}^{\infty}\int_{2t_{\text{mod}}(i)}^{2t_{\text{mod}}(i)+t_{pws}}\int_{2t_{\text{mod}}(j)+t_{\text{mod}}}^{2t_{\text{mod}}(j+1)} T_1T_3\overline{g(t-\theta)g(t-\theta')}H_1(\theta)H_3(\theta')\mathrm{d}\theta\mathrm{d}\theta' \\ &-2\sum_{i=0}^{\infty}\sum_{j=0}^{\infty}\int_{2t_{\text{mod}}(i)+t_{pws}}^{2t_{\text{mod}}(i)+t_{\text{mod}}}\int_{2t_{\text{mod}}(j)+t_{\text{mod}}}^{2t_{\text{mod}}(j+1)} T_2T_3\overline{g(t-\theta)g(t-\theta')}H_2(\theta)H_3(\theta')\mathrm{d}\theta\mathrm{d}\theta' \\ &= J_{1g}+J_{2g}+J_{3g}+J_{4g}-J_{5g}-J_{6g}.\end{aligned} \tag{11}$$

For the dispersion calculation take $g(t)$ stationary, with a normal distribution, the exponential autocorrelation function

$$\overline{g(t-\theta)g(t-\theta')} = \sigma_g^2\exp\left(-\frac{|\theta-\theta'|}{\tau_0}\right), \tag{12}$$

where σ_g^2 is the measuring path transmission coefficient fluctuations dispersion, τ_0—effective correlation time constant of the transmission coefficient fluctuations. Typically τ_0 is much greater than the radiometer accumulating filters time constants τ_1, τ_2 and τ_3.

Compute generalized integral, through which any of the six integrals in expression (11) can be expressed. For variable arguments during the generalized integral calculating accept the conditions $2t_{mod} \ge x, y \ge 0$; $2t_{mod} \ge z, v \ge 0$; $k,\ r \gg 2t_{mod}$. Then

$$\begin{aligned}&I(x,y,z,v,k,r) \\ &= \sum_{i=0}^{\infty}\sum_{j=0}^{\infty}\int_{2t_{\text{mod}}(i)+x}^{2t_{\text{mod}}(i)+y}\int_{2t_{\text{mod}}(j)+z}^{2t_{\text{mod}}(j)+v}\exp\left(-\frac{|\theta-\theta'|}{\tau_0}\right)\frac{1}{kr}\exp\left(-\frac{\theta}{k}-\frac{\theta'}{r}\right)\mathrm{d}\theta\mathrm{d}\theta' \\ &\cong \frac{(y-x)(v-z)}{(2t_{мод})^2 kr}\times\int_0^{\infty}\left[\int_0^{\theta}\exp\left(-\frac{\theta'}{\tau_0}\right)\exp\left(-\frac{\theta}{k}-\frac{\theta'}{r}\right)\mathrm{d}\theta' + \int_{\theta}^{\infty}\exp\left(-\frac{\theta'-\theta}{\tau_0}\right)\exp\left(-\frac{\theta}{k}-\frac{\theta'}{r}\right)\mathrm{d}\theta'\right]\mathrm{d}\theta \\ &= \frac{(y-x)(v-z)\tau_0}{(2t_{мод})^2}\times\frac{2rk+k\tau_0+r\tau_0}{(r+k)(k+\tau_0)(r+\tau_0)}.\end{aligned} \tag{13}$$

Using (13) calculate the integrals $J_{1g} - J_{6g}$ in (11) and obtain an expression for calculating the fluctuation dispersions of the transmission coefficient

$$\begin{aligned}\overline{\Delta q_g^2} = \frac{\sigma_g^2\tau_0}{4}\Bigg[&\frac{(T_3-T_2)^2}{(T_1-T_2)^2}\left(\frac{T_1^2}{\tau_1+\tau_0}+\frac{T_2^2}{\tau_2+\tau_0}\right)+\frac{T_2^2}{\tau_2+\tau_0}\times\frac{T_1+T_2-2T_3}{T_1-T_2}+\frac{T_3^2}{\tau_3+\tau_0} \\ &+2T_1T_2\frac{(T_3-T_2)(T_1-T_3)}{(T_1-T_2)^2}\times\frac{2\tau_1\tau_2+\tau_1\tau_0+\tau_2\tau_0}{(\tau_1+\tau_2)(\tau_1+\tau_0)(\tau_2+\tau_0)}-2T_1T_3\frac{T_3-T_2}{T_1-T_2} \\ &\times\frac{2\tau_1\tau_3+\tau_1\tau_0+\tau_3\tau_0}{(\tau_1+\tau_3)(\tau_1+\tau_0)(\tau_3+\tau_0)}-2T_2T_3\frac{T_1-T_3}{T_1-T_2}\times\frac{2\tau_2\tau_3+\tau_2\tau_0+\tau_3\tau_0}{(\tau_2+\tau_3)(\tau_2+\tau_0)(\tau_3+\tau_0)}.\end{aligned} \tag{14}$$

If the filters pulse characteristics are the same, obtain $\overline{\Delta q_g^2} = 0$. A zero value of obtained dispersion indicates on more less its value changing, than the dispersion caused by the signals noise nature. Given this dispersion of fluctuations of the transmission coefficient can be neglected. In fact an error of the dispersion determination can appear as results of the autocorrelation function approximate sampling (12), approximations for the exponential function. The received result is being coincided with the conclusions of other authors [10] [11]. These conclusions show that the zero method application allows minimizing the influence of fluctuations receivers increasing on the measurement results.

If pulse-width signal t_{pws} duration changing have happened at 1 time interval (discrete step), $Q_1(t)$ changing in the first modulation half-cycle (**Figure 1(a)**) equal to

$$\Delta Q_{1ds} = G_0\left(T_1 - T_2\right)\frac{1}{N}, \tag{15}$$

where N is quantity of discrete step to be placed on the half period t_{mod} duration. N characterizes measurements resolution.

The traditional sequence of operations is used to determine the sensitivity [12]. In the case of introduced null method measuring this traditional sequence of operations can be formulated as follows: the regulation of the pulse-width signal duration will be meaningful if this duration changing on one discrete step and associated with this volt-second areas changing of pulse signals at the receiver output in the first half-period modulation are equal to standard deviation from volt-second areas data equality. The standard deviation is caused by fluctuations and noise nature of the measuring signals. In accordance with this write

$$\frac{\Delta Q_{1ds}}{\sqrt{\overline{\Delta q^2}}} = \frac{1}{\sqrt{R}}, \tag{16}$$

where R is a number of accumulated values of the pulse-width signal duration digital codes. There are two stages of averaging the signal in radiometers which use this modification of zero receiving method. Firstly a low-frequency analog signal filtering at the output of the receiver is performed. Further, in the system of added noise signal duration automatic control, except the adjustment cycle, the accumulation of digital codes of this duration with following averaging is occurring. It is known from the theory of errors; the signal dispersion is reduced $\left(\sqrt{R}\right)^{-1}$ times.

After the replacement of (10), (15) into (16) obtain

$$2df\tau\left(T_1 - T_2\right)^2 R = N^2\left[T_3\left(T_1 + T_2 + T_3\right) - T_1T_2\right]. \tag{17}$$

Pulse-width signal duration is variated from 0 to t_{mod} by changing the antenna signal from minimum to maximum. Therefore for the minimum antenna signal ΔT_a, which can be detected, there is a proportion

$$\frac{dT_a}{N} = \Delta T_a \sim \Delta t = \frac{t_{\text{mod}}}{N}, \tag{18}$$

where dT_a—antenna signal measuring range, Δt—time discrete step duration, which varies by the duration t_{pws} sudden change. The value of the minimum detectable antenna signal ΔT_a characterizes the sensitivity.

Typically, ΔT_a and dT_a are defined by the device designing. In terms of these parameters, using (18) find N. N determines the number of order n of radiometer output digital code. Using (18) the formula of radiometer fluctuation sensitivity calculation is formed from (17)

$$\Delta T_a = \frac{dT_a}{\sqrt{2df\tau R}} \times \frac{\sqrt{T_3\left(T_1 + T_2 + T_3\right) - T_1T_2}}{T_1 - T_2}. \tag{19}$$

For the known radiometric receiver necessary sensitivity is achieved by selecting the τ and R.

Parameter R is related to measurement time t_{mes} by the ratio

$$R = \frac{t_{mes}}{2t_{\text{mod}}}.$$

5. Structured Modeling of the Radiometer Input Receiving Modules

The signal modulation before their arrival at the receiver is carried out in the radiometer input block. Three signals are subjected to modulation. Two signals are reference signals T_{ref} and T_{add}. They are produced by reference noise generator (NG) and additional reference noise generator (ANG) accordingly. The third signal is an antenna measurable signal (A) T_a. Combinations of these signals constitute the signals T_1, T_2, T_3 on time diagrams in **Figure 1**. Possible signals combinations T_{ref}, T_{add}, T_a for the level formations T_1, T_2, T_3 are given in **Table 1**. T_n–reduced to the input of the receiver self-noise effective temperature of the receiver and input block of the radiometer.

According to the received method functioning model (5) block diagrams of the three input blocks (for the respective positions of the **Table 1**) are shown in **Figure 2**. Modulator (M) alternately commutes to the receiver input or antenna signal T_a or noise reference generator signal T_{ref}. In the schemes in **Figure 2(a)** and **Figure 2(b)** additional reference signal T_{add} is introduced through the microwave switch (MWS) to the antenna or supporting duct through the directional coupler (DC). Additional modulator (M_1) is introduced into the scheme in **Figure 2(c)**.

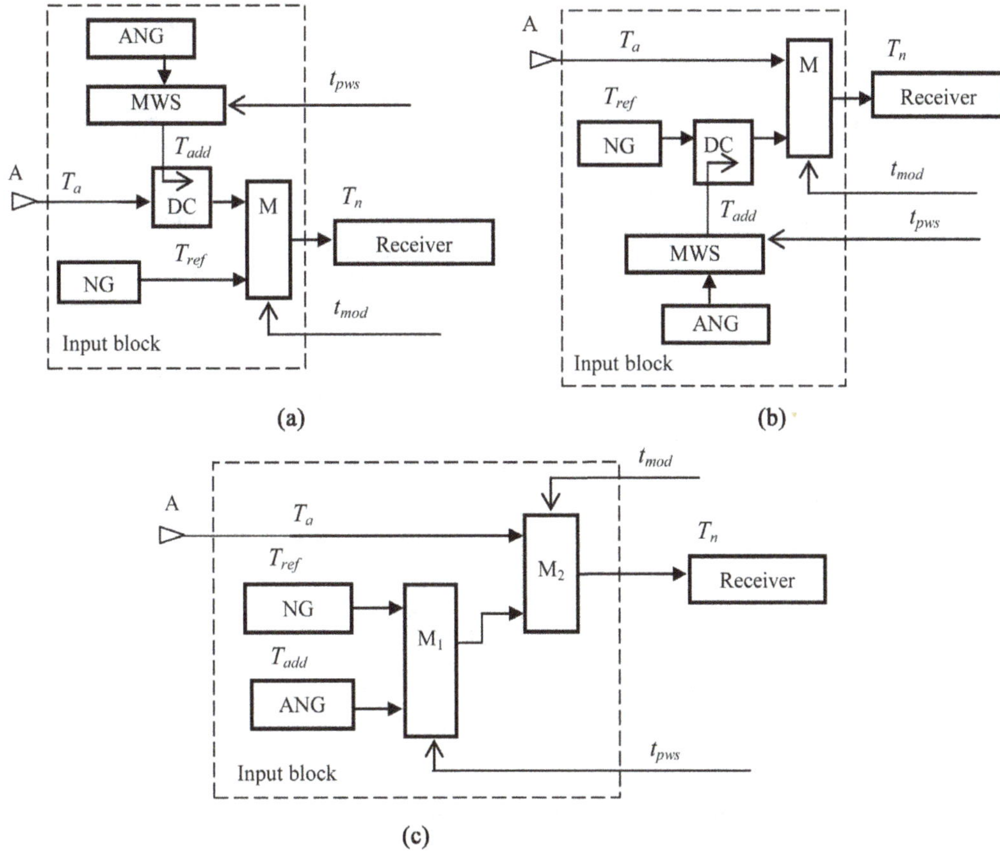

Figure 2. Input receiver block structure schemes of the modified zero radiometers.

Table 1. The modulated signals combinations for different measurement ranges.

Position	The modulated signals			
	T_1	T_2	T_3	Note
1	$T_a + T_{add} + T_n$	$T_a + T_n$	$T_{ref} + T_n$	$T_a < T_{ref}$
2	$T_{ref} + T_{add} + T_n$	$T_{ref} + T_n$	$T_a + T_n$	$T_a > T_{ref}$
3	$T_{add} + T_n$	$T_{ref} + T_n$	$T_a + T_n$	$T_{add} > T_a > T_{ref}$

Obtain the transfer characteristic of the radiometer for the circuit of input block in **Figure 2(a)** by the values of effective temperatures from Pos. 1 **Table 1** substituting in Equation (5)

$$t_{pws} = \frac{T_{ref} - T_a}{T_{add}} \times t_{\text{mod}}. \tag{20}$$

(t_{pws} duration shown without upper underlining and further averaging mark is ignored). From (20) follows that antenna signal can be determined through the pulse-width signal duration that controls the additional reference noise generator (ANG) modulation without signals change in the radiometer low-frequency part. t_{pws} duration is associated with antenna signal T_a by the linear law and does not depend on the transmission coefficient of the radiometer.

Obtain antenna signal from (20) $T_a = T_{ref} - T_{add}t_{pws}/t_{mod}$. Define the minimum and maximum of measured signals range limit by substituting two sample extremes of the duration t_{pws} (equal t_{mod} and 0) in the last equality. $T_{a,min} = T_{ref} - T_{add}$; $T_{a,max} = T_{ref}$. Where the effectives dynamic range is determined by the additional reference noise generator (ANG) signal $dT_a = T_{a,max} - T_{a,min} = T_{add}$.

The equation for sensitivity determining of radiometer with the input block **Figure 2(a)** can be found from (19) by substituting signals T_1, T_2, T_3 of the Pos. 1 **Table 1**

$$\Delta T_a = \frac{\sqrt{T_{ref}\left(T_{ref} + T_{add} + 4T_n\right) + 2T_n^2 - T_a\left(T_a + T_{add} - 2T_{ref}\right)}}{\sqrt{2df\tau R}}. \tag{21}$$

The radiometer measurement range dT_a with concerned input block is equal T_{add}. This fact has been taken into account in equation (21).

From (21) follows that the sensitivity depends on the specific antenna signal and is variable in measurement range. In [13], the attention was taken into the fluctuating sensitivity dependence on antenna signal. But the case of a large antenna signal was regarded in this article. Owing to super noiseless amplifier with noise floor of tens of Kelvin creation whole radiometric system self-noise were considerably reduced, which were comparable with measured signals.

Usable sensitivity occurs for the antenna signal $T_a = T_{ref} - T_{add}/2$ in the middle of measurement range

$$\Delta T_{a,\max} = \frac{\sqrt{2\left(T_{ref} + T_n\right)^2 + \frac{T_{add}^2}{4}}}{\sqrt{2df\tau R}}. \tag{22}$$

To achieve the necessary threshold of sensitivity during the designing of radiometer with the input block obtain the relation for product calculation from (22)

$$\tau R = \frac{\left(T_{ref} + T_n\right)^2 + \frac{T_{add}^2}{8}}{df\,\Delta T_{a,\max}^2}. \tag{23}$$

Consider τ and R determination strategy by example. Let it be required to secure the measurement dynamic range 0…300 K for a receiver with noise temperature T_n = 200 K and receiving band df = 100 MHz. It is required to provide a minimum signal-detection threshold 0.05 K ($\Delta T_{a,\max}$ = 0.05 K) in this range. The modulating frequency in radiometer is 1 kHz (t_{mod} = 500 ms).

Firstly determine the levels of reference signals $T_{ref} = T_{a,max}$ = 300 K, $T_{add} = T_{ref} - T_{a,min}$ = 300 K.

Then using (23) find $\tau R = \frac{\left(T_{ref} + T_n\right)^2 + \frac{T_{add}^2}{8}}{df\,\Delta T_{a,\max}^2} = 1.045$.

Choose $\tau = 30t_{mod}$ = 0.015 s to ensure the necessary dynamic properties of the regulatory system. Whence R = 1.045/τ = 69. Then, for the modulation period 1 ms, the measurement time will be 69 ms.

To provide the necessary sensitivity the required capacity of radiometer output digital code is determined by further calculations. Find the number of minimum antenna signal values on the measurement range $N = dT_a/\Delta T_{a,max}$ = 6000 (consider that for a given input block structure a measurement range $dT_a = T_{add}$). Then a rates number of radiometer output digital code $n = \log_2 N = \log_2 6000$. Get code size n = 13 by rounding up the whole.

Using similar calculations for other input blocks in **Figure 2(b)** and **Figure 2(c)**, it is useful to get the necessary data for a modified zero-radiometer designing.

As example of sensitivity calculations by Formula (29) determine $\sqrt{2df\tau R}$ which is situated in its denominator. $\sqrt{2df\tau R}$ characterizes radiometer receiver and signal processing at low frequency. Choose radiometer parameters as more typical: df = 100 MHz, modulation frequency 1 kHz, time constant of low frequency filter τ =30ms, R = 1000 which correspond to signal time storage equal to 1 second. $\sqrt{2df\tau R}$ is dimensionless quantity equal to 78383.7. The graph in **Figure 3** shows calculated by Equation (21) threshold sensitivity dependence of the modified radiometer for the most typical in remote sensing measurement ranges for a receiver with noise temperature T_n = 50 K. These calculations imply that the sensitivity within the range of measurements remains almost invariant. At the edges of measurement range the sensitivity takes the same minimum value and insignificantly increases in the middle of the range. The sensitivity depends on the upper limit of measurement range. The minimum threshold ΔT_a increases in the case of range expansion in the direction of higher measurement antenna temperatures measurement.

Found by similar way: transfer characteristic, measurement range, fluctuation sensitivity of radiometer with the input blocks are tabulated in **Table 2** and their schemes are shown in **Figure 2(b)**. Calculation of minimum detectable signal for both blocks gives the same results with the same measurement ranges. The dependence of fluctuation sensitivity on antenna signal diagrams for various full scales receivers with different noise temperatures are plotted in **Figure 4**. The sensitivity does not remain the same in the measurement range and changes during the antenna signal modifying with almost linear variation. ΔT_a reaches a peak at the maximum antenna signal and trough at the antenna signal equal to minimum value of the measurement range.

For all considered input units T_n augment leads to a proportional increase of the minimum detectable antenna signal and sensitivity deterioration occurs linearly with noise temperature increasing of the receiver and the input path.

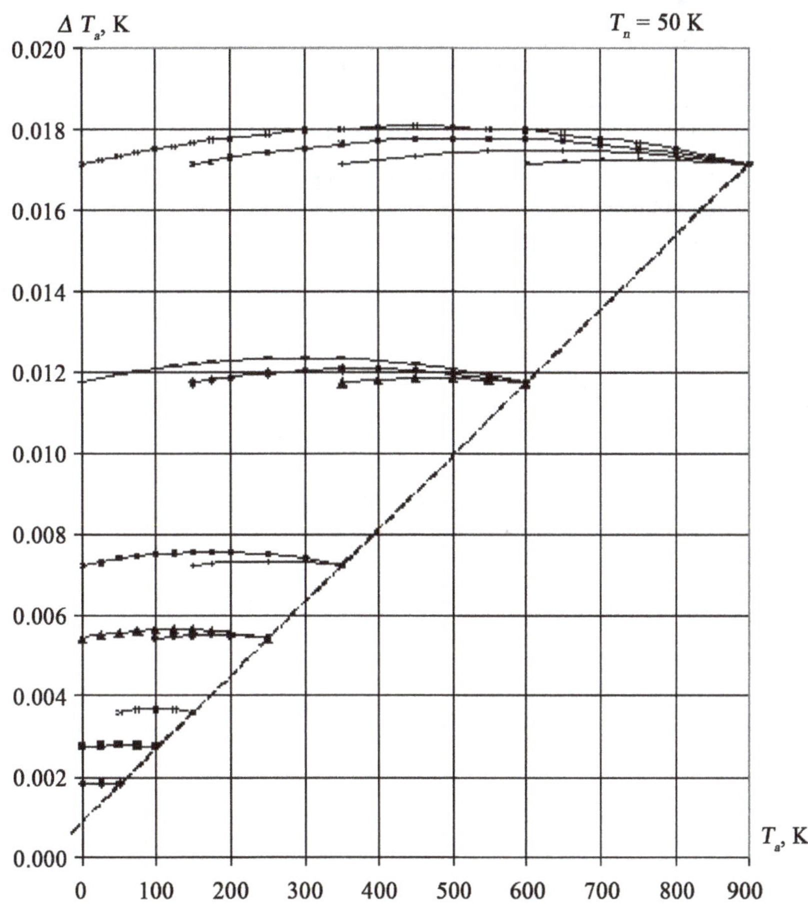

Figure 3. Fluctuation sensitivity responses of the radiometer with input block (**Figure 2(a)**) from antenna signal of different measuring range.

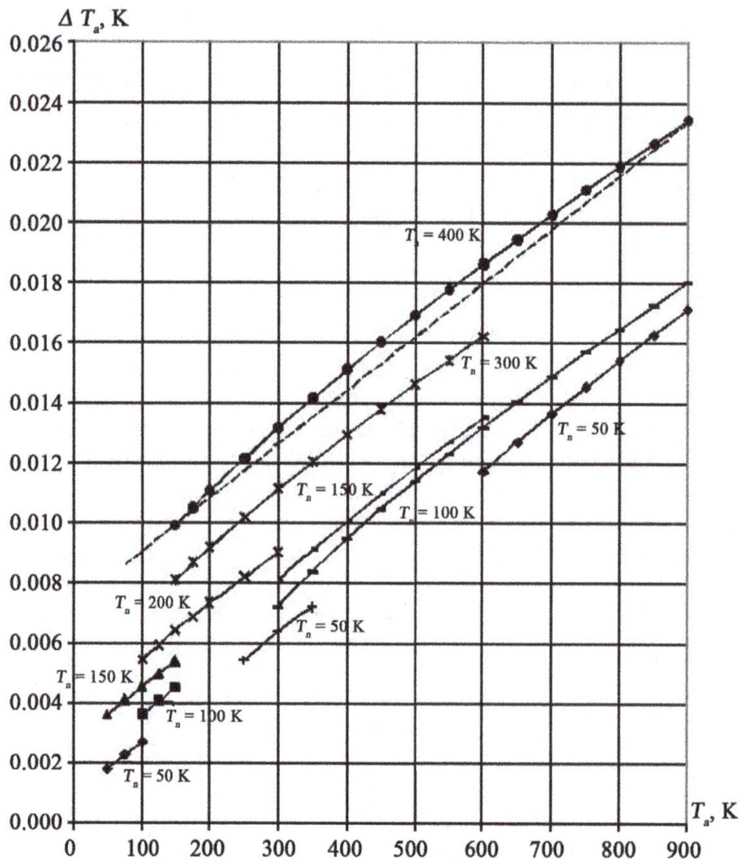

Figure 4. Fluctuation sensitivity responses of the radiometer with input block (**Figure 2(b)** and **Figure 2(c)**) from antenna signal of different measuring range and different receiver noise characteristics.

Table 2. The transfer characteristics, the measurement ranges, the radiometer sensitivity with the input blocks shown in **Figure 2(b)** and **Figure 2(c)**.

Input block	The transfer characteristics, the measurement ranges, the radiometer sensitivity
Input block by **Figure 2(b)** scheme	$t_{pws}=\frac{T_a-T_{ref}}{T_{add}}\times t_{\mathrm{mod}}$; $T_a=T_{ref}+T_{add}\frac{t_{pws}}{t_{\mathrm{mod}}}$ $T_{a,\min}=T_{ref}\left(t_{pws}=0\right)$; $T_{a,\max}=T_{ref}+T_{add}\quad\left(t_{pws}=t_{\mathrm{mod}}\right)$ $dT_a=T_{a,\max}-T_{a,\min}=T_{add}$ $\Delta T_a=\frac{\sqrt{T_a\left(2T_{ref}+T_{add}+T_a+4T_n\right)+2T_n^2-T_{ref}\left(T_{ref}+T_{add}\right)}}{\sqrt{2df\tau R}}$ $\Delta T_{a,\max}=\frac{\sqrt{2\left(T_{ref}+T_n\right)^2+2T_{add}^2+4T_{add}\left(T_{ref}+T_n\right)}}{\sqrt{2df\tau R}}$ $\tau R=\frac{\left(T_{ref}+T_n\right)^2+T_{add}^2+2T_{add}\left(T_{ref}+T_n\right)}{df\Delta T_{a,\max}^2}$
Input block by **Figure 2(c)** scheme	$t_{pws}=\frac{T_a-T_{ref}}{T_{add}-T_{ref}}\times t_{\mathrm{mod}}$. $T_a=T_{ref}+\left(T_{add}-T_{ref}\right)\times\frac{t_{pws}}{t_{\mathrm{mod}}}$. $T_{a,\min}=T_{ref}\quad T_{a,\max}=T_{add}$; $dT_a=T_{add}-T_{ref}$ $\Delta T_a=\frac{\sqrt{T_a\left(T_{add}+T_{ref}+T_a+4T_n\right)+2T_n^2-T_{add}T_{ref}}}{\sqrt{2df\tau R}}$, $\Delta T_{a,\max}=\frac{T_{add}+T_n}{\sqrt{df\tau R}}$; $\tau R=\frac{\left(T_{add}+T_n\right)^2}{df\Delta T_{a,\max}^2}$

6. Experimental Researches

The formulas obtained in section 5 for sensitivity calculations were tested in experimental researches. Sensitivity determine experiments were carried out for the considered input blocks with the radiometer at the wavelength 6.5 sm (total noise temperature of the whole system is equal to 600 K, receiving bandwidth—100 MHz, modulation frequency—1 kHz). Antenna signal changing was performed by antenna tilt variation. Directional diagram covered an area in which there were no sources of synthetic electromagnetic radiation. The sensitivity was determined for different antenna signals.

Radiometer was reconstructed and 8 series of 16 measurements with 5 minutes intervals were performed. Storage time was set equal to 1 second. Average values were calculated for each series. These values were plotted on a graph shown in **Figure 5**. The values of the minimum detectable antenna signal were imaged on ordinate axis; the values of the measured antenna signal were imaged on abscissa axis. Curves 1 and 2 on the graphs are constructed according to the obtained analytical dependences of the radiometer with the input blocks sensitivity calculation. Schemes of these blocks are shown in **Figure 2(a)** and **Figure 2(c)**, respectively. Experimentally obtained values of sensitivity are represented by vertical lines on the given graphs. The scatter of the sensitivity measured values over standard periods of time at a constant antenna signal was taken into account. Maximum spread between findings obtained by theoretical and experimental way was about 20% with standard deviation of 8% ... 10%.

7. Conclusions

On the basis of the combined pulse modulation and the original principle of signal processing, the algorithm of the following system functioning was developed. This algorithm showed that it was useful to carry out the auto-zero balance in the radiometer by changing of pulse-width signal duration. According to this algorithm, at the output of the radiometer after constant component exclusion in the first half-period of rectangular symmetrical modulation pulse volt-second areas, equalization is performed.

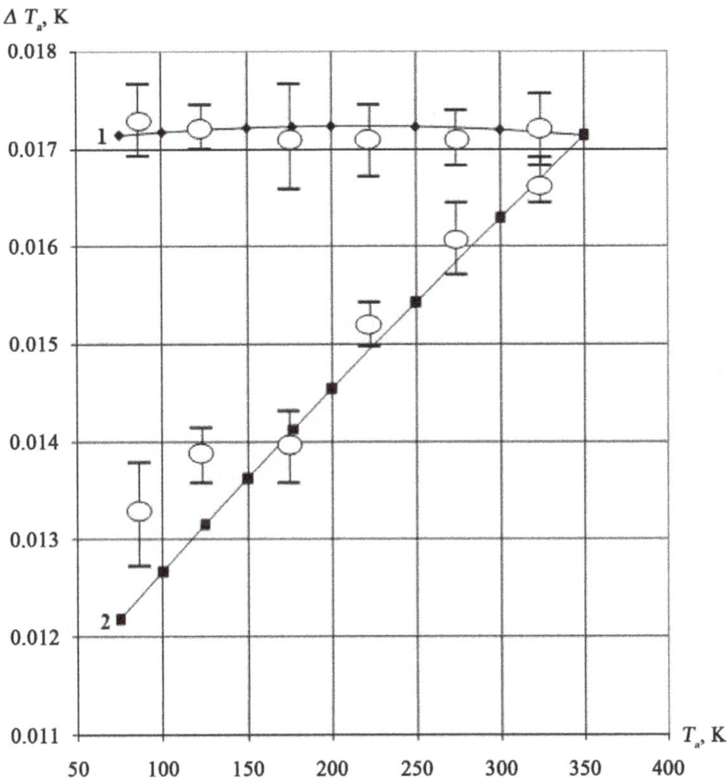

Figure 5. Radiometer fluctuation sensitivity with different receiving blocks at the input.

This procedure is equivalent to the signal energies' equalization at the radiometer receiver input at the different modulation half-periods. Zero voltage in the second modulation half period is an indicator of pulse volt-second areas' equality. As a result, a mathematical model is found. This model establishes a linear relation between antenna effective temperature and duration of reference noise signal modulated by pulse-width-law.

The analysis of the fluctuation sensitivity of this measuring method pointed to variable sensitivity, the dependence on the measured antenna signal. Formulas for fluctuation sensitivity calculation were obtained and experimental verification was carried out for three proposed schemes of the modified radiometer input blocks with different measured signals' range.

Acknowledgements

This work was supported by Russian Foundation for Basic Research, grants No. 13-07-98009, No. 15-07-04971.

References

[1] Sharkov, E.A. (2003) Passive Microwave Remote Sensing of the Earth: Physical Foundations. Springer/PRAXIS, Berlin.

[2] Astafieva, N.M., Raev, M.D. and Sharkov, E.A. (2006) Portret Zemli iz kosmosa. Globalnoe radioteplovoe pole. *Priroda*, **9**, 75-86.

[3] Dicke, R.H. (1946) The Measurement of Thermal Radiation at Microwave Frequencies. *Review of Scientific Instruments*, **17**, 268-275. http://dx.doi.org/10.1063/1.1770483

[4] Orhaur, T. and Waltman, W. (1962) Switched Load Radiometer. *Public National Radio Astronomy Observatory*, **1**, 179-204.

[5] Ryle, M. and Vonberg, D.D. (1948) An Investigation of Radio-Frequency Radiation from the Sun. *Proceeding of the Royal Society*, **193**, 98-119. http://dx.doi.org/10.1098/rspa.1948.0036

[6] Troitskiy, V.S., Lubina, A.G. and Zolotov, A.V. (1951) Sravnenie teplovih shumov nekotorix materialov nulevim metodom. *Dokladi Akademii Nauk SSSR*, **4**, 583-586.

[7] Hardy, W.N., Gray, K.W. and Love, A.W. (1974) An S-Band Radiometer Design with High Absolute Precision. *IEEE Transactions on Microwave Theory and Techniques*, **MTT-22**, 382-391.

[8] Lawrence, R.F., Harrington, R.F. and Higdon, N.S. (1982) Flight Test Evaluation of a Noise Injection Dicke Microwave Radiometer Employing Digital Signal Processing. *IEEE MTT-S International Microwave Symposium Digest*, New York, 15-17 June 1982, 90-92.

[9] de Maagt, P.J.I., Oerlemans, R.A.E., van Gestel, J.C.A.M. and Herben, M.H.B.J. (1992) A Novel Radiometer Receiver Stabilization Method. *International Journal of Infrared and Millimeter Waves*, **13**, 1075-1097. http://dx.doi.org/10.1007/BF01009052

[10] Brown, S.T., Desai, S., Wenwen, Lu. and Tanner, A.B. (2007) On the Long-Term Stability of Microwave Radiometers Using Noise Diodes for Calibration. *IEEE Transactions on Geoscience and Remote Sensing*, **45**, 1908-1920.

[11] Vlaby, F.T., Moore, R.K. and Fung, A.K. (1981) Microwave Remote Sensing. Artech House, Norwood.

[12] Sironi, G., Inzani, P., Limon, M. and Marchioni, C. (1990) Evaluation of Small Signals with a Differential Radiometer (with Application to Radio Observations at 2.5 Ghz). *Measurement Science and Technology*, **1**, 1119-1121. http://dx.doi.org/10.1088/0957-0233/1/10/025

[13] Esepkina, N.A., Korolkov, D.V. and Pariyskiy, U.N. (1973) Radioteleskopi i radiometri. Nauka, Moscow, 415 p.

Permissions

All chapters in this book were first published by Scientific Research Publishing; hereby published with permission under the Creative Commons Attribution License or equivalent. Every chapter published in this book has been scrutinized by our experts. Their significance has been extensively debated. The topics covered herein carry significant findings which will fuel the growth of the discipline. They may even be implemented as practical applications or may be referred to as a beginning point for another development.

The contributors of this book come from diverse backgrounds, making this book a truly international effort. This book will bring forth new frontiers with its revolutionizing research information and detailed analysis of the nascent developments around the world.

We would like to thank all the contributing authors for lending their expertise to make the book truly unique. They have played a crucial role in the development of this book. Without their invaluable contributions this book wouldn't have been possible. They have made vital efforts to compile up to date information on the varied aspects of this subject to make this book a valuable addition to the collection of many professionals and students.

This book was conceptualized with the vision of imparting up-to-date information and advanced data in this field. To ensure the same, a matchless editorial board was set up. Every individual on the board went through rigorous rounds of assessment to prove their worth. After which they invested a large part of their time researching and compiling the most relevant data for our readers.

The editorial board has been involved in producing this book since its inception. They have spent rigorous hours researching and exploring the diverse topics which have resulted in the successful publishing of this book. They have passed on their knowledge of decades through this book. To expedite this challenging task, the publisher supported the team at every step. A small team of assistant editors was also appointed to further simplify the editing procedure and attain best results for the readers.

Apart from the editorial board, the designing team has also invested a significant amount of their time in understanding the subject and creating the most relevant covers. They scrutinized every image to scout for the most suitable representation of the subject and create an appropriate cover for the book.

The publishing team has been an ardent support to the editorial, designing and production team. Their endless efforts to recruit the best for this project, has resulted in the accomplishment of this book. They are a veteran in the field of academics and their pool of knowledge is as vast as their experience in printing. Their expertise and guidance has proved useful at every step. Their uncompromising quality standards have made this book an exceptional effort. Their encouragement from time to time has been an inspiration for everyone.

The publisher and the editorial board hope that this book will prove to be a valuable piece of knowledge for researchers, students, practitioners and scholars across the globe.

List of Contributors

Ana P. de Moura, Larissa H. Oliveira, Paula F. S. Pereira, Elson Longo and José A. Varela
Chemistry Institute, State University of Sao Paulo-UNESP, Araraquara, Brazil

Içamira C. Nogueira
Department of Engineering Materials, Federal University of Sao Carlos, São Carlos, Brazil

Máximo S. Li
Institute of Physics of São Carlos, USP, São Carlos, Brazil

Ieda L. V. Rosa
Department of Chemistry, Federal University of Sao Carlos, São Carlos, Brazil

Wissam M. Alobaidi, Eric Sandgren and Hussain M. Al-Rizzo
Systems Engineering Department, Donaghey College of Engineering & Information Technology, University of Arkansas at Little Rock, Little Rock, Arkansas, USA

Entidhar A. Alkuam
Department of Physics and Astronomy, College of Arts, Letters, and Sciences, University of Arkansas at Little Rock, Little Rock, Arkansas, USA

Muhammadjon Gulomkodirovich Dadamirzaev
Namangan Engineering Pedagogical Institute, Namangan, Uzbekistan

Vittorio Romano and Rino Apicella
Dipartimento di Ingegneria Industriale, Università degli Studi di Salerno, Fisciano, Italy

Heesup Choi and Myungkwan Lim
Department of Civil Engineering, Kitami Institute of Technology, Hokkaido, Japan
Graduated School of Engineering, Hankyong National University, Ansung, Korea

Hyeonggil Choi, Ryoma Kitagaki and Takafumi Noguchi
Department of Architecture, The University of Tokyo, Tokyo, Japan

Obed Osorio-Esquivel
Department of Biochemical Engineering, National School of Biological Sciences, National Polytechnic Institute, Mexico City, Mexico
University of Chalcatongo, Avenida Universidad, Chalcatongo of Hidalgo, Mexico

Vianney Cortés-Viguri, Alicia Ortiz-Moreno and María Elena Sánchez-Pardo
Department of Biochemical Engineering, National School of Biological Sciences, National Polytechnic Institute, Mexico City, Mexico

Leticia Garduño-Siciliano
Department of Physiology, National School of Biological Sciences, National Polytechnic Institute, Mexico City, Mexico

Andre Ricard, Hayat Zerrouki and Jean-Philippe Sarrette
Laplace, Toulouse, France

Mamdouh Omran
Laboratory of Process Metallurgy Research Group, Process and Environmental Engineering Department, University of Oulu, Oulu, Finland
Central Metallurgical Research and Development Institute, Cairo, Egypt

Timo Fabritius
Laboratory of Process Metallurgy Research Group, Process and Environmental Engineering Department, University of Oulu, Oulu, Finland

Nagui Abdel-Khalek, Mahmoud Nasr and Ahmed Elmahdy
Central Metallurgical Research and Development Institute, Cairo, Egypt

Mortada El-Aref and Abd El-Hamid Elmanawi
Geology Department, Faculty of Science, Cairo University, Cairo, Egypt

Gafur Gulyamov, Muhammadjon Gulomkodirovich Dadamirzaev and Hasan Yusupovich Mavlyanov
Namangan Engineering Pedagogical Institute, Namangan, Uzbekistan
Namangan State University, Namangan, Uzbekistan

Mayur Shukla
Academy of Scientific and Innovative Research (AcSIR), CSIR-Central Glass and Ceramic Research Institute, Kolkata, India
CSIR-Central Glass and Ceramic Research Institute (CSIR-CGCRI), Kolkata, India

Sumana Ghosh, Nandadulal Dandapat, Ashis K. Mandal and Vamsi K. Balla
CSIR-Central Glass and Ceramic Research Institute (CSIR-CGCRI), Kolkata, India

Franz Müller, Thomas Scheller, Jinni Lee, Ralf Behr, Luis Palafox, Marco Schubert and Johannes Kohlmann
Physikalisch-Technische Bundesanstalt (PTB), Braunschweig, Germany

L. Mohamed, N. Ozawa, Y. Ono, T. Kamiya and Y. Kuwahara
Graduate School of Engineering, Shizuoka University, Hamamatsu-shi, Japan

G. Gulyamov and M. G. Dadamirzaev
Namangan Engineering Pedagogical Institute, Namangan, Uzbekistan

Eman Alzahrani, Abeer Sharfalddin and Mohamad Alamodi
Chemistry Department, Faculty of Science, Taif University, Taif, Kingdom of Saudi Arabia

S. B. Shruthi, Chandan Bhat, S. P. Bhaskar and G. Preethi
Chemical Engineering Department, Dayananda Sagar College of Engineering, Kumaraswamy Layout, Bengaluru, India

R. R. N. Sailaja
Resource Efficient Process Technology Application, The Energy and Resources Institute, Bengaluru, India

Prasanna K. Vuram1, C. Kabilan
Laboratory of Bioorganic Chemistry, Department of Biotechnology, Indian Institute of Technology Madras, Chennai, India

Anju Chadha
Laboratory of Bioorganic Chemistry, Department of Biotechnology, Indian Institute of Technology Madras, Chennai, India
National Centre for Catalysis Research, Indian Institute of Technology Madras, Chennai, India

Vinay Kumar Suryadevara, Suyog Patil and Paul Salama
Department of Electrical and Computer Engineering, Indiana University-Purdue University Indianapolis, Indianapolis, USA

James Rizkalla and Ahdy Helmy
Indiana University School of Medicine, Indianapolis, USA

Maher Rizkalla
Integrated Nanosystems Development Institute (INDI), Indiana University-Purdue University Indianapolis, Indianapolis, USA

Tawfik Elsayed Khattab
Department of Electrical Engineering, Engineering College, King Khalid University, Abha, Saudi Arabia

Peri A. Muradova and Yuriy N. Litvishkov
Institute of Catalysis and Inorganic Chemistry Named after M. Nagiyev of Azerbaijan National Academy of Sciences, Baku, Azerbaijan

Ivaylo Hinkov
University of Chemical Technology and Metallurgy, Sofia, Bulgaria

Samir Farhat, Alix Gicquel, François Silva, Amine Mesbahi and Ovidiu Brinza
Laboratoire des Sciences des Procédés et des Matériaux, CNRS, LSPM-UPR 3407, Université Paris 13, Villetaneuse, France

Cristian P. Lungu, Cornel Porosnicu and Alexandru Anghel
National Institute for Laser, Plasma and Radiation Physics, Bucharest, Romania

Ryosuke Nakata and Yusuke Asakuma
Department of Mechanical and System Engineering, University of Hyogo, Kobe, Japan

Bin Wu, Hongxi Yin, Anliang Liu and Chang Liu
Laboratory of Optical Communications and Photonic Technology, School of Information and Communication Engineering, Dalian University of Technology, Dalian, China

Jingchao Wang
Institute of China Electronic System Engineering Company, Beijing, China

T. Ohishi and S. Sugawara
Department of Applied Chemistry, Faculty of Engineering, Shibaura Institute of Technology, Tokyo, Japan

Alexander V. Filatov
Tomsk State University of Control Systems and Radio Engineering, Tomsk, Russia

www.ingramcontent.com/pod-product-compliance
Lightning Source LLC
Chambersburg PA
CBHW080508200326
41458CB00012B/4128

* 9 7 8 1 6 3 2 3 8 5 1 9 2 *